T0299716

Multiplicative Partial Differential Equations

This book presents an introduction to the theory of multiplicative partial differential equations (MPDEs). It is suitable for all types of basic courses on MPDEs. The authors' aim is to present a clear and well-organized treatment of the concept behind the development of mathematics and solution techniques. The text material of this book is presented in a highly readable, mathematically solid format. Many practical problems are illustrated displaying a wide variety of solution techniques.

Key features:
- The book includes new classification and canonical forms of second-order MPDEs.
- Authors propose a new technique to solving the multiplicative wave equation such as the method of separation of variables and the energy method.
- The proposed technique in the book can be used to give the basic properties of multiplicative elliptic problems, fundamental solutions, multiplicative integral representation of multiplicative harmonic functions, mean-value formulas, strong principle of maximum, multiplicative Poisson equation, multiplicative Green functions, method of separation of variables and theorems of Liouville and Harnack.

Svetlin G. Georgiev was born in 1974 in Rouse, Bulgaria. He is a mathematician who has worked in various areas of mathematics. He currently focuses on harmonic analysis, functional analysis, partial differential equations, ordinary differential equations, Clifford and quaternion analysis, integral equations and dynamic calculus on timescales.

Khaled Zennir was born in Skikda, Algeria in 1982. He received his PhD in mathematics in 2013 from Sidi Bel Abbès University, Algeria (assist. professor). He obtained his highest diploma in Algeria (habilitation, mathematics) from Constantine University, Algeria in 2015. He is now an associate professor at Qassim University, KSA. His research interests lie in nonlinear hyperbolic partial differential equations—global existence, blow-up and longtime behavior.

Advances in Applied Mathematics
Series Editor: Daniel Zwillinger

Introduction to Quantum Control and Dynamics
Domenico D'Alessandro

Handbook of Radar Signal Analysis
Bassem R. Mahafza, Scott C. Winton, Atef Z. Elsherbeni

Separation of Variables and Exact Solutions to Nonlinear PDEs
Andrei D. Polyanin, Alexei I. Zhurov

Boundary Value Problems on Time Scales, Volume I
Svetlin Georgiev, Khaled Zennir

Boundary Value Problems on Time Scales, Volume II
Svetlin Georgiev, Khaled Zennir

Observability and Mathematics
Fluid Mechanics, Solutions of Navier-Stokes Equations, and Modeling
Boris Khots

Handbook of Differential Equations, Fourth Edition
Daniel Zwillinger, Vladimir Dobrushkin

Experimental Statistics and Data Analysis for Mechanical and Aerospace Engineers
James Middleton

Advanced Engineering Mathematics with MATLAB°, Fifth Edition
Dean G. Duffy

Handbook of Fractional Calculus for Engineering and Science
Harendra Singh, H. M. Srivastava, Juan J. Nieto

Advanced Engineering Mathematics
A Second Course with MATLAB®
Dean G. Duffy

Quantum Computation
Helmut Bez and Tony Croft

Computational Mathematics
An Introduction to Numerical Analysis and Scientific Computing with Python
Dimitrios Mitsotakis

Delay Ordinary and Partial Differential Equations
Andrei D. Polyanin, Vsevolod G. Sorkin, Alexi I. Zhurov

Clean Numerical Simulation
Shijun Liao

Multiplicative Partial Differential Equations
Svetlin G. Georgiev and Khaled Zennir

*https://www.routledge.com/Advances-in-Applied-Mathematics/book-series/CRCADVAPP
MTH?pd=published,forthcoming&pg=1&pp=12&so=pub&view=list*

Multiplicative Partial Differential Equations

Svetlin G. Georgiev
Khaled Zennir

CRC Press
Taylor & Francis Group
Boca Raton London New York

CRC Press is an imprint of the
Taylor & Francis Group, an **informa** business

A CHAPMAN & HALL BOOK

First edition published 2024
by CRC Press
2385 NW Executive Center Drive, Suite 320, Boca Raton FL 33431

and by CRC Press
4 Park Square, Milton Park, Abingdon, Oxon, OX14 4RN

CRC Press is an imprint of Taylor & Francis Group, LLC

ISBN: 978-1-032-57503-2 (hbk)
ISBN: 978-1-032-57604-6 (pbk)
ISBN: 978-1-003-44011-6 (ebk)

DOI: 10.1201/9781003440116

Typeset in CMR10 font
by KnowledgeWorks Global Ltd.

Contents

Preface

This book presents an introduction to the theory of multiplicative partial differential equations (MPDEs). The book is suitable for all types of basic courses on MPDEs.

In Chapters 2 and 3, the classification and canonical forms of second-order MPDEs are considered. Chapter 4 is concerned with the multiplicative wave equation. They are investigated by even and odd dimensional multiplicative wave equations, method of separation of variables and energy method. It is introduced as the Riemann functions. Chapter 5 deals with the multiplicative heat equation. They are considered the weak and strong maximum principles, the Cauchy problem, the mean value formula, the method of separation of variables and the energy method. The multiplicative Laplace equation is introduced in Chapter 6. They are given the basic properties of multiplicative elliptic problems, the fundamental solutions, multiplicative integral representation of multiplicative harmonic functions, mean value formulas, strong principle of maximum, the multiplicative Poisson equation, multiplicative Green functions, method of separation of variables and theorems of Liouville and Harnack. Chapter 7 is focused on a multiplicative analogues of the classical Cauchy–Kovalevskaya Theorem. It is considered in the case of both MDEs and MPDEs.

The aim of this book is to present a clear and well-organized treatment of the concept behind the development of mathematics and solution techniques. The text of this book is presented in a highly readable, mathematically solid format. Many practical problems are illustrated displaying a wide variety of solution techniques.

Svetlin G. Georgiev and Khaled Zennir

Paris, France

1

General Introduction

A multiplicative partial differential equation (MPDE) describes a relation be-tween an unknown function and its multiplicative partial derivatives. The general form of a MPDE for a function $u(x_1, \ldots, x_n)$ is

$$F(x_1, \ldots, x_n, u, u^*_{x_1}, \ldots, u^*_{x_n}, \ldots) = 0_*,$$

or

$$F\left(x_1, \ldots, x_n, e^{x_1 \frac{u_{x_1}}{u}}, \ldots, e^{u^{x_n} \frac{u_{x_n}}{u}} \ldots, \right) = 1,$$

where x_i, $1 \le i \le n$, are multiplicative independent multiplicative variables, u is the unknown variable, $u^*_{x_i}$ denote the multiplicative partial derivative of u with respect to x_i, u_{x_i} denote the classical partial derivative of u with respect to x_i, $i \in \{1, \ldots, n\}$. The equation is, in general, supplemented by additional conditions such as initial conditions or boundary conditions.

The analysis of MPDEs has many facets. The classical approach is to develop methods for finding explicit solutions. The aim is to discover some of the solution properties before computing it, and sometimes even without a complete solution. There exist many equations that cannot be solved. All we can do in these cases is to obtain qualitative information on the solution. Furthermore, it is desired in many cases that the solution will be unique, and that it will be stable under small perturbations of the data. A theoretical understanding of the equation enables us to check whether these conditions are satisfied. As we will see in what follows, there are many ways to solve MPDEs, each way applicable to a certain class of MPDEs. Therefore, it is important to have an analysis of the equation before or during solving it.

The fundamental theoretical question is whether the problem consisting of the equation and its associated side conditions is well-posed.

Definition 1.1 A problem is called well-posed if it satisfies the following criteria.

1. *Existence:* The problem has a solution.

2. *Uniqueness:* There is no more than one solution.

3. *Stability:* A small change in the equation or in the side conditions gives rise to a small change in the solution.

DOI: 10.1201/9781003440116-1

If one or more conditions above does not hold, we say that the problem is ill-posed. MPDEs are often classified into different types. In fact, there exist several such classifications. The first classification is according to the order of the equation.

Definition 1.2 The order is defined to be the order of the highest multiplicative derivative in the equation. If the highest multiplicative derivative is of order m, then the equation is said to be of order m.

Other important classifications will be given in the next chapters.

Definition 1.3 A function in the set $\mathcal{C}_*^m$ that satisfies a MPDE of order m will be called a solution.

Definition 1.4 If each of the functions u_1, ..., u_l satisfies a MPDE and every multiplicative linear combination of them satisfies that equation too, this property is called multiplicative superposition principle. It allows the construction of complex solutions through multiplicative combinations of simple solutions.

In addition, we will use the superposition principle to obtain uniqueness of solutions of some PDEs.

2

Classification of Second-Order Multiplicative Partial Differential Equations

Definition 2.1 By a second-order multiplicative partial differential equation in n multiplicative variables $x_1, \ldots, x_n$, we mean any equation of the form

$$F(x_1, \ldots, x_n, u, u^*_{x_1}, \ldots, u^*_{x_n}, u^{**}_{x_1 x_1}, u^{**}_{x_1 x_2}, \ldots, u^{**}_{x_n x_n}) = 0_* \qquad (2.1)$$

or

$$F\left(x_1, \ldots, x_n, e^{x_1 \frac{u_{x_1}}{u}}, \ldots, e^{x_n \frac{u_{x_n}}{u}},\right.$$

$$e^{x_1 \left(\frac{u_{x_1}}{u} + x_1 \frac{u_{x_1 x_1} u - u_{x_1}^2}{u^2}\right)}, e^{x_1 x_2 \left(\frac{u_{x_1 x_2} u - u_{x_1} u_{x_2}}{u^2}\right)},$$

$$\left. \ldots, e^{x_n \left(\frac{u_{x_n}}{u} + x_n \frac{u_{x_n x_n} u - u_{x_n}^2}{u^2}\right)}\right) = 1.$$

Example 2.1 The equation

$$-_* \sum_{*i=1}^{n} u^{**}_{x_i x_i} = f(u)$$

will be called a multiplicative nonlinear Poisson equation. It is a second-order MPDE. In fact, we can rewrite it in the following manner:

$$
\begin{aligned}
f(u) &= -_* \sum_{*i=1}^{n} e^{x_i \left(\frac{u_{x_i}}{u} + x_i \frac{u_{x_i x_i} u - u_{x_i}^2}{u^2}\right)} \\
&= -_* e^{\sum_{i=1}^{n} x_i \left(\frac{u_{x_i}}{u} + x_i \frac{u_{x_i x_i} u - u_{x_i}^2}{u^2}\right)} \\
&= e^{-\sum_{i=1}^{n} x_i \left(\frac{u_{x_i}}{u} + x_i \frac{u_{x_i x_i} u - u_{x_i}^2}{u^2}\right)}.
\end{aligned}
$$

DOI: 10.1201/9781003440116-2

Example 2.2 The equation

$$u_{x_n}^* -_* \sum_{*i=1}^{n-1} (u^{\gamma*})_{x_i x_i}^{**} = 0_*$$

will be called multiplicative porous medium equation, where $\gamma > 0$ is a constant. It is a second-order MPDE. It can be rewritten in the form

$$0_* = 1$$

$$= e^{x_n \frac{u_{x_n}}{u}} -_* \sum_{*i=1}^{n-1} e^{x_i \left(\frac{\left(e^{(\log u)^{\gamma}} \right)_{x_i}}{\left(e^{(\log u)^{\gamma}} \right)} + x_i \frac{\left(e^{(\log u)^{\gamma}} \right)_{x_i x_i} \left(e^{(\log u)^{\gamma}} \right) - \left(e^{(\log u)^{\gamma}} \right)_{x_i}^2}{\left(e^{(\log u)^{2\gamma}} \right)} \right)}$$

$$= e^{x_n \frac{u_{x_n}}{u}} -_* e^{\sum_{i=1}^{n-1} x_i \left(\frac{\left(e^{(\log u)^{\gamma}} \right)_{x_i}}{\left(e^{(\log u)^{\gamma}} \right)} + x_i \frac{\left(e^{(\log u)^{\gamma}} \right)_{x_i x_i} \left(e^{(\log u)^{\gamma}} \right) - \left(e^{(\log u)^{\gamma}} \right)_{x_i}^2}{\left(e^{(\log u)^{2\gamma}} \right)} \right)}$$

$$= e^{x_n \frac{u_{x_n}}{u} - \sum_{i=1}^{n-1} x_i \left(\frac{\left(e^{(\log u)^{\gamma}} \right)_{x_i}}{\left(e^{(\log u)^{\gamma}} \right)} + x_i \frac{\left(e^{(\log u)^{\gamma}} \right)_{x_i x_i} \left(e^{(\log u)^{\gamma}} \right) - \left(e^{(\log u)^{\gamma}} \right)_{x_i}^2}{\left(e^{(\log u)^{2\gamma}} \right)} \right)},$$

or

$$x_n \frac{u_{x_n}}{u} - \sum_{i=1}^{n-1} x_i \left(\frac{\left(e^{(\log u)^{\gamma}} \right)_{x_i}}{\left(e^{(\log u)^{\gamma}} \right)} + x_i \frac{\left(e^{(\log u)^{\gamma}} \right)_{x_i x_i} \left(e^{(\log u)^{\gamma}} \right) - \left(e^{(\log u)^{\gamma}} \right)_{x_i}^2}{\left(e^{(\log u)^{2\gamma}} \right)} \right) = 0.$$

Definition 2.2 If equation (2.1) can be written in the form

$$\sum_{*i,j=1}^{n} a_{ij}(x_1, \ldots, x_n, u, u_{x_1}^*, \ldots, u_{x_n}^*) \cdot_* u_{x_i x_j}^{**} +_* \sum_{*i=1}^{n} b_i(x_1, \ldots, x_n, u) \cdot_* u_{x_i}^*$$

$$= f(x_1, \ldots, x_n, u, u_{x_1}^*, \ldots, u_{x_n}^*),$$

then we say that the equation is multiplicative quasilinear.

Example 2.3 The equation

$$u_{x_2}^* \cdot_* u_{x_1 x_1}^{**} -_* u_{x_1 x_2}^{**} \cdot_* u^{2*} = x_1^{2*} u^{2*}$$

is a multiplicative quasilinear second-order MPDE.

Definition 2.3 If equation (2.1) can be written in the form

$$\sum_{*i,j=1}^{n} a_{ij}(x_1, \ldots, x_n) \cdot_* u_{x_i x_j}^{**} + \sum_{*i=1}^{n} b_i(x_1, \ldots, x_n) \cdot_* u_{x_i}^*$$

$$= f(x_1, \ldots, x_n, u, u_{x_1}^*, \ldots, u_{x_n}^*),$$

then we say that the equation is multiplicative semilinear.

Example 2.4 The equation

$$u^{**}_{x_1x_1} -_* u^{**}_{x_1x_2} + u^{**}_{x_3x_3} = x_1^{2*} \cdot_* u^{4*}$$

is a multiplicative semiilinear second-order MPDE. It can be written in the form

$$x_1^{2*} \cdot_* u^{4*}$$

$$= e^{(\log x_1)^2} \cdot_* e^{(\log u)^4}$$

$$= e^{(\log x_1)^2 (\log u)^2}$$

$$= e^{x_1 \left(\frac{u_{x_1}}{u} + x_1 \frac{u_{x_1x_1} u - u_{x_1}^2}{u^2} \right)} -_* e^{x_1x_2 \left(\frac{u_{x_1x_2} u - u_{x_1} u_{x_2}}{u^2} \right)}$$

$$+_* e^{x_3 \left(\frac{u_{x_3}}{u} + x_3 \frac{u_{x_3x_3} u - u_{x_3}^2}{u^2} \right)}$$

$$= e^{x_1 \left(\frac{u_{x_1}}{u} + x_1 \frac{u_{x_1x_1} u - u_{x_1}^2}{u^2} \right) - x_1x_2 \left(\frac{u_{x_1x_2} u - u_{x_1} u_{x_2}}{u^2} \right) + x_3 \left(\frac{u_{x_3}}{u} + x_3 \frac{u_{x_3x_3} u - u_{x_3}^2}{u^2} \right)},$$

or

$$(\log x_1)^2 (\log u)^4$$

$$= x_1 \left(\frac{u_{x_1}}{u} + x_1 \frac{u_{x_1x_1} u - u_{x_1}^2}{u^2} \right) - x_1x_2 \left(\frac{u_{x_1x_2} u - u_{x_1} u_{x_2}}{u^2} \right)$$

$$+ x_3 \left(\frac{u_{x_3}}{u} + x_3 \frac{u_{x_3x_3} u - u_{x_3}^2}{u^2} \right).$$

Definition 2.4 If equation (2.1) can be written in the form

$$\sum_{*i,j=1}^{n} a_{ij}(x_1, \ldots, x_n) \cdot_* u^{**}_{x_ix_j} +_* \sum_{*i=1}^{n} b_i(x_1, \ldots, x_n) \cdot_* u^{*}_{x_i} = f(x_1, \ldots, x_n),$$

then we say that the equation is multiplicative linear. Moreover, if $f(x_1, \ldots, x_n) = 0$, then the equation is said to be multiplicative linear homogeneous second-order MPDE. Otherwise, the equation is said to be multiplicative linear nonhomogeneous second-order MPDE.

Example 2.5 The equation

$$\sum_{*i=1}^{n} u^{**}_{x_ix_i} = 0_*$$

is called multiplicative Laplace equation. It is a multiplicative linear homogeneous second-order MPDE. It can be rewritten in the form

$$0_* \quad = \quad 1$$

$$= \quad \sum_{*i=1}^{n} e^{x_i \left(\frac{u_{x_i}}{u} + x_i \frac{u_{x_i x_i} u - u_{x_i}^2}{u^2} \right)}$$

$$= \quad e^{\sum_{i=1}^{n} x_i \left(\frac{u_{x_i}}{u} + x_i \frac{u_{x_i x_i} u - u_{x_i}^2}{u^2} \right)},$$

or

$$\sum_{i=1}^{n} x_i \left(\frac{u_{x_i}}{u} + x_i \frac{u_{x_i x_i} u - u_{x_i}^2}{u^2} \right) = 0.$$

Example 2.6 The equation

$$u_{x_1 x_1}^{**} - {}_* \sum_{*i=2}^{n} u_{x_i x_i}^{**} = 0_*$$

will be called multiplicative wave equation. It is a multiplicative linear homogeneous second-order MPDE. It can be rewritten in the form

$$0_* \quad = \quad 1$$

$$= \quad e^{x_1 \left(\frac{u_{x_1}}{u} + x_1 \frac{u_{x_1 x_1} u - u_{x_1}^2}{u^2} \right)} - {}_* \sum_{*i=2}^{n} e^{x_i \left(\frac{u_{x_i}}{u} + x_i \frac{u_{x_i x_i} u - u_{x_i}^2}{u^2} \right)}$$

$$= \quad e^{x_1 \left(\frac{u_{x_1}}{u} + x_1 \frac{u_{x_1 x_1} u - u_{x_1}^2}{u^2} \right) - \sum_{i=2}^{n} x_i \left(\frac{u_{x_i}}{u} + x_i \frac{u_{x_i x_i} u - u_{x_i}^2}{u^2} \right)},$$

or

$$x_1 \left(\frac{u_{x_1}}{u} + x_1 \frac{u_{x_1 x_1} u - u_{x_1}^2}{u^2} \right) - \sum_{i=2}^{n} x_i \left(\frac{u_{x_i}}{u} + x_i \frac{u_{x_i x_i} u - u_{x_i}^2}{u^2} \right) = 0.$$

Example 2.7 The equation

$$u_{x_1}^* - \sum_{*i=2}^{n} u_{x_i x_i}^{**} = 0_*$$

will be called multiplicative heat (multiplicative diffusion) equation. It is a multiplicative linear homogeneous second-order MPDE. It can be rewritten in the form

$$0_* \quad = \quad 1$$

$$= \quad e^{x_1 \frac{u_{x_1}}{u}} - {}_* \sum_{*i=2}^{n} e^{x_i \left(\frac{u_{x_i}}{u} + x_i \frac{u_{x_i x_i} u - u_{x_i}^2}{u^2} \right)}$$

$$= \quad e^{x_1 \frac{u_{x_1}}{u}} - \sum_{i=2}^{n} x_i \left(\frac{u_{x_i}}{u} + x_i \frac{u_{x_i x_i} u - u_{x_i}^2}{u^2} \right),$$

or

$$x_1 \frac{u_{x_1}}{u} + \sum_{i=2}^{n} x_i \left(\frac{u_{x_i}}{u} + x_i \frac{u_{x_i x_i} u - u_{x_i}^2}{u^2} \right) = 0.$$

Example 2.8 The equation

$$u_{x_1}^* -_* \sum_{*i,j=2}^{n} a_{ij} \cdot_* u_{x_i x_j}^{**} +_* \sum_{*i=2}^{n} b_i \cdot_* u_{x_i}^* = x_1^{2*} +_* \cdots +_* x_n^{2*},$$

will be called multiplicative Kolmogorov equation, where a_{ij}, b_i, $1 \leq i,j \leq n$, are positive constants. It is a multiplicative linear nonhomogeneous second-order MPDE. It can be rewritten in the form

$$x_1^{2*} +_* \cdots +_* x_n^{2*} = e^{(\log x_1)^2} +_* \cdots +_* e^{(\log x_n)^2}$$

$$= e^{(\log x_1)^2 + \cdots + (\log x_n)^2}$$

$$= e^{x_1 \frac{u_{x_1}}{u}} -_* \sum_{*i,j=2,i\neq j}^{n} e^{x_i x_j \left(\frac{u_{x_i x_j} u - u_{x_i} u_{x_j}}{u^2} \right)}$$

$$-_* \sum_{*i=2}^{n} e^{x_i \left(\frac{u_{x_i}}{u} + x_i \frac{u_{x_i x_i} u - u_{x_i}^2}{u^2} \right)}$$

$$= e^{x_1 \frac{u_{x_1}}{u} - \sum_{i,j=2,i\neq j}^{n} x_i x_j \left(\frac{u_{x_i x_j} u - u_{x_i} u_{x_j}}{u^2} \right) - \sum_{i=2}^{n} x_i \left(\frac{u_{x_i}}{u} + x_i \frac{u_{x_i x_i} u - u_{x_i}^2}{u^2} \right)},$$

or

$$(\log x_1)^2 + \cdots + (\log x_n)^2$$

$$= x_1 \frac{u_{x_1}}{u} - \sum_{i,j=2,i\neq j}^{n} x_i x_j \left(\frac{u_{x_i x_j} u - u_{x_i} u_{x_j}}{u^2} \right) - \sum_{i=2}^{n} x_i \left(\frac{u_{x_i}}{u} + x_i \frac{u_{x_i x_i} u - u_{x_i}^2}{u^2} \right).$$

Definition 2.5 A second-order MPDE that is not multiplicative linear is said to be multiplicative nonlinear.

Example 2.9 The equation

$$u^{2*} \cdot_* u_{x_1}^{*3*} +_* u^{2*} \cdot_* u_{x_1 x_2}^{**} +_* u \cdot_* u_{x_2 x_2}^{**} = u^{3*} -_* u_{x_1}^*$$

is a multiplicative nonlinear second-order MPDE.

3

Classification and Canonical Forms for Multiplicative Linear and Multiplicative Quasilinear Second-Order Multiplicative Partial Differential Equations

3.1 Classification and Canonical Forms for Multiplicative Linear Second-Order Multiplicative Partial Differential Equations in Two Multiplicative Independent Multiplicative Variables

Let U be a domain in $\mathbb{R}_*^2$.

Definition 3.1 A multiplicative linear differential operator of second order for the function $u = u(x_1, x_2)$ is given by

$$L(u) = a \cdot_* u_{x_1 x_1}^{**} +_* e^2 \cdot_* b \cdot_* u_{x_1 x_2}^{**} +_* c \cdot_* u_{x_2 x_2}^{**},$$

where the coefficients a, b and c are supposed to be nonnegative continuously multiplicative differentiable and not simultaneously multiplicative vanishing functions of x_1 and x_2 in the domain U.

We consider the multiplicative differential operator

$$\tilde{L}(u) = L(u) +_* g(x_1, x_2, u, u_{x_1}^*, u_{x_2}^*) := L(u) +_* \cdots , \qquad (3.1)$$

where g is not necessarily multiplicative linear and does not contain second-order multiplicative derivatives.

Definition 3.2 The operator L is called the principal part of the operator $\tilde{L}$.

Our aim is to transform the operator (3.1) or the corresponding equation

$$L(u) +_* \cdots = 0_*$$

into a simple form, called the canonical form, by introducing new multiplicative independent multiplicative variables.

DOI: 10.1201/9781003440116-3

Let ξ_1 and ξ_2 be new multiplicative independent multiplicative variables which are connected with x_1 and x_2 in the following ways:

$$\xi_1 = \phi_1(x_1, x_2),$$

$$\xi_2 = \phi_2(x_1, x_2),$$

where $\phi_1, \phi_2 \in C^2_*(U)$. We will denote with $u(\xi_1, \xi_2)$ the transformed function $u(x_1, x_2)$ into the multiplicative variables ξ_1 and ξ_2. We have the following relations:

$$u^*_{x_1} = u^*_{\xi_1} \cdot_* \xi^*_{1x_1} +_* u^*_{\xi_2} \cdot_* \xi^*_{2x_1},$$

$$u^*_{x_2} = u^*_{\xi_1} \cdot_* \xi^*_{1x_2} +_* u^*_{\xi_2} \cdot_* \xi^*_{2x_2},$$

$$u^{**}_{x_1 x_1} = \left(u^*_{\xi_1} \cdot_* \xi^*_{1x_1} +_* u^*_{\xi_2} \cdot_* \xi^*_{2x_1}\right)^*_{x_1}$$

$$= \left(u^*_{\xi_1} \cdot_* \xi^*_{1x_1}\right)^*_{x_1} +_* \left(u^*_{\xi_2} \cdot_* \xi^*_{2x_1}\right)^*_{x_1}$$

$$= \left(u^*_{\xi_1}\right)^*_{x_1} \cdot_* \xi^*_{1x_1} +_* u^*_{\xi_1} \cdot_* \xi^{**}_{1x_1 x_1} +_* \left(u^*_{\xi_2}\right)^*_{x_1} \cdot_* \xi^*_{2x_1} +_* u^*_{\xi_2} \cdot_* \xi^{**}_{2x_1 x_1}$$

$$= \left(u^{**}_{\xi_1\xi_1} \cdot_* \xi^*_{1x_1} +_* u^{**}_{\xi_1\xi_2} \cdot_* \xi^*_{2x_1}\right) \cdot_* \xi^*_{1x_1} +_* u^*_{\xi_1} \cdot_* \xi^{**}_{1x_1 x_1}$$

$$+ \left(u^{**}_{\xi_1\xi_2} \cdot_* \xi^*_{1x_1} +_* u^{**}_{\xi_2\xi_2} \cdot_* \xi^*_{2x_1}\right) \cdot_* \xi^*_{2x_1} +_* u^*_{\xi_2} \cdot_* \xi^{**}_{2x_1 x_1}$$

$$= u^{**}_{\xi_1\xi_1} \cdot_* \left(\xi^*_{1x_1}\right)^{2*} +_* u^{**}_{\xi_1\xi_2} \cdot_* \xi^*_{1x_1} \cdot_* \xi^*_{2x_1} +_* u^*_{\xi_1} \cdot_* \xi^{**}_{1x_1 x_1}$$

$$+ u^{**}_{\xi_1\xi_2} \cdot_* \xi^*_{1x_1} \cdot_* \xi^*_{2x_1} +_* u^{**}_{\xi_2\xi_2} \cdot_* \left(\xi^*_{2x_1}\right)^{2*} +_* u^*_{\xi_2} \cdot_* \xi^{**}_{2x_1 x_1}$$

$$= u^{**}_{\xi_1\xi_1} \cdot_* \left(\xi^*_{1x_1}\right)^{2*} +_* e^2 \cdot_* u^{**}_{\xi_1\xi_2} \cdot_* \xi^*_{1x_1} \cdot_* \xi^*_{2x_1} +_* u^{**}_{\xi_2\xi_2} \cdot_* \left(\xi^*_{2x_1}\right)^{2*}$$

$$+ u^*_{\xi_1} \cdot_* \xi^{**}_{1x_1 x_1} +_* u^*_{\xi_2} \cdot_* \xi^{**}_{2x_1 x_1}$$

$$= u^{**}_{\xi_1\xi_1} \cdot_* \phi^{*2*}_{1x_1} +_* e^2 \cdot_* u^{**}_{\xi_1\xi_2} \cdot_* \phi^*_{1x_1} \cdot_* \phi^*_{2x_1} +_* u^{**}_{\xi_2\xi_2} \cdot_* \phi^{*2*}_{2x_1}$$

$$+_* u^*_{\xi_1} \cdot_* \phi^{**}_{1x_1 x_1} +_* u^*_{\xi_2} \cdot_* \phi^{**}_{2x_1 x_1},$$

$$u_{x_2x_2} = \left(u^*_{\xi_1} \cdot_* \xi^*_{1x_2} +_* u^*_{\xi_2} \cdot_* \xi^*_{2x_2} \right)^*_{x_2}$$

$$= \left(u^*_{\xi_1} \cdot_* \xi^*_{1x_2} \right)^*_{x_2} +_* \left(u^*_{\xi_2} \cdot_* \xi^*_{2x_2} \right)^*_{x_2}$$

$$= \left(u^*_{\xi_1} \right)^*_{x_2} \cdot_* \xi^*_{1x_2} +_* u^*_{\xi_1} \cdot_* \xi^{**}_{1x_2x_2} +_* \left(u^*_{\xi_2} \right)^*_{x_2} \cdot_* \xi^*_{2x_2} +_* u^*_{\xi_2} \cdot_* \xi^{**}_{2x_2x_2}$$

$$= \left(u^{**}_{\xi_1\xi_1} \cdot_* \xi^*_{1x_2} +_* u^{**}_{\xi_1\xi_2} \cdot_* \xi^*_{2x_2} \right) \cdot_* \xi^*_{1x_2} +_* u^*_{\xi_1} \cdot_* \xi^{**}_{1x_2x_2}$$

$$+ \left(u^{**}_{\xi_1\xi_2} \cdot_* \xi^*_{1x_2} +_* u^{**}_{\xi_2\xi_2} \cdot_* \xi^*_{2x_2} \right) \cdot_* \xi^*_{2x_2} +_* u^*_{\xi_2} \cdot_* \xi^{**}_{2x_2x_2}$$

$$= u^{**}_{\xi_1\xi_1} \cdot_* \left(\xi^*_{1x_2} \right)^{2*} +_* u^{**}_{\xi_1\xi_2} \cdot_* \xi^*_{1x_2} \cdot_* \xi^*_{2x_2} +_* u^*_{\xi_1} \cdot_* \xi^{**}_{1x_2x_2}$$

$$+ u^{**}_{\xi_1\xi_2} \cdot_* \xi^*_{1x_2} \cdot_* \xi^*_{2x_2} +_* u^{**}_{\xi_2\xi_2} \cdot_* \left(\xi^*_{2x_2} \right)^{2*} +_* u^*_{\xi_2} \cdot_* \xi^{**}_{2x_2x_2}$$

$$= u^{**}_{\xi_1\xi_1} \cdot_* \left(\xi^*_{1x_2} \right)^{2*} +_* e^2 \cdot_* u^{**}_{\xi_1\xi_2} \cdot_* \xi^*_{1x_2} \cdot_* \xi^*_{2x_2} +_* u^{**}_{\xi_2\xi_2} \cdot_* \left(\xi^*_{2x_2} \right)^{2*}$$

$$+_* u^*_{\xi_1} \cdot_* \xi^{**}_{1x_2x_2} +_* u^*_{\xi_2} \cdot_* \xi^{**}_{2x_2x_2}$$

$$= u^{**}_{\xi_1\xi_1} \cdot_* \phi^{*2*}_{1x_2} +_* e^2 \cdot_* u^{**}_{\xi_1\xi_2} \cdot_* \phi^*_{1x_2} \cdot_* \phi^*_{2x_2} +_* u^{**}_{\xi_2\xi_2} \cdot_* \phi^{*2*}_{2x_2}$$

$$+ u^*_{\xi_1} \cdot_* \phi^{**}_{1x_2x_2} +_* u^*_{\xi_2} \cdot_* \phi^{**}_{2x_2x_2},$$

$$u_{x_1x_2} = \left(u^*_{\xi_1} \cdot_* \xi^*_{1x_1} +_* u^*_{\xi_2} \cdot_* \xi^*_{2x_1} \right)^*_{x_2}$$

$$= \left(u^*_{\xi_1} \cdot_* \xi^*_{1x_1} \right)^*_{x_2} +_* \left(u^*_{\xi_2} \cdot_* \xi^*_{2x_1} \right)^*_{x_2}$$

$$= \left(u^*_{\xi_1} \right)^*_{x_2} \cdot_* \xi^*_{1x_1} +_* u^*_{\xi_1} \cdot_* \xi^{**}_{1x_1x_2} +_* \left(u^*_{\xi_2} \right)^*_{x_2} \cdot_* \xi^*_{2x_1} +_* u^*_{\xi_2} \cdot_* \xi^{**}_{2x_1x_2}$$

$$= \left(u^{**}_{\xi_1\xi_1} \cdot_* \xi^*_{1x_2} +_* u^{**}_{\xi_1\xi_2} \cdot_* \xi^*_{2x_2} \right) \cdot_* \xi^*_{1x_1} +_* u^*_{\xi_1} \cdot_* \xi^{**}_{1x_1x_2}$$

$$+ \left(u^{**}_{\xi_1\xi_2} \cdot_* \xi^*_{1x_2} +_* u^{**}_{\xi_2\xi_2} \cdot_* \xi^*_{2x_2} \right) \cdot_* \xi^*_{2x_1} +_* u^*_{\xi_2} \cdot_* \xi^{**}_{2x_1x_2}$$

$$= u^{**}_{\xi_1\xi_1} \cdot_* \xi^*_{1x_1} \cdot_* \xi^*_{1x_2} +_* u^{**}_{\xi_1\xi_2} \cdot_* \xi^*_{1x_1} \cdot_* \xi^*_{2x_2} +_* u^*_{\xi_1} \cdot_* \xi^{**}_{1x_1x_2}$$

$$+ u^{**}_{\xi_1\xi_2} \cdot_* \xi^*_{1x_2} \cdot_* \xi^*_{2x_1} +_* u^{**}_{\xi_2\xi_2} \cdot_* \xi^*_{2x_1} \cdot_* \xi^*_{2x_2} +_* u^*_{\xi_2} \cdot_* \xi^{**}_{2x_1x_2}$$

$$= \ u^{**}_{\xi_1\xi_1} \cdot_* \phi^*_{1x_1} \cdot_* \phi^*_{1x_2} +_* u^{**}_{\xi_1\xi_2} \cdot_* \left(\phi^*_{1x_1} \cdot_* \phi^*_{2x_2} +_* \phi^*_{1x_2} \cdot_* \phi^*_{2x_1} \right)$$

$$+ u^{**}_{\xi_2\xi_2} \cdot_* \phi^*_{2x_1} \cdot_* \phi^*_{2x_2} +_* u^*_{\xi_1} \cdot_* \phi^{**}_{1x_1x_2} +_* u^*_{\xi_2} \cdot_* \phi^{**}_{2x_1x_2}.$$

From here,

$$L(u) = a \cdot_* u^{**}_{x_1x_1} +_* e^2 \cdot_* b \cdot_* u^{**}_{x_1x_2} +_* c \cdot_* u^{**}_{x_2x_2}$$

$$= a \cdot_* \left(u^{**}_{\xi_1\xi_1} \cdot_* \phi^{*2*}_{1x_1} +_* e^2 \cdot_* u^{**}_{\xi_1\xi_2} \cdot_* \phi^*_{1x_1} \cdot_* \phi^*_{2x_1} +_* u^{**}_{\xi_2\xi_2} \cdot_* \phi^{*2*}_{2x_1} \right.$$

$$\left. +_* u^*_{\xi_1} \cdot_* \phi^{**}_{1x_1x_1} +_* u^*_{\xi_2} \cdot_* \phi^{**}_{2x_1x_1} \right)$$

$$+_* e^2 \cdot_* b \cdot_* \left(u^{**}_{\xi_1\xi_1} \cdot_* \phi^*_{1x_1} \cdot_* \phi^*_{1x_2} +_* u^{**}_{\xi_1\xi_2} \cdot_* \right.$$

$$\left(\phi^*_{1x_1} \cdot_* \phi^*_{2x_2} +_* \phi^*_{1x_2} \cdot_* \phi^*_{2x_1} \right)$$

$$\left. +_* u^{**}_{\xi_2\xi_2} \cdot_* \phi^*_{2x_1} \cdot_* \phi^*_{2x_2} +_* u^*_{\xi_1} \cdot_* \phi^{**}_{1x_1x_2} +_* u^*_{\xi_2} \cdot_* \phi^{**}_{2x_1x_2} \right)$$

$$+_* c \cdot_* \left(u^{**}_{\xi_1\xi_1} \cdot_* \phi^{*2*}_{1x_2} +_* e^2 \cdot_* u^{**}_{\xi_1\xi_2} \cdot_* \phi^*_{1x_2} \cdot_* \phi^*_{2x_2} +_* u^{**}_{\xi_2\xi_2} \cdot_* \phi^{*2*}_{2x_2} \right.$$

$$\left. +_* u^*_{\xi_1} \cdot_* \phi^{**}_{1x_2x_2} +_* u^*_{\xi_2} \cdot_* \phi^{**}_{2x_2x_2} \right)$$

$$= u^{**}_{\xi_1\xi_1} \cdot_* \left(a \cdot_* \phi^{*2*}_{1x_1} +_* e^2 \cdot_* b \cdot_* \phi^*_{1x_1} \cdot_* \phi^*_{1x_2} +_* c \cdot_* \phi^{*2*}_{1x_2} \right)$$

$$+_* e^2 \cdot_* u^{**}_{\xi_1\xi_2} \cdot_* \left(a \cdot_* \phi^*_{1x_1} \cdot_* \phi^*_{2x_1} +_* b \cdot_* \left(\phi^*_{1x_2} \cdot_* \phi^*_{2x_1} +_* \phi^*_{1x_1} \cdot_* \phi^*_{2x_2} \right) \right.$$

$$\left. +_* c \cdot_* \phi^*_{1x_2} \cdot_* \phi^*_{2x_2} \right)$$

$$+_* u^{**}_{\xi_2\xi_2} \cdot_* \left(a \cdot_* \phi^{*2*}_{2x_1} +_* e^2 \cdot_* b \cdot_* \phi^*_{2x_1} \cdot_* \phi^*_{2x_2} +_* c \cdot_* \phi^{*2*}_{2x_2} \right)$$

$$+_* u^*_{\xi_1} \cdot_* \left(a \cdot_* \phi^{**}_{1x_1x_1} +_* e^2 \cdot_* b \cdot_* \phi^{**}_{1x_1x_2} +_* c \cdot_* \phi^{**}_{1x_2x_2} \right)$$

$$+_* u^*_{\xi_2} \cdot_* \left(a \cdot_* \phi^{**}_{2x_1x_1} +_* e^2 \cdot_* b \cdot_* \phi^{**}_{2x_1x_2} +_* c \cdot_* \phi^{**}_{2x_2x_2} \right).$$

Let

$$\alpha = a \cdot_* \phi_{1x_1}^{*2*} +_* e^2 \cdot_* b \cdot_* \phi_{1x_1}^* \cdot_* \phi_{1x_2}^* +_* c \cdot_* \phi_{1x_2}^{*2*},$$

$$\beta = a \cdot_* \phi_{1x_1}^* \cdot_* \phi_{2x_1}^* +_* b \cdot_* \left(\phi_{1x_2}^* \cdot_* \phi_{2x_1}^* +_* \phi_{1x_1}^* \cdot_* \phi_{2x_2}^*\right) +_* c \cdot_* \phi_{1x_2}^* \cdot_* \phi_{2x_2}^*,$$

$$\gamma = a \cdot_* \phi_{2x_1}^{*2*} +_* e^2 \cdot_* b \cdot_* \phi_{2x_1}^* \cdot_* \phi_{2x_2}^* +_* c \cdot_* \phi_{2x_2}^{*2*},$$

$$\alpha_1 = a \cdot_* \phi_{1x_1x_1}^{**} +_* e^2 \cdot_* b \cdot_* \phi_{1x_1x_2}^{**} +_* c \cdot_* \phi_{1x_2x_2}^{**},$$

$$\gamma_1 = a \cdot_* \phi_{2x_1x_1}^{**} +_* e^2 \cdot_* b \cdot_* \phi_{2x_1x_2}^{**} +_* c \cdot_* \phi_{2x_2x_2}^{**}.$$

Then the multiplicative differential operator L assumes the following form:

$$L(u) = \alpha \cdot_* u_{\xi_1\xi_1}^{**} +_* e^2 \cdot_* \beta \cdot_* u_{\xi_1\xi_2}^{**} +_* \gamma \cdot_* u_{\xi_2\xi_2}^{**} +_* \alpha_1 \cdot_* u_{\xi_1}^* +_* \gamma_1 \cdot_* u_{\xi_2}^*,$$

which is called the canonical form of the operator L. We set

$$\Lambda(u) = \alpha \cdot_* u_{\xi_1\xi_1}^{**} +_* e^2 \cdot_* \beta \cdot_* u_{\xi_1\xi_2}^{**} +_* \gamma \cdot_* u_{\xi_2\xi_2}^{**}.$$

Proposition 3.1 *a, b, c and α, β, γ are related as follows:*

$$\alpha \cdot_* \gamma -_* \beta^{2*} = (a \cdot_* c -_* b^{2*}) \cdot_* \left(\phi_{1x_1}^* \cdot_* \phi_{2x_2}^* -_* \phi_{1x_2}^* \cdot_* \phi_{2x_1}^*\right)^{2*} \qquad (3.2)$$

Proof *We have*

$$\alpha \cdot_* \gamma -_* \beta^{2*} = \left(a \cdot_* \phi_{1x_1}^{*2*} +_* e^2 \cdot_* b \cdot_* \phi_{1x_1}^* \cdot_* \phi_{1x_2}^* +_* c \cdot_* \phi_{1x_2}^{*2*}\right)$$

$$\cdot_* \left(a \cdot_* \phi_{2x_1}^{*2*} +_* e^2 \cdot_* b \cdot_* \phi_{2x_1}^* \cdot_* \phi_{2x_2}^* +_* c \cdot_* \phi_{2x_2}^{*2*}\right)$$

$$-_* \left(a \cdot_* \phi_{1x_1}^* \cdot_* \phi_{2x_1}^* +_* b \cdot_* \left(\phi_{1x_2}^* \cdot_* \phi_{2x_1}^* +_* \phi_{1x_1}^* \cdot_* \phi_{2x_2}^*\right)\right.$$

$$\left. +_* c \cdot_* \phi_{1x_2}^* \cdot_* \phi_{2x_2}^*\right)^{2*}$$

$$= a^{2*} \cdot_* \phi_{1x_1}^{*2*} \cdot_* \phi_{2x_1}^{*2*} +_* e^2 \cdot_* a \cdot_* b \cdot_* \phi_{1x_1}^{*2*} \cdot_* \phi_{2x_1}^* \cdot_* \phi_{2x_2}^*$$

$$+_* a \cdot_* c \cdot_* \phi_{1x_1}^{*2*} \cdot_* \phi_{2x_2}^{*2*} +_* e^2 \cdot_* a \cdot_* b \cdot_* \phi_{1x_1}^* \cdot_* \phi_{1x_2}^* \cdot_* \phi_{2x_1}^{*2*}$$

$$+_* e^4 \cdot_* b^{2*} \cdot_* \phi_{1x_1}^* \cdot_* \phi_{1x_2}^* \cdot_* \phi_{2x_1}^* \cdot_* \phi_{2x_2}^* +_* e^2 \cdot_* b \cdot_* c \cdot_* \phi_{1x_1}^*$$

$$\cdot_* \phi_{1x_2}^* \cdot_* \phi_{2x_2}^{*2*} +_* a \cdot_* c \cdot_* \phi_{1x_2}^{*2*} \cdot_* \phi_{2x_1}^{*2*} +_* e^2 \cdot_* b \cdot_* c \cdot_* \phi_{1x_2}^{*2*} \cdot_* \phi_{2x_1}^* \cdot_* \phi_{2x_2}^*$$

$$+_*c^{2*}\cdot_*\phi_{1x_2}^{*2*}\cdot_*\phi_{2x_2}^{*2*}-_*a^{2*}\cdot_*\phi_{1x_1}^{*2*}\cdot_*\phi_{2x_1}^{*2*}-_*b^{2*}\cdot_*$$

$$\left(\phi_{1x_2}^{*2*}\cdot_*\phi_{2x_1}^{*2*}+_*e^2\cdot_*\phi_{1x_1}^{*}\cdot_*\phi_{1x_2}^{*}\cdot_*\phi_{2x_1}^{*}\cdot_*\phi_{2x_2}^{*}\cdot_*+_*\phi_{1x_1}^{*2*}\cdot_*\phi_{2x_2}^{*2*}\right)$$

$$-_*c^{2*}\cdot_*\phi_{1x_2}^{*2*}\cdot_*\phi_{2x_2}^{*2*}-_*e^2\cdot_*a\cdot_*b\cdot_*$$

$$\left(\phi_{1x_1}^{*}\cdot_*\phi_{1x_2}^{*}\cdot_*\phi_{2x_1}^{*2*}+_*\phi_{1x_1}^{*2*}\cdot_*\phi_{2x_1}^{*}\cdot_*\phi_{2x_2}^{*}\right)$$

$$-_*e^2\cdot_*a\cdot_*c\cdot_*\phi_{1x_1}^{*}\cdot_*\phi_{1x_2}^{*}\cdot_*\phi_{2x_1}^{*}\cdot_*\phi_{2x_2}^{*}-_*e^2\cdot_*b\cdot_*c\cdot_*$$

$$\left(\phi_{1x_2}^{*2*}\cdot_*\phi_{2x_1}^{*}\cdot_*\phi_{2x_2}^{*}+_*\phi_{1x_1}^{*}\cdot_*\phi_{1x_2}^{*}\cdot_*\phi_{2x_2}^{*2*}\right)$$

$$=\quad a\cdot_*c\cdot_*\left(\phi_{1x_1}^{*2*}\cdot_*\phi_{2x_2}^{*2*}+_*\phi_{1x_2}^{*2*}\cdot_*\phi_{2x_1}^{*2*}-_*e^2\cdot_*\phi_{1x_1}^{*}\cdot_*\phi_{1x_2}^{*}\cdot_*\phi_{2x_1}^{*}\cdot_*\phi_{2x_2}^{*}\right)$$

$$-_*b^{2*}\cdot_*\left(\phi_{1x_2}^{*2*}\cdot_*\phi_{2x_1}^{*2*}\cdot_*-_*e^2\cdot_*\phi_{1x_1}^{*}\cdot_*\phi_{1x_2}^{*}\cdot_*\phi_{2x_1}^{*}\cdot_*\phi_{2x_2}^{*}\right.$$

$$\left.+_*\phi_{1x_1}^{*2*}\cdot_*\phi_{2x_2}^{*2*}\right)$$

$$=\quad a\cdot_*c\cdot_*\left(\phi_{1x_1}^{*}\cdot_*\phi_{2x_2}^{*}-_*\phi_{1x_2}^{*}\cdot_*\phi_{2x_1}^{*}\right)^{2*}$$

$$-_*b^{2*}\cdot_*\left(\phi_{1x_1}^{*}\cdot_*\phi_{2x_2}^{*}-_*\phi_{1x_2}^{*}\cdot_*\phi_{2x_1}^{*}\right)^{2*}$$

$$=\quad\left(a\cdot_*c-_*b^{2*}\right)\cdot_*\left(\phi_{1x_1}^{*}\cdot_*\phi_{2x_2}^{*}-_*\phi_{1x_2}^{*}\cdot_*\phi_{2x_1}^{*}\right)^{2*},$$

which completes the proof.

Proposition 3.2 *(Identity for the Characteristic Multiplicative Quadratic Form) We have*

$$a\cdot_*l^{2*}+_*e^2\cdot_*b\cdot_*l\cdot_*m+_*c\cdot_*m^{2*}=\alpha\cdot_*\lambda^{2*}+_*e^2\cdot_*\beta\cdot_*\lambda\cdot_*\mu+_*\gamma\cdot_*\mu^{2*},$$

where

$$l\quad=\quad\lambda\cdot_*\phi_{1x_1}j+_*\mu\cdot_*\phi_{1x_2}^{*},$$

$$m\quad=\quad\lambda\cdot_*\phi_{2x_1}^{*}+_*\mu\cdot_*\phi_{2x_2}^{*}.$$

Proof *We have*

$$a\cdot_*l^{2*}+_*e^2\cdot_*b\cdot_*l\cdot_*m+_*c\cdot_*m^{2*}=a\cdot_*\left(\lambda\cdot_*\phi_{1x_1}^{*}+_*\mu\cdot_*\phi_{1x_2}^{*}\right)^{2*}$$

$$+_*e^2\cdot_*b\cdot_*\left(\lambda\cdot_*\phi_{1x_1}^{*}+_*\mu\cdot_*\phi_{1x_2}^{*}\right)\cdot_*\left(\lambda\cdot_*\phi_{2x_1}^{*}+_*\mu\cdot_*\phi_{2x_2}^{*}\right)$$

$$+_*c\cdot_*\left(\lambda\cdot_*\phi_{2x_1}^{*}+_*\mu\cdot_*\phi_{2x_2}^{*}\right)^{2*}$$

$$= \; a \cdot_* \left(\lambda^{*2*} \cdot_* \phi_{1x_1}^{*2*} +_* e^2 \cdot_* \lambda \cdot_* \mu \cdot_* \phi_{1x_1}^* \cdot_* \phi_{1x_2}^* +_* \mu^{2*} \cdot_* \phi_{1x_2}^{*2*} \right)$$

$$+_* e^2 \cdot_* b \cdot_* \left(\lambda^{2*} \cdot_* \phi_{1x_1}^* \cdot_* \phi_{2x_1}^* +_* \lambda \cdot_* \mu \cdot_* \phi_{1x_2}^* \cdot_* \phi_{2x_1}^* \right.$$

$$+_* \lambda \cdot_* \mu \cdot_* \phi_{1x_1}^* \cdot_* \phi_{2x_2}^* +_* \mu^{2*} \cdot_* \phi_{1x_2}^* \cdot_* \phi_{2x_2}^* \left. \right)$$

$$+_* c \cdot_* \left(\lambda^{2*} \cdot_* \phi_{2x_1}^{*2*} +_* e^2 \cdot_* \lambda \cdot_* \mu \cdot_* \phi_{2x_1}^* \cdot_* \phi_{2x_2}^* +_* \mu^{2*} \cdot_* \phi_{2x_2}^{*2*} \right) .$$

Also,

$$\alpha \cdot_* \lambda^{2*} +_* e^2 \cdot_* \beta \cdot_* \lambda \cdot_* \mu +_* \gamma \cdot_* \mu^{2*}$$

$$= \; \lambda^{2*} \cdot_* \left(a \cdot_* \phi_{1x_1}^{*2*} +_* e^2 \cdot_* b \cdot_* \phi_{1x_1}^* \cdot_* \phi_{1x_2}^* +_* c \cdot_* \phi_{1x_2}^{*2*} \right)$$

$$+_* e^2 \cdot_* \left(a \cdot_* \phi_{1x_1}^* \cdot_* \phi_{2x_1}^* +_* b \cdot_* \left(\phi_{1x_2}^* \cdot_* \phi_{2x_1}^* +_* \phi_{1x_1}^* \cdot_* \phi_{2x_2}^* \right) \right.$$

$$+_* c \cdot_* \phi_{1x_2}^* \cdot_* \phi_{2x_2}^* \left. \right) \cdot_* \lambda \cdot_* \mu$$

$$+_* \left(a \cdot_* \phi_{2x_1}^{*2*} +_* e^2 \cdot_* b \cdot_* \phi_{2x_1}^* \cdot_* \phi_{2x_2}^* +_* c \cdot_* \phi_{2x_2}^{*2*} \right) \cdot_* \mu^{2*}$$

$$= \; a \cdot_* \left(\lambda^{2*} \cdot_* \phi_{1x_1}^{*2*} +_* e^2 \cdot_* \lambda \cdot_* \mu \cdot_* \phi_{1x_1}^* \cdot_* \phi_{2x_1}^* +_* \mu^{2*} \cdot_* \phi_{2x_1}^{*2*} \right)$$

$$+_* e^2 \cdot_* b \cdot_* \left(\lambda^{2*} \cdot_* \phi_{1x_1}^* \cdot_* \phi_{1x_2}^* +_* \lambda \cdot_* \mu \cdot_* \left(\phi_{1x_2}^* \cdot_* \phi_{2x_1}^* +_* \phi_{1x_1}^* \cdot_* \phi_{2x_2}^* \right) \right.$$

$$+_* \mu^{2*} \cdot_* \phi_{2x_1}^* \cdot_* \phi_{2x_2}^* \left. \right)$$

$$+_* c \cdot_* \left(\lambda^{2*} \cdot_* \phi_{1x_2}^{*2*} +_* e^2 \cdot_* \lambda \cdot_* \mu \cdot_* \phi_{1x_2}^* \cdot_* \phi_{2x_2}^* +_* \mu^{2*} \cdot_* \phi_{2x_2}^{*2*} \right) ,$$

which completes the proof.

We will impose two conditions on the transformed coefficients α, β and γ so that to obtain a simple canonical form $\Lambda(u)$. We consider the following cases:

1. $\alpha = -_* \gamma$, $\beta = 0_*$ or $\alpha = \gamma = 0_*$.
2. $\alpha = \gamma$, $\beta = 0_*$.
3. $\beta = \gamma = 0_*$.

The transformations ϕ_1 and ϕ_2 satisfy one of the above cases. This depends on the algebraic character of the characteristic quadratic form:

$$Q(l, m) \; = \; a \cdot_* l^{2*} +_* e^2 \cdot_* b \cdot_* l \cdot_* m +_* c \cdot_* m^{2*}$$

$$= \; \alpha \cdot_* \lambda^{2*} +_* e^2 \cdot_* \beta \cdot_* \lambda \cdot_* \mu +_* \gamma \cdot_* \mu^{2*}.$$

Geometrically speaking, this depends on the character of the quadratic curve in the l, m-multiplicative plane, i.e., for fixed x_1 and x_2 such that $Q(l, m) = e$; this curve may be a multiplicative hyperbola, multiplicative ellipse or multiplicative parabola. From here and equation (3.2), we get to the following definition.

Definition 3.3 At a point (x_1, x_2), the operator $L(u)$ will be called as follows:

1. Multiplicative hyperbolic, if $a \cdot_* c -_* b^{2*} < 1$ or

$$1 > e^{\log a \log c} -_* e^{(\log b)^2}$$

$$= e^{\log a \log c - \log b^2},$$

or

$$\log a \log c - (\log b)^2 < 0.$$

2. Multiplicative elliptic, if $a \cdot_* c -_* b^{2*} > 1$, or

$$\log a \log c - (\log b)^2 > 0.$$

3. Multiplicative parabolic, if $a \cdot_* c -_* b^{2*} = 1$, or

$$\log a \log c - (\log b)^2 = 0.$$

Remark 3.1 Note that there are cases in which a second-order multiplicative linear multiplicative differential operator is multiplicative parabolic and at the same time the corresponding classical analogue is hyperbolic. We will see this in the following example.

Example 3.1 Consider the operator

$$L(u)(x_1, x_2) = u_{x_1 x_1}^{**} -_* e^2 \cdot_* u_{x_1 x_2}^{**} +_* u_{x_2 x_2}^{**}, \quad (x_1, x_2) \in \mathbb{R}_*^2. \quad (3.3)$$

Its classical analogue is

$$L_1(u) = u_{x_1 x_1} - e^2 u_{x_1 x_2} + u_{x_2 x_2}, \quad (x_1, x_2) \in \mathbb{R}^2. \quad (3.4)$$

Let

$$a(x_1, x_2) = e,$$

$$b(x_1, x_2) = e,$$

$$c(x_1, x_2) = e, \quad (x_1, x_2) \in \mathbb{R}_*^2,$$

and

$$a_1(x_1, x_2) \quad = \quad 1,$$

$$b_1(x_1, x_2) \quad = \quad \frac{e^2}{2},$$

$$c_1(x_1, x_2) \quad = \quad 1, \quad (x_1, x_2) \in \mathbb{R}^2.$$

Then,

$$\log(a(x_1, x_2)) \log(c(x_1, x_2)) - (\log(b(x_1, x_2)))^2 \quad = \quad \log e \log e - (\log e)^2$$

$$= \quad 1 - 1$$

$$= \quad 0, \quad (x_1, x_2) \in \mathbb{R}^2_*,$$

and

$$a_1(x_1, x_2) c_1(x_1, x_2) - (b_1(x_1, x_2))^2 \quad = \quad 1 \cdot 1 - \left(\frac{e^2}{2}\right)^2$$

$$= \quad 1 - \frac{e^4}{4}$$

$$< \quad 0, \quad (x_1, x_2) \in \mathbb{R}^2.$$

Thus, L is multiplicative parabolic in $\mathbb{R}^2_*$ and L_1 is hyperbolic in $\mathbb{R}^2$.

Example 3.2 Let

$$L(u) = e^{-*x_1} \cdot_* u^{**}_{x_1 x_1} +_* e^4 \cdot_* e^{-*(x_1 +_* x_2)/_* e^2} \cdot_* u^{**}_{x_1 x_2}$$

$$+_* e^{-*x_2} \cdot_* u^{**}_{x_2 x_2} +_* u^{*}_{x_1} +_* u^{*}_{x_2},$$

$(x_1, x_2) \in \mathbb{R}^2_*$. Here

$$a(x_1, x_2) \quad = \quad e^{-*x_1}$$

$$= \quad e^{\frac{1}{x_1}},$$

$$b(x_1, x_2) \quad = \quad e^4 \cdot_* e^{-*(x_1 +_* x_2)/_* e^2}$$

$$= \quad e^4 \cdot_* e^{-*((x_1 x_2)/_* e^2}$$

$$= e^4 \cdot_* e^{-}_* e^{\frac{\log(x_1 x_2)}{\log e^2}}$$

$$= e^4 \cdot_* e^{-}_* a^{\frac{\log(x_1 x_2)}{?}}$$

$$= e^4 \cdot_* e^{e^{-\frac{\log(x_1 x_2)}{2}}}$$

$$= e^{\log e^4 \log e^{e^{-\frac{\log(x_1 x_2)}{2}}}}$$

$$= e^{4e^{-\frac{\log(x_1 x_2)}{2}}},$$

$$c(x_1, x_2) \quad - \quad e^{-}_* x_2$$

$$= e^{\frac{1}{x_2}}, \quad (x_1, x_2) \in \mathbb{R}^2_*.$$

Then,

$$\log(a(x_1, x_2)) \log(c(x_1, x_2)) - (\log(b(x_1, x_2)))^2$$

$$= \log\left(e^{\frac{1}{x_1}}\right) \log\left(e^{\frac{1}{x_2}}\right) - \left(\log\left(e^{4e^{-\frac{\log(x_1 x_2)}{2}}}\right)\right)$$

$$= \frac{1}{x_1 x_2} - \left(4e^{-\frac{\log(x_1 x_2)}{2}}\right)^2$$

$$= \frac{1}{x_1 x_2} - 16e^{-2\frac{\log(x_1 x_2)}{2}}$$

$$= \frac{1}{x_1 x_2} - 16e^{-\log(x_1 x_2)}$$

$$= \frac{1}{x_1 x_2} - \frac{16}{x_1 x_2}$$

$$= -\frac{15}{x_1 x_2}$$

$$< 0, \quad (x_1, x_2) \in \mathbb{R}^2_*,$$

because $x_1 > 1$, $x_2 > 1$. Thus the considered operator is multiplicative hyperbolic.

Example 3.3 Let

$$L(u) = u^{**}_{x_1 x_1} +_* \sin_* x_1 \cdot_* \cos_* x_2 \cdot_* u^{**}_{x_1 x_2} +_* e^4 \cdot_* u^{**}_{x_2 x_2},$$

$(x_1, x_2) \in \mathbb{R}_*^2$. Here

$$
\begin{aligned}
a(x_1, x_2) &= e, \\[2mm]
b(x_1, x_2) &= e^{-2} \cdot_* \sin_* x_1 \cdot_* \cos_* x_2 \\[2mm]
&= e^{-2} \cdot_* e^{\sin(\log x_1)} \cdot_* e^{\cos(\log x_2)} \\[2mm]
&= e^{\log(e^{-2}) \log(e^{\sin(\log x_1)}) \log(e^{\log(\cos x_2)})} \\[2mm]
&= e^{\frac{\sin(\log x_1)\,\cos(\log x_2)}{2}}, \\[2mm]
c(x_1, x_2) &= e^4, \quad (x_1, x_2) \in \mathbb{R}_*^2.
\end{aligned}
$$

Then,

$$
\begin{aligned}
&\log(a(x_1, x_2)) \log(c(x_1, x_2)) - (\log(b(x_1, x_2)))^2 \\[2mm]
&= \log e \log\left(e^4\right) - \left(\log\left(e^{\frac{\sin(\log x_1)\,\cos(\log x_2)}{2}}\right)\right)^2 \\[2mm]
&= 4 - \left(\frac{\sin(\log x_1)\,\cos(\log x_2)}{2}\right)^2 \\[2mm]
&= 4 - \frac{(\sin(\log x_1))^2 (\cos(\log x_2))^2}{4} \\[2mm]
&> 0, \quad (x_1, x_2) \in \mathbb{R}_*^2.
\end{aligned}
$$

Hence, the operator L is multiplicative elliptic.

Example 3.4 Let

$$
\begin{aligned}
L(u) = {}&x_1^{2*} \cdot_* u_{x_1 x_1}^{**} +_* e^2 \cdot_* x_1 \cdot_* x_2 \cdot_* u_{x_1 x_2}^{**} +_* x_2^{2*} \cdot_* u_{x_2 x_2}^{**} \\
&+_* e^4 \cdot_* u_{x_1}^* -_* u_{x_2}^*,
\end{aligned}
$$

$(x_1, x_2) \in \mathbb{R}_*^2$. Here

$$
\begin{aligned}
a(x_1, x_2) &= x_1^{2*} \\[2mm]
&= e^{(\log x_1)^2}, \\[2mm]
b(x_1, x_2) &= x_1 \cdot_* x_2 \\[2mm]
&= e^{\log x_1 \, \log x_2}, \\[2mm]
c(x_1, x_2) &= x_2^{2*} \\[2mm]
&= e^{(\log x_2)^2}, \quad (x_1, x_2) \in \mathbb{R}_*^2.
\end{aligned}
$$

Then,

$$\log(a(x_1, x_2)) \log(c(x_1, x_2)) - (\log(b(x_1, x_2)))^2$$

$$= \log\left(e^{(\log x_1)^2}\right) \log\left(e^{(\log x_2)^2}\right) - \left(\log\left(e^{\log x_1 \log x_2}\right)\right)^2$$

$$= (\log x_1)^2 (\log x_2)^2 - (\log x_1 \log x_2)^2$$

$$= 0, \quad (x_1, x_2) \in \mathbb{R}_*^2.$$

Therefore, the operator L is multiplicative parabolic.

Exercise 3.1 Determine the operator L:

1. $L(u) = e^2 \cdot_* u_{x_1 x_1}^{**} -_* e^3 \cdot_* u_{x_1 x_2}^{**} +_* e^4 \cdot_* u_{x_2 x_2}^{**} +_* u_{x_1}^* -_* e^2 \cdot_* u_{x_2}^*,$
   $(x_1, x_2) \in \mathbb{R}_*^2.$

2. $L(u) = u_{x_1 x_1}^{**} -_* e^2 \cdot_* u_{x_1 x_2}^{**} +_* e^3 \cdot_* u_{x_2 x_2}^{**} + u, \quad (x_1, x_2) \in \mathbb{R}_*^2.$

3. $L(u) = u_{x_1 x_1}^{**} +_* e^4 \cdot_* u_{x_1 x_2}^{**} +_* u_{x_2 x_2}^{**}, \quad (x_1, x_2) \in \mathbb{R}_*^2.$

4. $L(u) = e^{-2} \cdot_* u_{x_1 x_1}^{**} +_* u_{x_1 x_2}^{**} +_* e^3 \cdot_* u_{x_2 x_2}^{**}, \quad (x_1, x_2) \in \mathbb{R}_*^2.$

5. $L(u) = e^{-3} \cdot_* u_{x_1 x_1}^{**} +_* e^2 \cdot_* u_{x_1 x_2}^{**} -_* e^4 \cdot_* u_{x_2 x_2}^{**} -_* u_{x_1}^*,$
   $(x_1, x_2) \in \mathbb{R}_*^2.$

6. $L(u) = u_{x_1 x_1}^{**} -_* u_{x_1 x_2}^{**} +_* u_{x_2 x_2}^{**}, \quad (x_1, x_2) \in \mathbb{R}_*^2.$

7. $L(u) = e^2 \cdot_* u_{x_1 x_1}^{**} +_* e^8 \cdot_* u_{x_1 x_2}^{**} +_* e^8 \cdot_* u_{x_2 x_2}^{**}, \quad (x_1, x_2) \in \mathbb{R}_*^2.$

Answer   *1. Multiplicative elliptic.*

  2. Multiplicative elliptic.

  3. Multiplicative hyperbolic.

  4. Multiplicative hyperbolic.

  5. Multiplicative elliptic.

  6. Multiplicative elliptic.

  7. Multiplicative parabolic.

Example 3.5 Let us consider the operator

$$L(u) = x_1 \cdot_* u_{x_1 x_1}^{**} -_* e^2 \cdot_* u_{x_1 x_2}^{**} +_* x_2 \cdot_* u_{x_2 x_2}^{**} +_* u_{x_1}^*, \quad (x_1, x_2) \in \mathbb{R}_*^2.$$

Here

$$a(x_1, x_2) = x_1,$$

$$b(x_1, x_2) = e^{-1},$$

$$c(x_1, x_2) = x_2, \quad (x_1, x_2) \in \mathbb{R}_*^2.$$

Then,

$$\log(a(x_1, x_2)) \log(c(x_1, x_2)) - (\log(b(x_1, x_2)))^2$$

$$= \log x_1 \log x_2 - (\log(e^{-1}))^2$$

$$= \log x_1 \log x_2 - (-1)^2$$

$$= \log x_1 \log x_2 - 1, \quad (x_1, x_2) \in \mathbb{R}_*^2.$$

Thus, we have the following cases:

1. If
$$\log x_1 \log x_2 < 1, \quad (x_1, x_2) \in \mathbb{R}_*^2,$$
then the considered equation is multiplicative hyperbolic.

2. If
$$\log x_1 \log x_2 = 1, \quad (x_1, x_2) \in \mathbb{R}_*^2,$$
then the considered equation is multiplicative parabolic.

3. If
$$\log x_1 \log x_2 > 1, \quad (x_1, x_2) \in \mathbb{R}_*^2,$$
then the considered equation is multiplicative elliptic.

Example 3.6 Consider the multiplicative analogue of the classical Tricomi operator:

$$L(u) = u_{x_1 x_1}^{**} +_* x_1 \cdot_* u_{x_2 x_2}^{**}, \quad (x_1, x_2) \in \mathbb{R}_*^2.$$

Here

$$a(x_1, x_2) = e,$$

$$b(x_1, x_2) = 1,$$

$$c(x_1, x_2) = x_1, \quad (x_1, x_2) \in \mathbb{R}_*^2.$$

Then,

$$\log(a(x_1, x_2)) \log(c(x_1, x_2)) - (\log(b(x_1, x_2)))^2 = \log e \log x_1 - (\log 1)^2$$

$$= \log x_1, \quad (x_1, x_2) \in \mathbb{R}_*^2.$$

Thus, we have the following cases:

1. If
$$x_1 < e,$$
then
$$\log x_1 < 1$$
and the considered equation is multiplicative hyperbolic.

2. If
$$x_1 = e,$$
then
$$\log x_1 = 1$$
and the considered equation is multiplicative parabolic.

3. If
$$x_1 > e,$$
then
$$\log x_1 > 1$$
and the considered equation is multiplicative elliptic.

The corresponding canonical forms of the multiplicative differential operator $L(u)$ are as follows:

1. $\alpha = -_*\gamma, \quad \beta = 0_*.$
$$\Lambda(u) +_* \cdots = \alpha(u^{**}_{\xi_1\xi_1} -_* u^{**}_{\xi_2\xi_2}).$$

When $\alpha = \gamma = 0_*$, we have
$$\Lambda(u) +_* \cdots = e^2 \cdot_* \beta \cdot_* u^{**}_{\xi_1\xi_2}.$$

2. $\alpha = \gamma, \quad \beta = 0_*.$
$$\Lambda(u) +_* \cdots = \alpha \cdot_* (u^{**}_{\xi_1\xi_1} +_* u^{**}_{\xi_2\xi_2}).$$

3. $\beta = \gamma = 0_*.$
$$\Lambda(u) +_* \cdots = \alpha \cdot_* u^{**}_{\xi_1\xi_1}.$$

The corresponding canonical forms of the considered multiplicative differential equations are as follows:

1. $\alpha = -_*\gamma, \quad \beta = 0_*.$
$$u^{**}_{\xi_1\xi_1} -_* u^{**}_{\xi_2\xi_2} +_* \cdots = 0_*.$$

When $\alpha = \gamma = 0_*$, we have
$$u^{**}_{\xi_1\xi_2} +_* \cdots = 0_*.$$

2. $\alpha = \gamma, \quad \beta = 0_*.$
$$u^{**}_{\xi_1\xi_1} +_* u^{**}_{\xi_2\xi_2} +_* \cdots = 0_*.$$

3. $\beta = \gamma = 0_*.$
$$u^{**}_{\xi_1\xi_1} +_* \cdots = 0_*.$$

For fixed $(x_1, x_2) \in \mathbb{R}_*^2$, such a canonical form can be obtained by the multiplicative linear transformation which takes Q into the corresponding canonical form.

If we assume that the operator L is of the same type in every point of the domain $U \subseteq \in \mathbb{R}_*^2$, we will search functions ϕ_1 and ϕ_2 which will transform $L(u)$ into a canonical form at every point of U. To find such functions, it depends on whether certain first-order systems of multiplicative linear multiplicative partial differential equations can be solved.

Without loss of generality, we suppose that $a \neq 0_*$ everywhere in the domain U.

3.1.1   The multiplicative hyperbolic case

We suppose that $L(u)$ is multiplicative hyperbolic in U and $\alpha = \gamma = 0_* = 1$. Then, using the definitions for α and γ, we obtain the following system:

$$a \cdot_* \phi_{1x_1}^{*2_*} +_* e^2 \cdot_* b \cdot_* \phi_{1x_1}^* \cdot_* \phi_{1x_2}^* +_* c \cdot_* \phi_{1x_2}^{*2_*} = 0_*,$$

$$a \cdot_* \phi_{2x_1}^{*2_*} +_* e^2 \cdot_* b \cdot_* \phi_{2x_1}^* \cdot_* \phi_{2x_2}^* +_* c \cdot_* \phi_{2x_2}^{*2_*} = 0_*;$$

or

$$a \cdot_* \left((\phi_{1x_1}^*)/_*(\phi_{1x_2}^*) \right)^{2_*} +_* e^2 \cdot_* b \cdot_* \left((\phi_{1x_1}^*)/_*(\phi_{1x_2}^*) \right) +_* c = 0_*,$$

$$a \cdot_* \left((\phi_{2x_1}^*)/_*(\phi_{2x_2}^*) \right)^{2_*} +_* e^2 \cdot_* b \cdot_* \left((\phi_{2x_1}^*)/_*(\phi_{2x_2}^*) \right) +_* c = 0_*.$$

From the above system we see that if there exist such functions ϕ_1 and ϕ_2, then $((\phi_{1x_1}^*)/_*(\phi_{1x_2}^*))$ and $(\phi_{2x_1}^*)/_*(\phi_{2x_2}^*)$ satisfy the multiplicative quadratic equation:

$$a \cdot_* p^{2_*} +_* e^2 \cdot_* b \cdot_* p +_* c = 0_*, \tag{3.5}$$

or

$$
\begin{aligned}
0_* &= 1 \\
&= e^0 \\
&= a \cdot_* e^{(\log p)^2} +_* e^2 \cdot_* e^{\log b \log p} +_* c \\
&= e^{\log a \log e^{(\log p)^2}} +_* e^{\log e^2 \log e^{\log b \log p}} +_* e^{\log c} \\
&= e^{\log a (\log p)^2} +_* e^{2 \log b \log p} +_* e^{\log c} \\
&= e^{\log a (\log p)^2 + 2 \log b \log p + \log c},
\end{aligned}
$$

or

$$\log a (\log p)^2 + 2 \log b \log p + \log c = 0,$$

where p is unknown. Since $L(u)$ is multiplicative hyperbolic in U,

$$\log a \log c - (\log b)^2 < 0$$

and then equation (3.5) has two roots p_1 and p_2. Thus in the multiplicative hyperbolic case, we obtain the canonical form

$$e^2 \cdot_* \beta \cdot_* u^{**}_{\xi_1 \xi_2} +_* \cdots$$

by determining the functions ϕ_1 and ϕ_2 from the multiplicative differential equations:

$$\phi^*_{1x_1} -_* p_1 \cdot_* \phi^*_{1x_2} = 0_*,$$

$$\phi^*_{2x_1} -_* p_2 \cdot_* \phi^*_{2x_2} = 0_*.$$

These two first-order multiplicative linear homogeneous multiplicative partial differential equations yield two families of multiplicative curves:

$$\phi_1 = \text{const}, \quad \phi_2 = \text{const}.$$

These two families can be defined as the families of solutions of the multiplicative ordinary differential equations:

$$(d_* x_2)/_*(d_* x_1) = -_* p_1,$$

$$(d_* x_2)/_*(d_* x_1) = -_* p_2,$$

and since p_1 and p_2 are roots of equation (3.5), we have

$$a \cdot_* ((d_* x_2)/_*(d_* x_1))^{*2*} -_* e^2 \cdot_* b \cdot_* ((d_* x_2)/_*(d_* x_1)) +_* c = 0_*. \qquad (3.6)$$

Here x_2 is considered as a function of x_1 along the multiplicative curves of the family.

We have that

$$p_{1,2} = (b \pm_* (b^{2*} -_* a \cdot_* c)^{\frac{1}{2}*})/_* a$$

and let us set

$$p_1 = (b +_* (b^{2*} -_* a \cdot_* c)^{\frac{1}{2}*})/_* a,$$

$$p_2 = (b -_* (b^{2*} -_* a \cdot_* c)^{\frac{1}{2}*})/_* a.$$

Then,

$$x_2 +_* \int_* (b +_* (b^{2*} -_* a \cdot_* c)^{\frac{1}{2}*})/_* a \cdot_* d_* x_1 = \text{const},$$

$$x_2 +_* \int_* (b -_* (b^{2*} -_* a \cdot_* c)^{\frac{1}{2}*})/_* a \cdot_* d_* x_2 = \text{const},$$

$$p_1 -_* p_2 = (e^2 \cdot_* (b^{2*} -_* a \cdot_* c))/_* a.$$

Definition 3.4 The multiplicative curves

$$\xi_1 = \phi_1 \cdot_* \left(x_2 +_* (b +_* \int_* (b^{2*} -_* a \cdot_* c))/_* a \right) \cdot_* d_* x_1,$$

$$\xi_2 = \phi_2 \cdot_* \left(x_2 +_* \int_* (b -_* (b^{2*} -_* a \cdot_* c))/_* a \right) \cdot_* d_* x_1,$$

(3.7)

are called the characteristic multiplicative curves of the multiplicative linear multiplicative hyperbolic differential operator $L(u)$.

Remark 3.2 For convenience, in practice, we very often take

$$\xi_1 \;=\; x_2 +_* \int_* ((b +_* (b^{2*} -_* a \cdot_* c))/_* a) \cdot_* d_* x_1,$$

$$\xi_2 \;=\; x_2 +_* \int_* ((b -_* (b^{2*} -_* a \cdot_* c))/_* a) \cdot_* d_* x_1.$$

Definition 3.5 The multiplicative curves

$$\xi_1 = \phi_1(x_1, x_2) = \text{const}, \quad (x_1, x_2) \in \mathbb{R}_*^2,$$

$$\xi_2 = \phi_2(x_1, x_2) = \text{const}, \quad (x_1, x_2) \in \mathbb{R}_*^2,$$

which satisfy the system (3.7), are called the characteristic multiplicative curves of the multiplicative linear multiplicative hyperbolic operator $L(u)$.

In the case when $\alpha = -_* \gamma, \beta = 0_*$, the functions ϕ_1 and ϕ_2 satisfy the system:

$$a \cdot_* \phi_{1x_1}^{*2*} +_* e^2 \cdot_* b \cdot_* \phi_{1x_1}^* \cdot_* \phi_{1x_2}^* +_* c \cdot_* \phi_{1x_2}^{*2*}$$
$$= -_*(a \cdot_* \phi_{2x_1}^{*2*} +_* e^2 \cdot_* b \cdot_* \phi_{2x_1}^* \cdot_* \phi_{2x_2}^* +_* c \cdot_* \phi_{2x_2}^{*2*})$$

$$a \cdot_* \phi_{1x_1}^* \cdot_* \phi_{2x_1}^* +_* b \cdot_* (\phi_{1x_2}^* \cdot_* \phi_{2x_1}^* +_* \phi_{1x_2}^* \cdot_* \phi_{2x_1}^*) +_* c \cdot_* \phi_{1x_2}^* \cdot_* \phi_{2x_2}^* = 0_*$$

or

$$a \cdot_* (\phi_{1x_1}^{*2*} +_* \phi_{2x_1}^{*2*}) +_* e^2 \cdot_* b \cdot_* (\phi_{1x_1}^* \cdot_* \phi_{1x_2}^* +_* \phi_{2x_1}^* \cdot_* \phi_{2x_2}^*)$$
$$+_* c \cdot_* (\phi_{1x_2}^{*2*} +_* \phi_{2x_2}^{*2*}) = 0_*,$$

$$a \cdot_* \phi_{1x_1}^* \cdot_* \phi_{2x_1}^* +_* b \cdot_* (\phi_{1x_2}^* \cdot_* \phi_{2x_1}^* +_* \phi_{1x_1}^* \cdot_* \phi_{2x_2}^*)$$
$$+_* c \cdot_* \phi_{1x_2}^* \cdot_* \phi_{2x_2}^* = 0_*.$$

(3.8)

Thus we have the canonical form:

$$\alpha \cdot_* (u_{\xi_1\xi_1}^{**} -_* u_{\xi_2\xi_2}^{**}) +_* \cdots = 0_*.$$

Definition 3.6 Equation (3.6) is called the characteristic equation.

Now we will simplify equation (3.8). Let us assume that $\phi_3, \phi_4 \in \mathcal{C}^1_*(U)$ and

$$\phi_1 = \phi_3 +_* \phi_4,$$

$$\phi_2 = \phi_3 -_* \phi_4.$$

Then,

$$\phi^*_{1x_1} = \phi^*_{3x_1} +_* \phi^*_{4x_1},$$

$$\phi^*_{1x_2} = \phi^*_{3x_2} +_* \phi^*_{4x_2},$$

$$\phi^*_{2x_1} = \phi^*_{3x_1} -_* \phi^*_{4x_1},$$

$$\phi^*_{2x_2} = \phi^*_{3x_2} -_* \phi^*_{4x_2},$$

and

$$
\begin{aligned}
\phi^{*2*}_{1x_1} +_* \phi^{*2*}_{2x_1} &= \left(\phi^*_{3x_1} +_* \phi^*_{4x_1}\right)^{2*} +_* \left(\phi^*_{3x_1} -_* \phi^*_{4x_1}\right)^{2*} \\
&= \phi^{*2*}_{3x_1} +_* e^2 \cdot_* \phi^*_{3x_1} \cdot_* \phi^*_{4x_1} +_* \phi^{*2*}_{4x_1} +_* \phi^{*2*}_{3x_1} \\
&\quad -_* e^2 \cdot_* \phi^*_{3x_1} \cdot_* \phi^*_{4x_1} +_* \phi^{*2*}_{4x_1} \\
&= e^2 \cdot_* \left(\phi^{*2*}_{3x_1} +_* \phi^{*2*}_{4x_1}\right), \\[4pt]
\phi^{*2*}_{1x_2} +_* \phi^{*2*}_{2x_2} &= \left(\phi^*_{3x_2} +_* \phi^*_{4x_2}\right)^{2*} +_* \left(\phi^*_{3x_2} -_* \phi^*_{4x_2}\right)^{2*} \\
&= \phi^{*2*}_{3x_2} +_* e^2 \cdot_* \phi^*_{3x_2} \cdot_* \phi^*_{4x_2} +_* \phi^{*2*}_{4x_2} +_* \phi^{*2*}_{3x_2} \\
&\quad -_* e^2 \cdot_* \phi^*_{3x_2} \cdot_* \phi^*_{4x_2} +_* \phi^{*2*}_{4x_2} \\
&= e^2 \cdot_* \left(\phi^{*2*}_{3x_2} +_* \phi^{*2*}_{4x_2}\right),
\end{aligned}
$$

and

$$
\begin{aligned}
\phi^*_{1x_1} \cdot_* \phi^*_{1x_2} +_* \phi^*_{2x_1} \cdot_* \phi^*_{2x_2} &= \left(\phi^*_{3x_1} +_* \phi^*_{4x_1}\right) \cdot_* \left(\phi^*_{3x_2} +_* \phi^*_{4x_2}\right) \\
&\quad +_* \left(\phi^*_{3x_1} -_* \phi^*_{4x_1}\right) \cdot_* \left(\phi^*_{3x_2} -_* \phi^*_{4x_2}\right) \\
&= \phi^*_{3x_1} \cdot_* \phi^*_{3x_2} +_* \phi^*_{3x_1} \cdot_* \phi^*_{4x_2} +_* \phi^*_{4x_1} \cdot_* \phi^*_{3x_2} \\
&\quad +_* \phi^*_{4x_1} \cdot_* \phi^*_{4x_2} +_* \phi^*_{3x_1} \cdot_* \phi^*_{3x_2} -_* \phi^*_{3x_1} \cdot_* \phi^*_{4x_2} \\
&\quad -_* \phi^*_{4x_1} \cdot_* \phi^*_{3x_2} +_* \phi^*_{4x_1} \cdot_* \phi^*_{4x_2} \\
&= e^2 \cdot_* \left(\phi^*_{3x_1} \cdot_* \phi^*_{3x_2} +_* \phi^*_{4x_1} \cdot_* \phi^*_{4x_2}\right);
\end{aligned}
$$

and

$$a \cdot_* \left(\phi_{1x_1}^{*2*} +_* \phi_{2x_1}^{*2*} \right) +_* e^2 \cdot_* b \cdot_* \left(\phi_{1x_1}^* \cdot_* \phi_{1x_2}^* +_* \phi_{2x_1}^* \cdot_* \phi_{2x_2}^* \right)$$
$$+_* c \cdot_* \left(\phi_{1x_2}^{*2*} +_* \phi_{2x_2}^{*2*} \right)$$
$$= e^2 \cdot_* a \cdot_* \left(\phi_{3x_1}^{*2*} +_* \phi_{4x_1}^{*2*} \right) +_* e^4 \cdot_* b \cdot_* \left(\phi_{3x_1}^* \cdot_* \phi_{3x_2}^* +_* \phi_{4x_1}^* \cdot_* \phi_{4x_2}^* \right)$$
$$+_* e^2 \cdot_* c \cdot_* \left(\phi_{3x_2}^{*2*} +_* \phi_{4x_2}^{*2*} \right),$$

from where

$$a \cdot_* \left(\phi_{3x_1}^{*2*} +_* \phi_{4x_1}^{*2*} \right) +_* e^2 \cdot_* b \cdot_* \left(\phi_{3x_1}^* \cdot_* \phi_{3x_2}^* +_* \phi_{4x_1}^* \cdot_* \phi_{4x_2}^* \right)$$
$$+_* c \cdot_* \left(\phi_{3x_2}^{*2*} +_* \phi_{4x_2}^{*2*} \right) = 0_*. \tag{3.9}$$

Moreover,

$$\begin{aligned}
\phi_{1x_1}^* \cdot_* \phi_{2x_1}^* &= \left(\phi_{3x_1}^* +_* \phi_{4x_1}^* \right) \cdot_* \left(\phi_{3x_1}^* -_* \phi_{4x_1}^* \right) \\
&= \phi_{3x_1}^{*2*} -_* \phi_{4x_1}^{*2*}, \\
\phi_{1x_2}^* \cdot_* \phi_{2x_2}^* &= \left(\phi_{3x_2}^* +_* \phi_{4x_2}^* \right) \cdot_* \left(\phi_{3x_2}^* -_* \phi_{4x_2}^* \right) \\
&= \phi_{3x_2}^{*2*} -_* \phi_{4x_2}^{*2*},
\end{aligned}$$

$$\begin{aligned}
\phi_{1x_2}^* \cdot_* \phi_{2x_1}^* +_* \phi_{1x_1}^* \cdot_* \phi_{2x_2}^* &= \left(\phi_{3x_2}^* +_* \phi_{4x_2}^* \right) \cdot_* \left(\phi_{3x_1}^* -_* \phi_{4x_1}^* \right) \\
&\quad +_* \left(\phi_{3x_1}^* +_* \phi_{4x_1}^* \right) \cdot_* \left(\phi_{3x_2}^* -_* \phi_{4x_2}^* \right) \\
&= \phi_{3x_1}^* \cdot_* \phi_{3x_2}^* +_* \phi_{3x_1}^* \cdot_* \phi_{4x_2}^* -_* \phi_{3x_2}^* \cdot_* \phi_{4x_1}^* \\
&\quad -_* \phi_{4x_1}^* \cdot_* \phi_{4x_2}^* +_* \phi_{3x_1}^* \cdot_* \phi_{3x_2}^* -_* \phi_{3x_1}^* \cdot_* \phi_{4x_2}^* \\
&\quad +_* \phi_{4x_1}^* \cdot_* \phi_{3x_2}^* -_* \phi_{4x_1}^* \cdot_* \phi_{4x_2}^* \\
&= e^2 \cdot_* \left(\phi_{3x_1}^* \cdot_* \phi_{3x_2}^* -_* \phi_{4x_1}^* \cdot_* \phi_{4x_2}^* \right),
\end{aligned}$$

and

$$a \cdot_* \phi_{1x_1}^* \cdot_* \phi_{2x_1}^* +_* b \cdot_* \left(\phi_{1x_2}^* \cdot_* \phi_{2x_1}^* +_* \phi_{1x_1}^* \cdot_* \phi_{2x_2}^* \right) +_* c \cdot_* \phi_{1x_2}^* \cdot_* \phi_{2x_2}^*$$
$$= a \cdot_* \left(\phi_{3x_1}^{*2*} -_* \phi_{4x_1}^{*2*} \right) +_* e^2 \cdot_* b \cdot_* \left(\phi_{3x_1}^* \cdot_* \phi_{3x_2}^* -_* \phi_{4x_1}^* \cdot_* \phi_{4x_2}^* \right)$$
$$+ c \cdot_* \left(\phi_{3x_2}^{*2*} -_* \phi_{4x_2}^{*2*} \right).$$

Using the last identity and equations (3.8), (3.9), we obtain

$$a \cdot_* \left(\phi_{3x_1}^{*2*} +_* \phi_{4x_1}^{*2*} \right) +_* e^2 \cdot_* b \cdot_* \left(\phi_{3x_1}^* \cdot_* \phi_{3x_2}^* +_* \phi_{4x_1}^* \cdot_* \phi_{4x_2}^* \right)$$

$$+_* c \cdot_* \left(\psi_{3x_2}^{*2*} +_* \psi_{4x_2}^{*2-*} \right) = 0_*$$

$$a \cdot_* \left(\phi_{3x_1}^{*2*} -_* \phi_{4x_1}^{2*} \right) +_* e^2 \cdot_* b \cdot_* \left(\phi_{3x_1}^* \cdot_* \phi_{3x_2}^* -_* \phi_{4x_1}^* \cdot_* \phi_{4x_2}^* \right)$$

$$+_* c \cdot_* \left(\phi_{3x_2}^{2*} -_* \phi_{4x_2}^{*2*} \right) = 0_*,$$

from where

$$a \cdot_* \phi_{3x_1}^{*2*} +_* e^2 \cdot_* b \cdot_* \phi_{3x_1}^* \cdot_* \phi_{3x_2}^* +_* c \cdot_* \phi_{3x_2}^{*2*} \quad = \quad 0_*$$

$$a \cdot_* \phi_{4x_1}^{*2*} +_* e^2 \cdot_* b \cdot_* \phi_{4x_1}^* \cdot_* \phi_{4x_2}^* +_* c \cdot_* \phi_{4x_2}^{*2*} \quad = \quad 0_*.$$

In this way, we have reduced the case $\alpha = -_*\gamma$, $\beta = 0_*$ to the case $\alpha = \gamma = 0_*$.

Example 3.7 We will find the canonical form of the multiplicative analogue of the classical Tricomi equation:

$$u_{x_1 x_1}^{**} +_* x_1 \cdot_* u_{x_2 x_2}^{**} = 0_*, \quad x_1 < 0_*.$$

The characteristic equation is

$$(d_* x_2)^{2*} +_* x_1 \cdot_* (d_* x_1)^{2*} = 0_*, \quad x_1 < 0_*,$$

whereupon

$$((d_* x_2)/_*(d_* x_1))^{2*} +_* x_1 = 0_*, \quad x_1 < 1,$$

or

$$(d_* x_2)/_*(d_* x_1) = \pm_*(-_* x_1)^{\frac{1}{2}*}, \quad x_1 < 1,$$

and

$$d_* x_2 = \pm_*(-_* x_1)^{\frac{1}{2}*} \cdot_* d_* x_1, \quad x_1 < 1,$$

and

$$\int_* d_* x_2 = \pm_* \int_* (-_* x_1)^{\frac{1}{2}*} \cdot_* d_* x_1 +_* c, \quad x_1 < 1,$$

or

$$x_2 = \mp_*((-_* x_1)^{\frac{3}{2}*}/_*(\frac{3}{2}_*)) +_* c, \quad x_1 < 1,$$

or

$$x_2 \pm_* \left(\frac{2}{3_*} \right) \cdot_* (-_* x_1)^{\frac{3}{2}*} = c.$$

We set

$$\xi_1 \;=\; x_2 +_* \left(\frac{2}{3_*}\right) \cdot_* (-_*x_1)^{\frac{3}{2}}{}_*,$$

$$\xi_2 \;=\; x_2 -_* \left(\frac{2}{3_*}\right) \cdot_* (-_*x_1)^{\frac{3}{2}}{}_*, \quad x_1 < 1.$$

Then,

$$\xi^*_{1x_1} \;=\; -_*(-_*x_1)^{\frac{1}{2}}{}_*,$$

$$\xi^*_{1x_2} \;=\; e,$$

$$\xi^*_{2x_1} \;=\; (-_*x_1)^{\frac{1}{2}}{}_*,$$

$$\xi^*_{2x_2} \;=\; e, \quad x_1 < 1.$$

Hence,

$$u^*_{x_1} \;=\; u^*_{\xi_1} \cdot_* xi^*_{1x_1} +_* u^*_{\xi_2} \cdot_* \xi^*_{2x_1}$$

$$=\; -_*(-_*x_1)^{\frac{1}{2}}{}_* \cdot_* u^*_{\xi_1} +_* (-_*x_1)^{\frac{1}{2}}{}_* \cdot_* u^*_{\xi_2},$$

$$u^{**}_{x_1x_1} \;=\; \left(e/_*\left(e^2 \cdot_* (-_*x_1)^{\frac{1}{2}}{}_*\right)\right) \cdot_* u^*_{\xi_1} -_* (-_*x_1)^{\frac{1}{2}}{}_* \cdot_*$$

$$\left(u^{**}_{\xi_1\xi_1} \cdot_* \xi^*_{1x_1} +_* u^{**}_{\xi_1\xi_2} \cdot_* \xi^*_{2x_1}\right)$$

$$-_* \left(e/_*\left(e^2 \cdot_* (-_*x_1)^{\frac{1}{2}}{}_*\right)\right) \cdot_* u^*_{\xi_2} +_* (-_*x_1)^{\frac{1}{2}}{}_* \cdot_*$$

$$\left(u^{**}_{\xi_1\xi_2} \cdot_* \xi^*_{1x_1} +_* u^{**}_{\xi_2\xi_2} \cdot_* \xi^*_{2x_1}\right)$$

$$=\; \left(e/_*\left(e^2 \cdot_* (-_*x_1)^{\frac{3}{2}}{}_*\right)\right) \cdot_* u^*_{\xi_1} -_* \left(e/_*\left(e^2 \cdot_* (-_*x_1)^{\frac{3}{2}}{}_*\right)\right) \cdot_* u^*_{\xi_2}$$

$$-_*(-_*x_1)^{\frac{1}{2}}{}_* \cdot_* \left(-_*(-_*x_1)^{\frac{1}{2}}{}_* \cdot_* u^{**}_{\xi_1\xi_1} +_* (-_*x_1)^{\frac{1}{2}}{}_* \cdot_* u^{**}_{\xi_1\xi_2}\right)$$

$$+_*(-_*x_1)^{\frac{1}{2}}{}_* \cdot_* \left(-_*(-_*x_1)^{\frac{1}{2}}{}_* \cdot_* u^{**}_{\xi_1\xi_2} +_* (-_*x_1)^{\frac{1}{2}}{}_* \cdot_* u^{**}_{\xi_2\xi_2}\right)$$

$$=\; \left(e/_*\left(e^2 \cdot_* (-_*x_1)^{\frac{3}{2}}{}_*\right)\right) \cdot_* u^*_{\xi_1} -_* \left(e/_*\left(e^2 \cdot_* (-_*x_1)^{\frac{3}{2}}{}_*\right)\right) \cdot_* u^*_{\xi_2}$$

$$-_*x_1 \cdot_* u^{**}_{\xi_1\xi_1} +_* x_1 \cdot_* u^{**}_{\xi_1\xi_2} +_* x_1 \cdot_* u^{**}_{\xi_1\xi_2} -_* x_1 \cdot_* u^{**}_{\xi_2\xi_2}$$

$$
= \left(e/_*\left(e^2\cdot_*(-_*x_1)^{\frac{1}{2}}_*\right)\right)\cdot_* u^*_{\xi_1} -_* \left(e/_*\left(e^2\cdot_*(-_*x_1)^{\frac{1}{2}}_*\right)\right)\cdot_* u^*_{\xi_2}
$$

$$
-_*x_1\cdot_* u^{**}_{\xi_1\xi_1} +_* e^2\cdot_* x_1\cdot_* u^{**}_{\xi_1\xi_2} -_* x_1\cdot_* u^{**}_{\xi_2\xi_2},
$$

$$
u^*_{x_2} = u^*_{\xi_1}\cdot_* \xi^*_{1x_2} +_* u^*_{\xi_2}\cdot_* \xi^*_{2x_2}
$$

$$
= u^*_{\xi_1} +_* u^*_{\xi_2},
$$

$$
u^{**}_{x_2x_2} = u^{**}_{\xi_1\xi_1}\cdot_* \xi^*_{1x_2} +_* u^{**}_{\xi_1\xi_2}\cdot_* \xi^*_{2x_2} +_* u^{**}_{\xi_1\xi_2}\cdot_* \xi^*_{1x_2} +_* u^{**}_{\xi_2\xi_2}\cdot_* \xi^*_{2x_2}
$$

$$
= u^{**}_{\xi_1\xi_1} +_* e^2\cdot_* u^{**}_{\xi_1\xi_2} +_* u^{**}_{\xi_2\xi_2}.
$$

From here,

$$
u^{**}_{x_1x_1} +_* x_1\cdot_* u^{**}_{x_2x_2} = \left(e/_*\left(e^2\cdot_*(-_*x_1)^{\frac{1}{2}}_*\right)\right)\cdot_* u^*_{\xi_1}
$$

$$
-_* \left(e/_*\left(e^2\cdot_*(-_*x_1)^{\frac{1}{2}}_*\right)\right)\cdot_* u^*_{\xi_2} -_* x_1\cdot_* u^{**}_{\xi_1\xi_1}
$$

$$
+_* e^2\cdot_* x_1\cdot_* u^{**}_{\xi_1\xi_2} -_* x_1\cdot_* u^{**}_{\xi_2\xi_2} +_* x_1\cdot_* u^{**}_{\xi_1\xi_1} +_* e^2\cdot_* x_1\cdot_* u^{**}_{\xi_1\xi_2}
$$

$$
+_* x_1\cdot_* u^{**}_{\xi_2\xi_2}
$$

$$
= \left(e/_*\left(e^2\cdot_*(-_*x_1)^{\frac{1}{2}}_*\right)\right)\cdot_* u^*_{\xi_1} -_* \left(e/_*\left(e^2\cdot_*(-_*x_1)^{\frac{1}{2}}_*\right)\right)
$$

$$
\cdot_* u^*_{\xi_2} +_* e^4\cdot_* x_1\cdot_* u^{**}_{\xi_1\xi_2}, \quad x_1 < 1.
$$

Therefore,

$$
0_* = u^{**}_{x_1x_1} +_* x_1\cdot_* u^{**}_{x_2x_2}
$$

$$
= \left(e/_*\left(e^2\cdot_*(-_*x_1)^{\frac{1}{2}}_*\right)\right)\cdot_* u^*_{\xi_1} -_* \left(e/_*\left(e^2\cdot_*(-_*x_1)^{\frac{1}{2}}_*\right)\right)
$$

$$
\cdot_* u^*_{\xi_2} +_* e^4\cdot_* x_1\cdot_* u^{**}_{\xi_1\xi_2}
$$

$$
= e^{-8}\cdot_*(-_*x_1)^{\frac{3}{2}}_*\cdot_* u^{**}_{\xi_1\xi_2} +_* u^*_{\xi_1} -_* u^*_{\xi_2}
$$

$$
= e^6\cdot_*(\xi_2 -_* \xi_1)\cdot_* u^{**}_{\xi_1\xi_2} +_* u^*_{\xi_1} -_* u^*_{\xi_2},
$$

i.e.,

$$e^6 \cdot_* (\xi_2 -_* \xi_1) \cdot_* u^{**}_{\xi_1 \xi_2} +_* u^*_{\xi_1} -_* u^*_{\xi_2} = 0_*$$

is the canonical form of the considered equation.

Example 3.8 We will find the canonical form of the equation:

$$x_2^{5*} \cdot_* u^{**}_{x_1 x_1} -_* x_2 \cdot_* u^{**}_{x_2 x_2} +_* e^2 \cdot_* u^*_{x_2} = 0_*, \quad x_2 \neq 0_*.$$

The characteristic equation is

$$x_2^{5*} \cdot_* (d_* x_2)^{2*} -_* x_2 \cdot_* (d_* x_1)^{2*} = 0_*, \quad x_2 \neq 0_*,$$

whereupon

$$x_2^{2*} \cdot_* d_* x_2 = \pm_* d_* x_1.$$

Hence,

$$\int_* x_2^{2*} \cdot_* d_* x_2 = \pm_* \int_* d_* x_1 +_* c, \quad x_2 \neq 0_*,$$

or

$$\left(x_2^{3*}\right) /_* e^3 = \pm_* x_1 +_* c, \quad x_2 \neq 0_*,$$

or

$$x_2^{3*} \pm_* e^3 \cdot_* x_1 = c, \quad x_2 \neq 0_*.$$

We set

$$\xi_1 = x_2^{3*} +_* e^3 \cdot_* x_1,$$

$$\xi_2 = x_2^{3*} -_* e^3 \cdot_* x_1, \quad x_2 \neq 0_*.$$

Then,

$$\xi^*_{1x_1}(x_1, x_2) = e^3,$$

$$\xi^*_{1x_2}(x_1, x_2) = e^3 \cdot_* x_2^{2*},$$

$$\xi^*_{2x_1}(x_1, x_2) = e^{-3},$$

$$\xi^*_{2x_2}(x_1, x_2) = e^3 \cdot_* x_2^{2*}, \quad x_2 \neq 0_*.$$

From here,

$$u^*_{x_1} = u^*_{\xi_1} \cdot_* \xi^*_{1x_1} +_* u^*_{\xi_2} \cdot_* \xi^*_{2x_1}$$

$$= e^3 \cdot_* u^*_{\xi_1} -_* e^3 \cdot_* u^*_{\xi_2},$$

$$u^{**}_{x_1 x_1} = e^3 \cdot_* \left(u^{**}_{\xi_1 \xi_1} \cdot_* \xi^*_{1 x_1} +_* u^{**}_{\xi_1 \xi_2} \cdot_* \xi^*_{2 x_1} \right)$$

$$-_* e^3 \cdot_* \left(u^{**}_{\xi_1 \xi_2} \cdot_* \xi^*_{1 x_1} +_* u^{**}_{\xi_2 \xi_2} \cdot_* \xi^*_{2 x_2} \right)$$

$$= c^3 \cdot_* \left(c^3 \cdot_* u^{**}_{\xi_1 \xi_1} -_* e^3 \cdot_* u_{\xi_1 \xi_2} \right) -_* e^3 \cdot_* \left(e^3 \cdot_* u^{**}_{\xi_1 \xi_2} -_* e^3 \cdot_* u^{**}_{\xi_2 \xi_2} \right)$$

$$= e^9 \cdot_* u^{**}_{\xi_1 \xi_1} -_* e^{18} \cdot_* u^{**}_{\xi_1 \xi_2} +_* e^9 \cdot_* u^{**}_{\xi_2 \xi_2},$$

$$u^*_{x_2} = u^*_{\xi_1} \cdot_* \xi^*_{1 x_2} +_* u^*_{\xi_2} \cdot_* \xi^*_{2 x_2}$$

$$= e^3 \cdot_* x^{2*}_2 \cdot_* u^*_{\xi_1} +_* e^3 \cdot_* x^{2*}_2 \cdot_* u^*_{\xi_2},$$

$$u^{**}_{x_2 x_2} = e^6 \cdot_* x_2 \cdot_* u^{**}_{\xi_1} +_* e^3 \cdot_* x^{2*}_2 \cdot_* \left(u^{**}_{\xi_1 \xi_1} \cdot_* \xi^*_{1 x_2} +_* u^{**}_{\xi_1 \xi_2} \cdot_* \xi^*_{2 x_2} \right)$$

$$+_* e^6 \cdot_* x_2 \cdot_* u^*_{\xi_2} +_* e^3 \cdot_* x^{2*}_2 \cdot_* \left(u^{**}_{\xi_1 \xi_2} \cdot_* \xi^*_{1 x_2} +_* u^{**}_{\xi_2 \xi_2} \cdot_* \xi^*_{2 x_2} \right)$$

$$= e^6 \cdot_* x_2 \cdot_* u^*_{\xi_1} +_* e^6 \cdot_* x_2 \cdot_* u^*_{\xi_2} +_* e^3 \cdot_* x^{2*}_2 \cdot_*$$

$$\left(e^3 \cdot_* x^{2*}_2 \cdot_* u^{**}_{\xi_1 \xi_1} +_* e^3 \cdot_* x^{2*}_2 \cdot_* u^{**}_{\xi_1 \xi_2} \right)$$

$$+_* e^3 \cdot_* x^{2*}_2 \cdot_* \left(e^3 \cdot_* x^{2*}_2 \cdot_* u^{**}_{\xi_1 \xi_2} +_* e^3 \cdot_* x^{2*}_2 \cdot_* u^{**}_{\xi_2 \xi_2} \right)$$

$$= e^6 \cdot_* x_2 \cdot_* u^*_{\xi_1} +_* e^6 \cdot_* x_2 \cdot_* u^*_{\xi_2} +_* e^9 \cdot_* x^{4*}_2 \cdot_* u^{**}_{\xi_1 \xi_1}$$

$$+_* e^{18} \cdot_* x^{4*}_2 \cdot_* u^{**}_{\xi_1 \xi_2} +_* e^9 \cdot_* x^{4*}_2 \cdot_* u^{**}_{\xi_2 \xi_2}.$$

Therefore,

$$0_* = x^{5*}_2 \cdot_* u^{**}_{x_1 x_1} -_* x_2 \cdot_* u^{**}_{x_2 x_2} +_* e^2 \cdot_* u^*_{x_2}$$

$$= e^9 \cdot_* x^{5*}_2 \cdot_* u^{**}_{\xi_1 \xi_1} -_* e^{18} \cdot_* x^{5*}_2 \cdot_* u^{**}_{\xi_1 \xi_2} +_* e^9 \cdot_* x^{5*}_2 \cdot_* u^{**}_{\xi_2 \xi_2}$$

$$-_* e^6 \cdot_* x^{2*}_2 \cdot_* u^*_{\xi_1} -_* e^6 \cdot_* x^{2*}_2 \cdot_* u^*_{\xi_2} -_* e^9 \cdot_* x^{5*}_2 \cdot_* u^{**}_{\xi_1 \xi_1}$$

$$-_* e^{18} \cdot_* x^{5*}_2 \cdot_* u^{**}_{\xi_1 \xi_2} -_* e^9 \cdot_* x^{5*}_2 \cdot_* u^{**}_{\xi_2 \xi_2}$$

$$+_* e^6 \cdot_* x^{2*}_2 \cdot_* u^*_{\xi_1} +_* e^6 \cdot_* x^{2*}_2 \cdot_* u^*_{\xi_2}$$

$$= -_* e^{36} \cdot_* x^{5*}_2 \cdot_* u^{**}_{\xi_1 \xi_2},$$

whereupon

$$u^{**}_{\xi_1 \xi_2} = 0_*$$

is the canonical form of the considered equation.

Example 3.9 We will find the canonical form of the equation:

$$u^{**}_{x_1 x_1} -_* e^2 \cdot_* \sin_* x_1 \cdot_* u^{**}_{x_1 x_2} -_* (\cos_* x_1)^{2*} \cdot_* u^{**}_{x_2 x_2} -_* \cos_* x_1 \cdot_* u^*_{x_2} = 0_*.$$

The characteristic equation is

$$(d_* x_2)^{2*} +_* e^2 \cdot_* \sin_* x_1 \cdot_* d_* x_1 \cdot_* d_* x_2 -_* (\cos_* x_1)^{2*} \cdot_* (d_* x_1)^{2*} = 0_*,$$

whereupon

$$((d_* x_2)/_*(d_* x_1))^{2*} +_* e^2 \cdot_* \sin_* x_1 \cdot_* ((d_* x_2)/_*(d_* x_1)) -_* (\cos_* x_1)^{2*} = 0_*.$$

Hence,

$$((d_* x_2)/_*(d_* x_1))_{1,2} = \left(\left(-_* \sin_* x_1 \pm_* \left((\sin_* x_1)^{2*} +_* (\cos_* x_1)^{2*} \right) \right) \right) /_* e$$

$$= -_* \sin_* x_1 \pm_* e,$$

or

$$d_* x_2 = (-_* \sin_* x_1 \pm_* e) \cdot_* d_* x_1,$$

from where

$$\int d_* x_2 = \int_* (-_* \sin_* x_1 \pm_* e) \cdot_* d_* x_1 +_* c$$

$$= \cos_* x_1 \pm_* x_1 +_* c$$

and

$$x_2 -_* \cos_* x_1 \mp_* x_1 = c.$$

We set

$$\xi_1 = x_2 -_* \cos_* x_1 -_* x_1,$$

$$\xi_2 = x_2 -_* \cos_* x_1 +_* x_1.$$

Then,

$$\xi^*_{1 x_1} = \sin_* x_1 -_* e,$$

$$\xi^*_{1 x_2} = e,$$

$$\xi^*_{2 x_1} = \sin_* x_1 +_* e,$$

$$\xi^*_{2 x_2} = 1.$$

From here,

$$u^*_{x_1} = u^*_{\xi_1} \cdot_* \xi^*_{1 x_1} +_* u^*_{\xi_2} \cdot_* \xi^*_{2 x_1}$$

$$= (\sin_* x_1 -_* e) \cdot_* u^*_{\xi_1} +_* (\sin_* x_1 +_* 1) u^*_{\xi_2},$$

$$u^{**}_{x_1 x_1} = \cos_* x_1 \cdot_* u^*_{\xi_1} +_* \cos_* x_1 \cdot_* u^*_{\xi_2} +_* (\sin_* x_1 -_* e)$$

$$\cdot_* \left(u^{**}_{\xi_1 \xi_1} \cdot_* \xi^*_{1x_1} +_* u^{**}_{\xi_1 \xi_2} \cdot_* \xi^*_{2x_1} \right)$$

$$+_* (\sin_* x_1 +_* e) \cdot_* \left(u^{**}_{\xi_1 \xi_2} \cdot_* \xi^*_{1x_1} +_* u^{**}_{\xi_2 \xi_2} \cdot_* \xi^*_{2x_1} \right)$$

$$= \cos_* x_1 \cdot_* u^*_{\xi_1} +_* \cos_* x_1 \cdot_* u^*_{\xi_2}$$

$$+_* (\sin_* x_1 -_* e) \cdot_* \left((\sin_* x_1 -_* e) \cdot_* u^{**}_{\xi_1 \xi_1} +_* (\sin_* x_1 +_* e) \cdot_* u^{**}_{\xi_1 \xi_2} \right)$$

$$+_* (\sin_* x_1 +_* e) \cdot_* \left((\sin_* x_1 -_* e) \cdot_* u^{**}_{\xi_1 \xi_2} +_* (\sin_* x_1 +_* e) \cdot_* u^{**}_{\xi_2 \xi_2} \right)$$

$$= \cos_* x_1 \cdot_* u^*_{\xi_1} +_* \cos_* x_1 \cdot_* u^*_{\xi_2} +_* (\sin_* x_1 -_* e)^{2*} \cdot_* u^{**}_{\xi_1 \xi_1}$$

$$+_* e^2 \cdot_* \left((\sin_* x_1)^{2*} -_* e \right) \cdot_* u^{**}_{\xi_1 \xi_2} +_* (\sin_* x_1 +_* e)^{2*} \cdot_* u^{**}_{\xi_2 \xi_2},$$

$$u^{**}_{x_1 x_2} = (\sin_* x_1 -_* e) \cdot_* \left(u^{**}_{\xi_1 \xi_1} \cdot_* \xi^*_{1x_2} +_* u^{**}_{\xi_1 \xi_2} \cdot_* \xi^*_{2x_2} \right)$$

$$+_* (\sin_* x_1 +_* e) \cdot_* \left(u^{**}_{\xi_1 \xi_2} \cdot_* \xi^*_{1x_2} +_* u^{**}_{\xi_2 \xi_2} \cdot_* \xi^*_{2x_2} \right)$$

$$= (\sin_* x_1 -_* e) \cdot_* \left(u^{**}_{\xi_1 \xi_1} +_* u^{**}_{\xi_1 \xi_2} \right)$$

$$+_* (\sin_* x_1 +_* e) \cdot_* \left(u^{**}_{\xi_1 \xi_2} +_* u^{**}_{\xi_2 \xi_2} \right)$$

$$= (\sin_* x_1 -_* e) \cdot_* u^{**}_{\xi_1 \xi_1} +_* e^2 \cdot_* \sin_* x_1 u^{**}_{\xi_1 \xi_2}$$

$$+_* (\sin_* x_1 +_* e) \cdot_* u^{**}_{\xi_2 \xi_2},$$

$$u^*_{x_2} = u^*_{\xi_1} \cdot_* \xi^*_{1x_2} +_* u^*_{\xi_2} \cdot_* \xi^*_{2x_2}$$

$$= u^*_{\xi_1} +_* u^*_{\xi_2},$$

$$u^{**}_{x_2 x_2} = u^{**}_{\xi_1 \xi_1} \cdot_* \xi^*_{1x_1} +_* u^{**}_{\xi_1 \xi_2} \cdot_* \xi^*_{2x_1} +_* u^{**}_{\xi_1 \xi_2} \cdot_* \xi^*_{1x_1} +_* u^{**}_{\xi_2 \xi_2} \cdot_* \xi^*_{2x_1}$$

$$= u^{**}_{\xi_1 \xi_1} +_* e^2 \cdot_* u^{**}_{\xi_1 \xi_2} +_* u^{**}_{\xi_2 \xi_2}.$$

Therefore,

$$
\begin{aligned}
0_* \;=\; & u^{**}_{x_1 x_1} -_* e^2 \sin_* x_1 \cdot_* u^{**}_{x_1 x_2} -_* (\cos_* x_1)^{2*} \cdot_* u^{**}_{x_2 x_2} -_* \cos_* x_1 \cdot_* u^{*}_{x_2} \\[4pt]
=\; & \cos_* x_1 \cdot_* u^{*}_{\xi_1} +_* \cos_* x_1 \cdot_* u^{*}_{\xi_2} +_* (\sin_* x_1 -_* e)^{2*} \cdot_* u^{**}_{\xi_1 \xi_1} \\[4pt]
& +_* e^2 \cdot_* ((\sin_* x_1)^{2*} -_* e) \cdot_* u^{**}_{\xi_1 \xi_2} +_* (\sin_* x_1 +_* e)^{2*} \cdot_* u^{**}_{\xi_2 \xi_2} \\[4pt]
& -_* e^2 \cdot_* \sin_* x_1 \cdot_* (\sin_* x_1 -_* e) \cdot_* u^{**}_{\xi_1 \xi_1} -_* e^4 \cdot_* (\sin_* x_1)^{2*} \cdot_* u^{**}_{\xi_1 \xi_2} \\[4pt]
& -_* e^2 \cdot_* \sin_* x_1 \cdot_* (\sin_* x_1 +_* e) \cdot_* u^{**}_{\xi_2 \xi_2} -_* (\cos_* x_1)^{2*} \cdot_* u^{**}_{\xi_1 \xi_1} \\[4pt]
& -_* e^2 \cdot_* (\cos_* x_1)^{2*} \cdot_* u^{**}_{\xi_1 \xi_2} -_* (\cos_* x_1)^{2*} \cdot_* u^{**}_{\xi_2 \xi_2} -_* \cos_* x_1 u^{*}_{\xi_1} \\[4pt]
& -_* \cos_* x_1 u^{*}_{\xi_2} -_* e^4 \cdot_* u^{**}_{\xi_1 \xi_2}, \\[4pt]
=\; & u^{**}_{\xi_1 \xi_2},
\end{aligned}
$$

i.e.,

$$
u^{**}_{\xi_1 \xi_2} = 0_*
$$

is the canonical form of the considered equation.

Exercise 3.2 Find the canonical form of the following equations:

1. $u^{**}_{x_1 x_1} -_* e^5 \cdot_* u^{**}_{x_1 x_2} +_* e^6 \cdot_* u^{**}_{x_2 x_2} = 0_*.$

2. $u^{**}_{x_1 x_1} +_* e^5 \cdot_* u^{**}_{x_1 x_2} +_* e^4 \cdot_* u^{**}_{x_2 x_2} = 0_*.$

3. $x_1^{2*} \cdot_* u^{**}_{x_1 x_1} -_* x_2^{2*} \cdot_* u^{**}_{x_2 x_2} -_* e^2 \cdot_* x_2 \cdot_* u^{*}_{x_2} = 0_*, \quad x_1 \ne 0,$
   $x_2 \ne 0.$

Answer   *1.*

$$
\begin{aligned}
u^{**}_{\xi_1 \xi_2} &= 0_*, \\[6pt]
x_2 +_* e^3 \cdot_* x_1 &= \xi_1, \\[6pt]
x_2 +_* e^2 \cdot_* x_1 &= \xi_2.
\end{aligned}
$$

2.

$$
\begin{aligned}
u^{**}_{\xi_1 \xi_2} &= 0_*, \\[6pt]
x_2 +_* e^4 \cdot_* x_1 &= \xi_1, \\[6pt]
x_2 +_* x_1 &= \xi_2.
\end{aligned}
$$

3.

$$u^{**}_{\xi_1\xi_2} +_* \left(e/_*\left(e^2 \cdot_* \xi_1\right)\right) \cdot_* u^*_{\xi_2} = 0,$$

$$x_2/_* x_1 = \xi_1,$$

$$x_1 \cdot_* x_2 = \xi_2.$$

Exercise 3.3 Consider the equation

$$u^{**}_{x_1 x_1} -_* u^{**}_{x_2 x_2} +_* e^2 \cdot_* u^*_{x_1} +_* e^2 \cdot_* u^*_{x_2} = 0_*.$$

1. Find the canonical form.
2. Find the general solution.
3. Find the solution for which

$$u(1, x_2) = e,$$

$$u^*_{x_1}(1, x_2) = -_* e.$$

Solution   1. The characteristic equation is

$$(d_* x_2)^{2*} -_* (d_* x_1)^{2*} = 0_*,$$

whereupon

$$d_* x_2 = \pm_* d_* x_1$$

and

$$x_2 \pm_* x_1 = c.$$

We set

$$\xi_1 = x_2 +_* x_1,$$

$$\xi_2 = x_2 -_* x_1.$$

Then,

$$\xi^*_{1x_1} = e,$$

$$\xi^*_{1x_2} = e,$$

$$\xi^*_{2x_1} = -_* e,$$

$$\xi_{2x_2}^* \;=\; e,$$

$$u_{x_1}^* \;=\; u_{\xi_1}^* \;\cdot_* \xi_{1x_1}^* +_* u_{\xi_2}^* \cdot_* \xi_{2x_1}^*$$

$$=\; u_{\xi_1}^* -_* u_{\xi_2}^*,$$

$$u_{x_1 x_1}^{**} \;=\; u_{\xi_1 \xi_1}^{**} \cdot_* \xi_{1x_1}^* +_* u_{\xi_1 \xi_2}^{**} \cdot_* \xi_{2x_1}^* -_* u_{\xi_1 \xi_2}^{**} \cdot_* \xi_{1x_1}^* -_* u_{\xi_2 \xi_2}^{**} \cdot_* \xi_{2x_1}^*$$

$$=\; u_{\xi_1 \xi_1}^{**} -_* u_{\xi_1 \xi_2}^{**} -_* u_{\xi_1 \xi_2}^{**} +_* u_{\xi_2 \xi_2}^{**}$$

$$=\; u_{\xi_1 \xi_1}^{**} -_* e^2 \cdot_* u_{\xi_1 \xi_2}^{**} +_* u_{\xi_2 \xi_2}^{**},$$

$$u_{x_2}^* \;=\; u_{\xi_1}^* \cdot_* \xi_{1x_2}^* +_* u_{\xi_2}^* \cdot_* \xi_{2x_2}^*$$

$$=\; u_{\xi_1}^* +_* u_{\xi_2}^*,$$

$$u_{x_2 x_2}^{**} \;=\; u_{\xi_1 \xi_1}^{**} \cdot_* \xi_{1x_2}^* +_* u_{\xi_1 \xi_2}^{**} \cdot_* \xi_{2x_2}^* +_* u_{\xi_1 \xi_2}^{**} \cdot_* \xi_{1x_2}^* +_* u_{\xi_2 \xi_2}^{**} \cdot_* \xi_{2x_2}^*$$

$$=\; u_{\xi_1 \xi_1}^{**} +_* u_{\xi_1 \xi_2}^{**} +_* u_{\xi_1 \xi_2}^{**} +_* u_{\xi_2 \xi_2}^{**}$$

$$=\; u_{\xi_1 \xi_1}^{**} +_* e^2 \cdot_* u_{\xi_1 \xi_2}^{**} +_* u_{\xi_2 \xi_2}^{**}.$$

Hence,

$$0_* \;=\; u_{x_1 x_1}^{**} -_* u_{x_2 x_2}^{**} +_* e^2 \cdot_* u_{x_1}^* +_* e^2 \cdot_* u_{x_2}^*$$

$$=\; u_{\xi_1 \xi_1}^{**} -_* e^2 \cdot_* u_{\xi_1 \xi_2}^{**} +_* u_{\xi_2 \xi_2}^{**} -_* u_{\xi_1 \xi_1}^{**}$$

$$-_* e^2 \cdot_* u_{\xi_1 \xi_2}^{**} -_* u_{\xi_2 \xi_2}^{**} +_* e^2 \cdot_* u_{\xi_1}^*$$

$$-_* e^2 \cdot_* u_{\xi_2}^* +_* e^2 \cdot_* u_{\xi_1}^* +_* e^2 \cdot_* u_{\xi_2}^*$$

$$-_* e^4 \cdot_* u_{\xi_1 \xi_2}^{**} +_* e^4 \cdot_* u_{\xi_1}^*,$$

whereupon

$$u_{\xi_1 \xi_2}^{**} -_* u_{\xi_1}^* = 0_* \qquad\qquad (3.10)$$

is the canonical form.

2. Fix ξ_1 and set

$$v = u_{\xi_1}^*.$$

Then equation (3.10) takes the form

$$v_{\xi_2} -_* v = 0_*,$$

from where

$$v = f_1(\xi_1) \cdot_* e^{\xi_2}.$$

Hence,

$$u_{\xi_1}^* = f_1(\xi_1) \cdot_* e^{\xi_2}$$

and

$$u = e^{\xi_2} \cdot_* f(\xi_1) +_* g(\xi_2),$$

where f and g are $\mathcal{C}_*^2$-functions. Therefore,

$$u(x_1, x_2) = f(x_1 +_* x_2) \cdot_* e^{x_2 -_* x_1} +_* g(x_2 -_* x_1) \qquad (3.11)$$

is the general solution.

3. Using equation (3.11), we get

$$u(1, x_2) = f(x_2) \cdot_* e^{x_2} +_* g(x_2),$$

$$u_{x_1}^*(x_1, x_2) = f^*(x_1 +_* x_2) \cdot_* e^{x_2 -_* x_1}$$

$$-_* f(x_1 +_* x_2) \cdot_* e^{x_2 -_* x_1} -_* g^*(x_2 -_* x_1),$$

$$u_{x_1}^*(1, x_2) = f^*(x_2) \cdot_* e^{x_2} -_* f(x_2) \cdot_* e^{x_2} -_* g^*(x_2),$$

i.e., we obtain the system

$$f(x_2) \cdot_* e^{x_2} +_* g(x_2) = e,$$

$$\qquad (3.12)$$

$$f^*(x_2) \cdot_* e^{x_2} -_* f(x_2) \cdot_* e^{x_2} -_* g^*(x_2) = -_* e.$$

We multiplicative differentiate the first equation of the last system with respect to x_2 and we obtain

$$f^*(x_2) \cdot_* e^{x_2} +_* f(x_2) \cdot_* e^{x_2} +_* g^*(x_2) = 0_*,$$

$$f^*(x_2) \cdot_* e^{x_2} -_* f(x_2) \cdot_* e^{x_2} -_* g^*(x_2) = -_* e.$$

whereupon

$$e^2 \cdot_* f^*(x_2) \cdot_* e^{x_2} = -_* e$$

or

$$f^*(x_2) = -_* (e /_* e^2) \cdot_* e^{-_* x_2}$$

or

$$f(x_2) = e^{\frac{1}{2}} \cdot_* e^{-_* x_2} +_* c.$$

From here and from the first equation of (3.12), we find

$$g(x_2) \;=\; e -_* f(x_2) \cdot_* e^{x_2}$$

$$=\; e -_* \left(e^{\frac{1}{2}} \cdot_* e^{-*x_2} +_* c \right) \cdot_* e^{x_2}$$

$$=\; e^{\frac{1}{2}} -_* c \cdot_* e^{x_2}.$$

Therefore,

$$u(x_1, x_2) \;=\; \left(e^{\frac{1}{2}} \cdot_* e^{-*x_1 -* x_2} +_* c \right) \cdot_* e^{x_2 -* x_1} +_* e^{\frac{1}{2}} -_* c \cdot_* e^{x_2 -* x_1}$$

$$=\; e^{\frac{1}{2}} \cdot_* e^{-*e^2 \cdot_* x_1} +_* e^{\frac{1}{2}}.$$

Exercise 3.4 Consider the equation

$$x_1^{2*} \cdot_* u^*_{x_1 x_1} -_* e^2 \cdot_* x_1 \cdot_* x_2 \cdot_* u^*_{x_1 x_2}$$
$$-_* e^3 \cdot_* x_2^{2*} \cdot_* u^*_{x_2 x_2} = 0_*, \quad x_1 > 0_*, \quad x_2 > 0_*.$$

1. Find the canonical form.
2. Find the general solution.
3. Find the solution $u(x_1, x_2)$ for which

$$u(x_1, e) \;=\; 1,$$

$$u^*_{x_2}(x_1, e) \;=\; (e^4 /_* (e^3 \cdot_* x_1^{\frac{5}{4}*}).$$

Solution   1. The characteristic equation is

$$x_1^{2*} \cdot_* (d_* x_2)^2 +_* e^2 \cdot_* x_1 \cdot_* x_2 \cdot_* d_* x_1 \cdot_* d_* x_2$$
$$-_* e^3 \cdot_* x_2^{2*} \cdot_* (d_* x_1)^{2*} = 0_*,$$

whereupon

$$x_1^{2*} \cdot_* ((d_* x_2)/_* (d_* x_1))^{2*} +_* e^2 \cdot_* x_1 \cdot_* x_2 \cdot_* ((d_* x_2)/_* (d_* x_1))$$
$$-_* e^3 \cdot_* x_2^{2*} = 0_*$$

and

$$((d_* x_2)/_* (d_* x_1))_{1,2} \;=\; (-_* x_1 \cdot_* x_2 \pm_* (x_1^{2*} \cdot_* x_2^{2*}$$
$$+_* e^3 \cdot_* x_1^{2*} \cdot_* x_2^{2*})^{\frac{1}{2}*}) /_* (x_1^{2*})$$

$$=\; (-_* x_1 \cdot_* x_2 \pm_* e^2 \cdot_* x_1 \cdot_* x_2) /_* (x_1^{2*}),$$

$$(d_* x_2)/_* (d_* x_1) \;=\; x_2 /_* x_1$$

and

$$(d_* x_2)/_*(d_* x_1) = -_* e^3 (x_2/_* x_1).$$

Hence,

$$x_2/_* x_1 \quad = \quad c,$$

$$x_1^{3*} \cdot_* x_2 \quad = \quad c.$$

We set

$$\xi_1 \quad = \quad x_2/_* x_1,$$

$$\xi_2 \quad = \quad x_1^{3*} \cdot_* x_2.$$

Then,

$$\xi_{1 x_1}^* \quad = \quad -_*(x_2/_*(x_1^{2*})),$$

$$\xi_{1 x_2}^* \quad = \quad e/_* x_1,$$

$$\xi_{2 x_1}^* \quad = \quad e^3 \cdot_* x_1^{2*} \cdot_* x_2,$$

$$\xi_{2 x_2}^* \quad = \quad x_1^{3*},$$

$$u_{x_1}^* \quad = \quad u_{\xi_1}^* \cdot_* \xi_{1 x_1}^* +_* u_{\xi_2}^* \cdot_* \xi_{2 x_1}^*$$

$$\quad = \quad -_*(x_2/_*(x_1^{2*})) \cdot_* u_{\xi_1}^* +_* e^3 \cdot_* x_1^{2*} \cdot_* x_2 \cdot_* u_{\xi_2}^*,$$

$$u_{x_1 x_1}^{**} \quad = \quad ((e^2 \cdot_* x_2)/_*(x_1^{3*})) \cdot_* u_{\xi_1}^* -_* (x_2/_*(x_1^{2*}))$$

$$\cdot_* \left(u_{\xi_1 \xi_1}^{**} \cdot_* \xi_{1 x_1}^* +_* u_{\xi_1 \xi_2}^{**} \cdot_* \xi_{2 x_1}^* \right)$$

$$+_* e^6 \cdot_* x_1 \cdot_* x_2 \cdot_* u_{\xi_2}^* +_* e^3 \cdot_* x_1^{2*} \cdot_* x_2$$

$$\cdot_* \left(u_{\xi_1 \xi_2}^{**} \cdot_* \xi_{1 x_1}^* +_* u_{\xi_2 \xi_2}^{**} \cdot_* \xi_{2 x_1}^* \right)$$

$$\quad = \quad ((e^2 \cdot_* x_2)/_*(x_1^{3*})) \cdot_* u_{\xi_1}^* -_* (x_2/_*(x_1^{2*})) \cdot_*$$

$$\left(-_*(x_2/_*(x_1^{2*})) \cdot_* u_{\xi_1 \xi_1}^{**} +_* e^3 \cdot_* x_1^{2*} \cdot_* x_2 \cdot_* u_{\xi_1 \xi_2}^{**} \right)$$

$$+_* e^6 \cdot_* x_1 \cdot_* x_2 \cdot_* u^*_{\xi_2}$$

$$+_* e^3 \cdot_* x_1^{2*} \cdot_* x_2 \cdot_* \left(-_*(x_2/_*(x_1^{2*})) \cdot_* u^{**}_{\xi_1\xi_2}\right.$$

$$\left.+_* e^3 \cdot_* x_1^{2*} \cdot_* x_2 \cdot_* u^{**}_{\xi_2\xi_2}\right)$$

$$= \ ((e^2 \cdot_* x_2)/_*(x_1^{3*})) \cdot_* u^*_{\xi_1} +_* e^6 \cdot_* x_1 \cdot_* x_2 \cdot_* u^*_{\xi_2}$$

$$+_* ((x_2^{2*})/_*(x_1^{4*})) \cdot_* u^{**}_{\xi_1\xi_1}$$

$$-_* e^6 \cdot_* x_2^{2*} \cdot_* u^{**}_{\xi_1\xi_2} +_* e^9 \cdot_* x_1^{4*} \cdot_* x_2^{2*} \cdot_* u^{**}_{\xi_2\xi_2},$$

$$u^{**}_{x_1 x_2} = \ -_*(e/_*(x_1^{2*})) \cdot_* u^*_{\xi_1} -_* (x_2/_*(x_1^{2*}))$$

$$\cdot_* \left(u^{**}_{\xi_1\xi_1} \cdot_* \xi^*_{1x_2} +_* u^{**}_{\xi_1\xi_2} \cdot_* \xi^*_{2x_2}\right)$$

$$+_* e^3 \cdot_* x_1^{2*} \cdot_* u^*_{\xi_2} +_* e^3 \cdot_* x_1^{2*} \cdot_* x_2$$

$$\cdot_* \left(u^{**}_{\xi_1\xi_2} \cdot_* \xi^*_{1x_2} +_* u^{**}_{\xi_2\xi_2} \cdot_* \xi^*_{2x_2}\right)$$

$$= \ -_*(e/_*(x_1^{2*})) \cdot_* u^*_{\xi_1} +_* e^3 \cdot_* x_1^{2*} \cdot_* u^*_{\xi_2} -_* (x_2/_*(x_1^{2*}))$$

$$\cdot_* \left((e/_* x_1) \cdot_* u^{**}_{\xi_1\xi_1} +_* x_1^{3*} \cdot_* u^{**}_{\xi_1\xi_2}\right)$$

$$+_* e^3 \cdot_* x_1^{2*} \cdot_* x_2 \cdot_* \left((e/_* x_1) \cdot_* u^{**}_{\xi_1\xi_2} +_* x_1^{3*} \cdot_* u^{**}_{\xi_2\xi_2}\right)$$

$$= \ -_*(e/_*(x_1^{2*})) \cdot_* u^*_{\xi_1} +_* e^3 \cdot_* x_1^{2*} \cdot_* u^*_{\xi_2} -_* (x_2/_*(x_1^{3*}))$$

$$\cdot_* u^{**}_{\xi_1\xi_1} +_* e^2 \cdot_* x_1 \cdot_* x_2 \cdot_* u^{**}_{\xi_1\xi_2}$$

$$+_* e^3 \cdot_* x_1^{5*} \cdot_* x_2 \cdot_* u^{**}_{\xi_2\xi_2},$$

$$u^*_{x_2} = \ u^*_{\xi_1} \cdot_* \xi^*_{1x_2} +_* u^*_{\xi_2} \cdot_* \xi^*_{2x_2}$$

$$= \ (e/_* x_1) \cdot_* u^*_{\xi_1} +_* x_1^{3*} \cdot_* u^*_{\xi_2},$$

$$u^*_{x_2 x_2} = (e/_* x_1) \cdot_* \left(u^{**}_{\xi_1 \xi_1} \cdot_* \xi^*_{1 x_2} +_* u^{**}_{\xi_1 \xi_2} \cdot_* \xi^*_{2 x_2} \right)$$

$$+_* x_1^{3*} \cdot_* \left(u^{**}_{\xi_1 \xi_2} \cdot_* \xi^*_{1 x_2} +_* u^{**}_{\xi_2 \xi_2} \cdot_* \xi^*_{2 x_2} \right)$$

$$= (e/_* x_1) \cdot_* \left((e/_* x_1) \cdot_* u^{**}_{\xi_1 \xi_1} +_* x_1^{3*} \cdot_* u^{**}_{\xi_1 \xi_2} \right)$$

$$+_* x_1^{3*} \cdot_* \left((e/_* x_1) \cdot_* u^{**}_{\xi_1 \xi_2} +_* x_1^{3*} \cdot_* u^{**}_{\xi_2 \xi_2} \right)$$

$$= (e/_* (x_1^{2*})) \cdot_* u^{**}_{\xi_1 \xi_1} +_* e^2 \cdot_* x_1^{2*} \cdot_* u^{**}_{\xi_1 \xi_2} +_* x_1^{6*} \cdot_* u^{**}_{\xi_2 \xi_2}.$$

Therefore,

$$0_* = x_1^{2*} \cdot_* u^{**}_{x_1 x_1} -_* e^2 \cdot_* x_1 \cdot_* x_2 \cdot_* u^{**}_{x_1 x_2} -_* e^3 x_2^{2*} \cdot_* u^{**}_{x_2 x_2}$$

$$= x_1^{2*} \cdot_* \left(((e^2 \cdot_* x_2)/_* (x_1^{3*})) \cdot_* u^*_{\xi_1} +_* e^6 \cdot_* x_1 \cdot_* x_2 \cdot_* u^*_{\xi_2} \right.$$

$$+_* ((x_2^{2*})/_* (x_1^{4*})) \cdot_* u^{**}_{\xi_1 \xi_1} -_* e^6 \cdot_* x_2^{2*}$$

$$\left. \cdot_* u^{**}_{\xi_1 \xi_2} +_* e^9 \cdot_* x_1^{4*} \cdot_* x_2^{2*} \cdot_* u^{**}_{\xi_2 \xi_2} \right) -_* e^2 \cdot_* x_1 \cdot_* x_2 \cdot_*$$

$$\left(-_* (e/_* (x_1^{2*})) \cdot_* u^*_{\xi_1} +_* e^3 \cdot_* x_1^{2*} \cdot_* u^*_{\xi_2} -_* (x_2/_* (x_1^{3*})) \cdot_* u^{**}_{\xi_1 \xi_1} \right.$$

$$\left. +_* e^2 \cdot_* x_1 \cdot_* x_2 \cdot_* u^{**}_{\xi_1 \xi_2} +_* e^3 \cdot_* x_1^{5*} \cdot_* x_2 \cdot_* u^{**}_{\xi_2 \xi_2} \right)$$

$$-_* e^3 \cdot_* x_2^{2*} \cdot_* \left((e/_* (x_1^{2*})) \cdot_* u^{**}_{\xi_1 \xi_1} \right.$$

$$\left. +_* e^2 \cdot_* x_1^{2*} \cdot_* u^{**}_{\xi_1 \xi_2} +_* x_1^{6*} \cdot_* u^{**}_{\xi_2 \xi_2} \right)$$

$$= e^2 \cdot_* (x_2/_* x_1) \cdot_* u^*_{\xi_1} +_* e^6 \cdot_* x_1^{3*} \cdot_* x_2 \cdot_* u^*_{\xi_2} +_* ((x_2^{2*})/_* (x_1^{2*}))$$

$$\cdot_* u^{**}_{\xi_1 \xi_1} -_* e^6 \cdot_* x_1^{2*} \cdot_* x_2^{2*} \cdot_* u^{**}_{\xi_1 \xi_2}$$

$$+_* e^9 \cdot_* x_1^{6*} \cdot_* x_2^{2*} \cdot_* u^{**}_{\xi_2 \xi_2} +_* e^2 \cdot_* (x_2/_* x_1) \cdot_* u^*_{\xi_1}$$

$$-_* e^6 \cdot_* x_1^{3*} \cdot_* x_2 \cdot_* u^*_{\xi_2} +_* e^2 \cdot_* ((x_2^{2*})/_* (x_1^{2*})) \cdot_* u^{**}_{\xi_1 \xi_1}$$

$$-{}_*e^4 \cdot{}_* x_1^{2*} \cdot{}_* x_2^{2*} u_{\xi_1\xi_2}^{**} -{}_* e^6 \cdot{}_* x_1^{6*} \cdot{}_* x_2^{2*} \cdot{}_* u_{\xi_2\xi_2}^{**}$$

$$-{}_*e^3 \cdot{}_* ((x_2^{2*})/{}_*(x_1^{2*})) \cdot{}_* u_{\xi_1\xi_1}^{**} -{}_* e^6 \cdot{}_* x_1^{2*} \cdot{}_* x_2^{2*} \cdot{}_* u_{\xi_1\xi_2}^{**}$$

$$-{}_*e^3 \cdot{}_* x_1^{6*} \cdot{}_* x_2^{2*} \cdot{}_* u_{\xi_2\xi_2}^{**}$$

$$= \ e^4 \cdot{}_* (x_2/{}_*x_1) \cdot{}_* u_{\xi_1}^{*} -{}_* e^{16} \cdot{}_* x_1^{2*} \cdot{}_* x_2^{2*} \cdot{}_* u_{\xi_1\xi_2}^{**}$$

$$= \ e^4 \cdot{}_* u_{\xi_1\xi_2}^{**} -{}_* (e/{}_*(x_1^{3*} \cdot{}_* x_2)) \cdot{}_* u_{\xi_1}^{*},$$

i.e.,

$$e^4 \cdot{}_* u_{\xi_1\xi_2}^{**} -{}_* (e/{}_*\xi_2) \cdot{}_* u_{\xi_1}^{*} = 0_* \tag{3.13}$$

is the canonical form of the considered equation.

2. Fix ξ_1 and set

$$v = u_{\xi_1}^{*}.$$

Then equation (3.13) takes the form

$$e^4 \cdot{}_* v_{\xi_2}^{*} -{}_* (e/{}_*\xi_2) \cdot{}_* v = 0_*$$

or

$$e^4 \cdot{}_* ((v_{\xi_2}^{*})/{}_*v) = (e/{}_*\xi_2)$$

or

$$v = x_2^{\frac{1}{4}*} \cdot{}_* f_1^{*}(\xi_1),$$

and

$$u_{\xi_1}^{*} = \xi_2^{\frac{1}{4}*} \cdot{}_* f_1^{*}(\xi_1)$$

and

$$u = \xi_2^{\frac{1}{4}*} \cdot{}_* f(\xi_1) +{}_* g(\xi_2),$$

i.e.,

$$u(x_1, x_2) = (x_1^{3*} \cdot{}_* x_2)^{\frac{1}{4}*} \cdot{}_* f(x_2/{}_*x_1) +{}_* g(x_1^{3*} \cdot{}_* x_2) \tag{3.14}$$

is the general solution of the considered equation, where f and g are C_*^2-functions.

3. From equation (3.14), we get

$$u(x_1, 1) \ = \ (x_1^{3*})^{\frac{4}{3}*} \cdot{}_* f(e/{}_*x_1) +{}_* g(x_1^{3*}),$$

$$u(x_1, x_2) \ = \ e^{\frac{1}{4}} \cdot{}_* ((x_1^{3*})/{}_*(x_2^{3*})) \cdot{}_* f(x_2/{}_*x_1) +{}_* (x_2/{}_*x_1)^{\frac{1}{4}*}$$

$$\cdot{}_*f^{*}(x_2/{}_*x_1) +{}_* x_1^{3*} \cdot{}_* g^{*}(x_1^{3*} \cdot{}_* x_2),$$

$$u(x_1, e) = e^{\frac{1}{4}} \cdot_* (x_1^{3*})^{\frac{1}{4}*} \cdot_* f\left(e/_* x_1\right) +_* \left(e/_*(x_1)^{\frac{1}{4}*}\right)$$

$$\cdot_* f^*\left(e/_* x_1\right) +_* x_1^{3*} \cdot_* g^*(x_1^{3*}).$$

In this way we get the system

$$(x_1^{3*})^{\frac{1}{4}*} \cdot_* f\left(e/_* x_1\right) +_* g(x_1^{3*}) = 0_*$$

$$e^{\frac{1}{4}} \cdot_* (x_1^{3*})^{\frac{1}{4}*} \cdot_* f\left(e/_* x_1\right) +_* \left(e/_*(x_1)^{\frac{1}{4}*}\right) \cdot_* f^*\left(e/_* x_1\right) \quad (3.15)$$

$$+_* x_1^{3*} \cdot_* g^*(x_1^{3*}) = \left(e^4/_*(e^3 \cdot_* x_1^{\frac{1}{4}*} \cdot_* x_1\right).$$

We multiplicative differentiate the first equation of the last system and we get

$$\left(e^3/_*(e^4 \cdot_* x_1^{\frac{1}{4}*}\right) \cdot_* f\left(e/_* x_1\right) -_* \left(e/_*(x_1^{\frac{5}{4}*})\right) \cdot_* f^*\left(e/_* x_1\right)$$

$$+_* e^3 \cdot_* x_1^{2*} \cdot_* g^*(x_1^{3*}) = 0_*$$

$$e^{\frac{1}{4}} \cdot_* (x_1^{3*})^{\frac{1}{4}*} \cdot_* f\left(e/_* x_1\right) +_* \left(e/_*(x_1^{\frac{1}{4}*})\right) \cdot_* f^*\left(e/_* x_1\right)$$

$$+_* x_1^{3*} \cdot_* g^*(x_1^{3*}) = \left(e^4/_*(e^3 \cdot_* x_1^{\frac{1}{4}*} \cdot_* x_1\right),$$

whereupon

$$e^{\frac{3}{4}} \cdot_* (x_1^{3*})^{\frac{1}{4}*} \cdot_* f\left(e/_* x_1\right) -_* \left(e/_*(x_1^{\frac{1}{4}*})\right) \cdot_* f^*\left(e/_* x_1\right)$$

$$+_* e^3 \cdot_* x_1^{3*} \cdot_* g^*(x_1^{3*}) = 0_*,$$

$$e^{\frac{1}{4}} \cdot_* (x_1^{3*})^{\frac{1}{4}*} \cdot_* f\left(e/_* x_1\right) +_* \left(e/_*(x_1^{\frac{1}{4}*})\right) \cdot_* f^*\left(e/_* x_1\right)$$

$$+_* x_1^{3*} \cdot_* g^*(x_1^{3*}) = \left(e^4/_*(e^3 \cdot_* x_1^{\frac{1}{4}*} \cdot_* x_1\right))$$

and

$$f^*\left(e/_* x_1\right) = e/_* x_1,$$

i.e.,

$$f^*(z) = z.$$

Therefore,

$$f(z) = e^{\frac{1}{2}} \cdot_* z^{2*} +_* c.$$

Hence, from the first equation of (3.15), we obtain

$$(x_1^{3*})^{\frac{1}{4}*} \cdot_* \left(e/_*(e^2 \cdot_* x_1^{2*}) +_* c\right) +_* g(x_1^{3*}) = 0_*.$$

From here,

$$g(z) = -_*c \cdot_* z^{\frac{1}{4}*} -_* (e/_*(e^2 \cdot_* z^{\frac{5}{12}*})).$$

Consequently,

$$
\begin{aligned}
u(x_1, x_2) \ &= \ (x_1^{3*} \cdot_* x_2)^{\frac{1}{4}*} \cdot_* f\left(x_2/_*x_1\right) +_* g(x_1^{3*} \cdot_* x_2) \\[2mm]
&= \ (x_1^{3*} \cdot_* x_2)^{\frac{1}{4}*} \cdot_* \left(e^{\frac{1}{2}} \cdot_* ((x_2^{2*})/_*(x_1^{2*})) +_* c\right) \\[2mm]
&\quad -_*c \cdot_* (x_1^{3*} \cdot_* x_2)^{\frac{1}{4}*} -_* (\frac{e}{7}_* (e^2 \cdot_* x_1^{\frac{5}{4}*} \cdot_* x_2^{\frac{5}{12}*}) \\[2mm]
&= \ e^{\frac{1}{2}} \cdot_* ((x_2^{\frac{9}{4}*})/_*(x_1^{\frac{5}{4}*})) -_* (e/_*(e^2 \cdot_* x_1^{\frac{5}{4}*} \cdot_* x_2^{\frac{5}{12}*}).
\end{aligned}
$$

Exercise 3.5 Consider the equation

$$u_{x_1 x_1}^{**} +_* e^6 \cdot_* u_{x_1 x_2}^{**} -_* e^{16} \cdot_* u_{x_2 x_2}^{**} = 0_*.$$

1. Find the canonical form.
2. Find the general solution.
3. Find the solution $u(x_1, x_2)$ for which

$$u(-_*x_1, e^2 \cdot_* x_1) \ = \ x_1,$$

$$u(x_1, 1) \ = \ e^2 \cdot_* x_1.$$

Solution   1. The characteristic equation is

$$(d_*x_2)^{2*} -_* e^6 \cdot_* d_*x_1 \cdot_* d_*x_2 -_* e^{16} \cdot_* (d_*x_1)^2 = 0_*,$$

whereupon

$$((d_*x_2)/_*(d_*x_1))^{2*} -_* e^6 \cdot_* ((d_*x_2)/_*(d_*x_1)) -_* e^{16} = 0_*$$

and

$$
\begin{aligned}
(d_*x_2)/_*(d_*x_1) \ &= \ (e^3 \pm_* (e^9 +_* e^{16})^{\frac{1}{2}*})/_*e \\[2mm]
&= \ e^3 \pm_* e^5, \\[2mm]
d_*x_2 \ &= \ e^8 \cdot_* d_*x_1, \\[2mm]
d_*x_2 \ &= \ -_*e^2 \cdot_* d_*x_1.
\end{aligned}
$$

Therefore,

$$x_2 -_* e^8 \cdot_* x_1 \;\; = \;\; c,$$

$$x_2 +_* e^2 \cdot_* x_1 \;\; = \;\; c.$$

We set

$$\xi_1 \;\; = \;\; x_2 -_* e^8 \cdot_* x_1,$$

$$\xi_2 \;\; = \;\; x_2 +_* e^2 \cdot_* x_1.$$

Then,

$$\xi^*_{1x_1} \;\; = \;\; -_* e^8,$$

$$\xi^*_{1x_2} \;\; = \;\; e,$$

$$\xi^*_{2x_1} \;\; = \;\; e^2,$$

$$\xi^*_{2x_2} \;\; = \;\; e,$$

$$u^*_{x_1} \;\; = \;\; u^*_{\xi_1} \cdot_* \xi^*_{1x_1} +_* u^*_{\xi_2} \cdot_* \xi^*_{2x_1}$$

$$= \;\; -_* e^8 \cdot_* u^*_{\xi_1} +_* e^2 \cdot_* u^*_{\xi_2},$$

$$u^{**}_{x_1 x_1} \;\; = \;\; -_* e^8 \cdot_* \left(u^{**}_{\xi_1 \xi_1} \cdot_* \xi^*_{1x_1} +_* u^{**}_{\xi_1 \xi_2} \cdot_* \xi^*_{2x_1} \right)$$

$$+_* e^2 \cdot_* \left(u^{**}_{\xi_1 \xi_2} \cdot_* \xi^*_{1x_1} +_* u^{**}_{\xi_2 \xi_2} \cdot_* \xi^*_{2x_1} \right)$$

$$= \;\; -_* e^8 \cdot_* \left(-_* e^8 \cdot_* u^{**}_{\xi_1 \xi_1} +_* e^2 \cdot_* u^{**}_{\xi_1 \xi_2} \right)$$

$$+_* e^2 \cdot_* \left(-_* e^8 \cdot_* u^{**}_{\xi_1 \xi_1} +_* e^2 \cdot_* u^{**}_{\xi_2 \xi_2} \right)$$

$$= \;\; e^{64} \cdot_* u^{**}_{\xi_1 \xi_1} -_* e^{32} \cdot_* u^{**}_{\xi_1 \xi_2} +_* e^4 \cdot_* u^{**}_{\xi_2 \xi_2},$$

$$u^{**}_{x_1 x_2} \;\; = \;\; -_* e^8 \cdot_* \left(u^{**}_{\xi_1 \xi_1} \cdot_* \xi^*_{1x_2} +_* u^{**}_{\xi_1 \xi_2} \cdot_* \xi^*_{2x_2} \right)$$

$$+_* e^2 \cdot_* \left(u^{**}_{\xi_1 \xi_2} \cdot_* \xi^*_{1x_2} +_* u^{**}_{\xi_2 \xi_2} \cdot_* \xi^*_{2x_2} \right)$$

$$= \;\; -_* e^8 \cdot_* \left(u^{**}_{\xi_1 \xi_1} +_* u^{**}_{\xi_1 \xi_2} \right)$$

$$+_* e^2 \cdot_* \left(u^{**}_{\xi_1 \xi_2} +_* u^{**}_{\xi_2 \xi_2} \right)$$

$$= \;\; -_* e^8 \cdot_* u^{**}_{\xi_1 \xi_1} -_* e^6 \cdot_* u^{**}_{\xi_1 \xi_2} +_* e^2 \cdot_* u^{**}_{\xi_2 \xi_2},$$

$$u_{x_2}^* = u_{\xi_1}^* \cdot_* \xi_{1x_2}^* +_* u_{\xi_2}^* \cdot_* \xi_{2x_2}^*$$

$$= u_{\xi_1}^* +_* u_{\xi_2}^*,$$

$$u_{x_2 x_2}^* = u_{\xi_1 \xi_1}^{**} \cdot_* \xi_{1x_2}^* +_* u_{\xi_1 \xi_2}^{**} \cdot_* \xi_{2x_2}^* +_* u_{\xi_1 \xi_2}^{**} \cdot_* \xi_{1x_2}^* +_* u_{\xi_2 \xi_2}^{**} \cdot_* \xi_{2x_2}^*$$

$$= u_{\xi_1 \xi_1}^{**} +_* u_{\xi_1 \xi_2}^{**} +_* u_{\xi_1 \xi_2}^{**} +_* u_{\xi_2 \xi_2}^{**}$$

$$= u_{\xi_1 \xi_1}^{**} +_* e^2 \cdot_* u_{\xi_1 \xi_2}^{**} +_* u_{\xi_2 \xi_2}^{**}.$$

Therefore,

$$0_* = u_{x_1 x_1}^{**} +_* e^6 \cdot_* u_{x_1 x_2}^{**} -_* e^{16} \cdot_* u_{x_2 x_2}^{**}$$

$$= e^{64} \cdot_* u_{\xi_1 \xi_1}^{**} -_* e^{32} \cdot_* u_{\xi_1 \xi_2}^{**} +_* e^4 \cdot_* u_{\xi_2 \xi_2}^{**}$$

$$-_* e^{48} \cdot_* u_{\xi_1 \xi_1}^{**} -_* e^{36} \cdot_* u_{\xi_1 \xi_2}^{**} +_* e^{12} \cdot_* u_{\xi_2 \xi_2}^{**}$$

$$= -_* e^{16} \cdot_* u_{\xi_1 \xi_1}^{**} -_* e^{32} \cdot_* u_{\xi_1 \xi_2}^{**} -_* e^{16} \cdot_* u_{\xi_2 \xi_2}^*$$

$$= -_* e^{100} \cdot_* u_{\xi_1 \xi_2}^{**},$$

i.e.,

$$u_{\xi_1 \xi_2}^{**} = 0_* \tag{3.16}$$

is the canonical form of the considered equation.

2. Fix ξ_2 and set

$$v = u_{\xi_2}^*.$$

Then equation (3.16) takes the form

$$v_{\xi_1}^* = 0_*,$$

from where

$$v = f_1(\xi_2)$$

and

$$u_{\xi_2}^* = f_1(\xi_2).$$

Hence,

$$u = f(\xi_2) +_* g(\xi_1),$$

i.e.,

$$u(x_1, x_2) = f(x_2 +_* e^2 \cdot_* x_1) +_* g(x_2 -_* e^8 \cdot_* x_1) \tag{3.17}$$

is the general solution of the considered equation, where f and g are C_*^2-functions.

3. From equation (3.17), we get

$$u(-_*x_1, 2x_1) = f(e^2 \cdot_* x_1 -_* e^2 \cdot_* x_1) +_* g(e^2 \cdot_* x_1 +_* e^8 \cdot_* x_1)$$

$$= f(0_*) +_* g(e^{10} \cdot_* x_1),$$

$$u(x_1, 1) = f(e^2 \cdot_* x_1) +_* g(-_* e^8 \cdot_* x_1).$$

In this way we get the system

$$f(1) +_* g(e^{10} \cdot_* x_1) = x_1$$

$$f(e^2 \cdot_* x_1) +_* g(-_* e^8 \cdot_* x_1) = e^2 \cdot_* x_1, \tag{3.18}$$

whereupon

$$g(e^{10} \cdot_* x_1) = x_1 -_* f(0_*)$$

and

$$g(z) = e^{\frac{1}{10}} \cdot_* z -_* f(0_*).$$

From here and from the second equation of (3.18), we find

$$f(e^2 \cdot_* x_1) = e^2 \cdot_* x_1 -_* g(-_* e^8 \cdot_* x_1)$$

$$= e^2 \cdot_* x_1 +_* (e^8 \cdot_* x_1)/_* e^{10} +_* f(0_*)$$

$$= e^{\frac{14}{5}} \cdot_* x_1 +_* f(0_*),$$

i.e.,

$$f(z) = e^{\frac{7}{5}} \cdot_* z +_* f(0_*).$$

Consequently,

$$u(x_1, x_2) = e^{\frac{7}{5}} \cdot_* (x_2 +_* e^2 \cdot_* x_1)$$

$$+_* f(0_*) +_* (x_2 -_* e^8 \cdot_* x_1)/_* e^{10} -_* f(0_*)$$

$$= e^{\frac{3}{2}} \cdot_* x_2 +_* e^2 \cdot_* x_1.$$

Exercise 3.6 Consider the equation

$$u^{**}_{x_1 x_1} +_* e^2 \cdot_* u^{**}_{x_1 x_2} -_* e^3 \cdot_* u^{**}_{x_2 x_2} = 0_*.$$

1. Find the canonical form.
2. Find the general solution.

3. Find the solution $u(x_1, x_2)$ for which

$$u(x_1, 1) \;=\; e^3 \cdot_* x_1^{2*},$$

$$u_{x_2}^*(x_1, 1) \;=\; 1.$$

Answer   1.

$$u_{\xi_1 \xi_2}^{**} \;=\; 0_*,$$

$$\xi_1 \;=\; x_2 -_* e^3 \cdot_* x_1,$$

$$\xi_2 \;=\; x_2 +_* x_1.$$

2.

$$u(x_1, x_2) = f(x_2 -_* e^3 \cdot_* x_1) +_* g(x_2 +_* x_1),$$
where f and g are C_*^2-functions.

3.

$$u(x_1, x_2) = e^3 \cdot_* x_1^{2*} +_* x_2^{2*}.$$

Exercise 3.7 Consider the equation

$$x_1 \cdot_* u_{x_1 x_1}^{**} +_* (x_1 +_* x_2) \cdot_* u_{x_1 x_2}^{**} +_* x_2 \cdot_* u_{x_2 x_2}^{**} = 0_*, \qquad x_1 > 0_*, x_2 > 0.$$

1. Find the canonical form.
2. Find the general solution.
3. Find a solution $u(x_1, x_2)$ for which

$$u\left(x_1, \frac{1}{x_1}\right) \;=\; x_1^{3*},$$

$$u_{x_1}^*\left(x_1, (e/_* x_1)\right) \;=\; e^2 \cdot_* x_1^{2*}.$$

Answer   1.

$$-_* \xi_1 \cdot_* u_{\xi_1 \xi_2}^{**} +_* u_{\xi_2}^* \;=\; 0_*,$$

$$\xi_1 \;=\; x_2 -_* x_1,$$

$$\xi_2 \;=\; x_2 /_* x_1.$$

2.

$$u(x_1, x_2) = (x_2 -_* x_1) \cdot_* f(x_2 /_* x_1) +_* g(x_2 -_* x_1),$$
where f and g are C_*^2-functions,

3.

$$u(x_1, x_2) = (x_1^{2*}) /_* x_2.$$

3.1.2 The multiplicative elliptic case

We suppose that $L(u)$ is multiplicative elliptic in U. In this case, we have

$$a \cdot_* c -_* b^{2*} > 0_*.$$

Also,

$$\alpha = \gamma$$

and

$$\beta = 0_*.$$

Then the multiplicative linear independent variables

$$\xi_1 = \phi_1(x_1, x_2)$$

and

$$\xi_2 = \phi_2(x_1, x_2)$$

satisfy the system

$$a \cdot_* \phi_{1x_1}^{*2*} +_* e^2 \cdot_* b \cdot_* \phi_{1x_1}^* \cdot_* \phi_{1x_2}^* +_* c \cdot_* \phi_{1x_2}^{*2*}$$

$$= a \cdot_* \phi_{2x_1}^{*2*} +_* e^2 \cdot_* b \cdot_* \phi_{2x_1}^* \cdot_* \phi_{2x_2}^* +_* c \cdot_* \phi_{2x_2}^{*2*}$$

$$a \cdot_* \phi_{1x_1}^* \cdot_* \phi_{2x_1}^* +_* b \cdot_* \left(\phi_{1x_2}^* \cdot_* \phi_{2x_1}^* +_* \phi_{1x_1}^* \cdot_* \phi_{2x_2}^* \right)$$

$$+_* c \cdot_* \phi_{1x_2}^* \cdot_* \phi_{2x_2}^* = 0_*$$

or

$$a \cdot_* \left(\phi_{1x_1}^* -_* \phi_{2x_1}^* \right) \cdot_* \left(\phi_{1x_1}^* +_* \phi_{2x_1}^* \right) +_* e^2 \cdot_* b \cdot_* \left(\phi_{1x_1}^* \right.$$

$$\left. \cdot_* \phi_{1x_2}^* -_* \phi_{2x_1}^* \cdot_* \phi_{2x_2}^* \right)$$

$$+_* c \cdot_* \left(\phi_{1x_2}^* -_* \phi_{2x_2}^* \right) \cdot_* \left(\phi_{1x_2}^* +_* \phi_{2x_2}^* \right) = 0_*, \qquad (3.19)$$

$$a \cdot_* \phi_{1x_1}^* \phi_{2x_1}^* +_* b \cdot_* \left(\phi_{1x_2}^* \cdot_* \phi_{2x_1}^* +_* \phi_{1x_1}^* \cdot_* \phi_{2x_2}^* \right)$$

$$+_* c \cdot_* \phi_{1x_2}^* \cdot_* \phi_{2x_2}^* = 0_*.$$

Next, we rewrite equation (3.19) in the following form:

$$a \cdot_* \left(\phi_{1x_1}^{*2*} +_* (i \cdot_* \phi_{2x_1}^*)^{2*} \right) +_* e^2 \cdot_* b \cdot_* \left(\phi_{1x_1}^* \cdot_* \phi_{1x_2}^* -_* \phi_{2x_1}^* \cdot_* \phi_{2x_2}^* \right)$$

$$+_* c \cdot_* \left(\phi_{1x_2}^{*2*} +_* (i \cdot_* \phi_{2x_2})^{2*} \right) = 0_*$$

$$e^2 \cdot_* i \cdot_* a \cdot_* \phi_{1x_1}^* \cdot_* \phi_{2x_1}^* +_* e^2 \cdot_* i \cdot_* b \cdot_* \left(\phi_{1x_2}^* \cdot_* \phi_{2x_1}^* +_* \phi_{1x_1}^* \cdot_* \phi_{2x_2}^* \right)$$

$$+_* e^2 \cdot_* i \cdot_* c \cdot_* \phi_{1x_2}^* \cdot_* \phi_{2x_2}^* = 0_*$$

or

$$a \cdot_* \left(\phi_{1x_1}^* +_* i \cdot_* \phi_{2x_1}^*\right)^{*2*} +_* e^2 \cdot_* b \cdot_* \left(\phi_{1x_1}^* +_* i \cdot_* \phi_{2x_1}^*\right) \cdot_* \left(\phi_{1x_2}^* +_* i \cdot_* \phi_{2x_2}^*\right)$$

$$+_* c \cdot_* \left(\phi_{x_2}^* +_* i \cdot_* \phi_{2x_2}^*\right)^{2*} = 0_*.$$

(3.20)

Let

$$\phi_3 = \phi_{1x_1}^* +_* i \cdot_* \phi_{2x_1}^*,$$

$$\phi_4 = \phi_{1x_2}^* +_* i \cdot_* \phi_{2x_2}^*.$$

Then, using equation (3.20), we have

$$a \cdot_* \phi_3^{2*} +_* e^2 \cdot_* b \cdot_* \phi_3 \cdot_* \phi_4 +_* c \cdot_* \phi_4^{2*} = 0_*$$

and

$$a \cdot_* \left(\phi_3/_*\phi_4\right)^{2*} +_* e^2 \cdot_* b(\phi_3/_*\phi_4) +_* c = 0_*.$$

Consequently,

$$\left(\phi_3/_*\phi_4\right)_{1,2} = \left(-_* b \pm_* i \cdot_* (a \cdot_* c -_* b^{2*})^{\frac{1}{2}*}\right)/_*a$$

or

$$\phi_3 = \left(-_*(b/_*a) \pm_* i \cdot_* ((a \cdot_* c -_* b^{2*})/_*a)\right) \cdot_* \phi_4,$$

or

$$\phi_{1x_1}^* +_* i \cdot_* \phi_{2x_1}^* = \left(-_*(b/_*a) \pm_* i \cdot_* ((a \cdot_* c -_* b^{2*})^{\frac{1}{2}*})/_*a\right)$$

$$\cdot_* \left(\phi_{1x_2}^* +_* i \cdot_* \phi_{2x_2}^*\right).$$

Let

$$\phi_{1x_1}^* +_* i \cdot_* \phi_{2x_1}^* = \left(-_*(b/_*a) +_* i \cdot_* ((a \cdot_* c -_* b^{2*})/_*a)\right) \cdot_* \left(\phi_{1x_2}^* +_* i \cdot_* \phi_{2x_2}^*\right).$$

Then,

$$\phi_{1x_1}^* +_* i \cdot_* \phi_{2x_1}^* = -_*(b/_*a) \cdot_* \phi_{1x_2}^* +_* i \cdot_* ((a \cdot_* c -_* b^{2*})/_*a) \cdot_* \phi_{1x_2}^*$$

$$-_*(b/_*a) \cdot_* i \cdot_* \phi_{2x_2}^*$$

$$-_*((a \cdot_* c -_* b^{2*})/_*a) \cdot_* \phi_{2x_2}^*.$$

From here,

$$\phi_{1x_1}^* = -_*(b/_*a) \cdot_* \phi_{1x_2}^* -_* ((a \cdot_* c -_* b^{2*})/_*a) \cdot_* \phi_{2x_2}^*,$$

$$\phi_{2x_1}^* = ((a \cdot_* c -_* b^{2*})/_*a) \cdot_* \phi_{1x_2}^* -_* (b/_*a) \cdot_* \phi_{2x_2}^*.$$

From the first equation of the last system, we get

$$\phi^*_{2x_2} = -_*((a \cdot_* \phi^*_{1x_1} +_* b \cdot_* \phi^*_{1x_2})/_*(a \cdot_* c -_* b^{2*})).$$

Then, using its second equation, we find

$$
\begin{aligned}
\phi^*_{2x_1} &= ((a \cdot_* c -_* b^{2*})/_* a) \cdot_* \phi^*_{1x_2} \\
&\quad -_*(b/_* a) \cdot_* \left(-_*((a \cdot_* \phi^*_{1x_1} +_* b \cdot_* \phi^*_{1x_2})/_*(a \cdot_* c -_* b^{2*})^{\frac{1}{2}}_*)\right) \\
&= (((a \cdot_* c -_* b^{2*})^{\frac{1}{2}}_*)/_* a) \cdot_* \phi^*_{1x_2} +_* ((\cdot_* a \cdot_* b \cdot_* \phi^*_{1x_1} +_* b^{2*} \cdot_* \phi^*_{1x_2})/_* \\
&\quad (a \cdot_* (a \cdot_* c -_* b^{2*})^{\frac{1}{2}}_*)) \\
&= (a \cdot_* c -_* b^{2*} +_* b^{2*} \cdot_* \phi^*_{1x_2} +_* a \cdot_* b \cdot_* \phi^*_{1x_1})/_* \\
&\quad (a \cdot_* ((a \cdot_* c -_* b^{2*})^{\frac{1}{2}}_*)) \\
&= (a \cdot_* c \cdot_* \phi^*_{1x_2} +_* a \cdot_* b \cdot_* \phi_1 x^*_1)/_*(a \cdot_* (a \cdot_* c -_* b^{2*})^{\frac{1}{2}}_*) \\
&= (b \cdot_* \phi^*_{1x_1} +_* c \cdot_* \phi^*_{1x_2})/_*((a \cdot_* c -_* b^{2*})^{\frac{1}{2}}_*),
\end{aligned}
$$

i.e., we obtain the system

$$
\begin{aligned}
\phi^*_{2x_1} &= (b \cdot_* \phi^*_{1x_1} +_* c \cdot_* \phi^*_{1x_2})/_*((a \cdot_* c -_* b^{2*})^{\frac{1}{2}}_*) \\
\phi^*_{2x_2} &= -_*((a \cdot_* \phi^*_{1x_1} +_* b \cdot_* \phi^*_{1x_2})/_*((a \cdot_* c -_* b^{2*})^{\frac{1}{2}}_*)).
\end{aligned}
\tag{3.21}
$$

Definition 3.7 Equation (3.21) will be called multiplicative Beltrami differential equations.

From these multiplicative Beltrami differential equations, by eliminating one of the unknowns, for instance, ϕ_2, we get

$$
\begin{aligned}
(\partial_*/_*(\partial_* x_1)) \cdot_* \left((a \cdot_* \phi^*_{1x_1} +_* b \cdot_* \phi^*_{1x_2})/_*(a \cdot_* c -_* b^{2*})^{\frac{1}{2}}_*\right) \\
+_*(\partial_*/_*(\partial_* x_2)) \cdot_* \left((b \cdot_* \phi^*_{1x_1} +_* c \cdot_* \phi^*_{1x_2})/_*((a \cdot_* c -_* b^{2*})^{\frac{1}{2}}_*)\right) = 0_*.
\end{aligned}
\tag{3.22}
$$

Using equation (3.21), we find

$$
\begin{aligned}
\phi^*_{2x_1} \cdot_* \phi^*_{1x_2} -_* \phi^*_{1x_1} \cdot_* \phi^*_{2x_2} &= ((b \cdot_* \phi^*_{1x_1} +_* c \cdot_* \phi^*_{1x_2}) \\
/_*((a \cdot_* c -_* b^{2*})^{\frac{1}{2}}_*))/_*(\phi^*_{1x_2}) & \\
+_*((a \cdot_* \phi^*_{1x_1} +_* b \cdot_* \phi^*_{1x_2})/_*((a \cdot_* c -_* b^{2*})^{\frac{1}{2}}_*)) \cdot_* \phi^*_{1x_1} & \\
&= (e/_*((a \cdot_* c -_* b^{2*})^{\frac{1}{2}}_*)) \cdot_* \left(a \cdot_* \phi^{*2*}_{1x_1} +_* e^2 \cdot_* b \cdot_* \phi^*_{1x_1} \cdot_* \phi^*_{1x_2} +_* c \cdot_* \phi^{*2*}_{1x_2}\right).
\end{aligned}
$$

If we assume that

$$ a \cdot_* \phi_{1x_1}^{*2*} +_* e^2 \cdot_* b \cdot_* \phi_{1x_1}^* \cdot_* \phi_{1x_2}^* +_* c \cdot_* \phi_{1x_2}^{*2*} = 0_*, $$

then from the first equation of (3.19), we get

$$ a \cdot_* \phi_{2x_1}^{*2*} +_* e^2 \cdot_* b \cdot_* \phi_{2x_1}^* \cdot_* \phi_{2x_2}^* +_* c \cdot_* \phi_{2x_2}^{*2*} = 0_*. $$

Therefore, $(\phi_{1x_1}^*)/_*(\phi_{1x_2}^*)$ and $(\phi_{2x_1}^*)/_*(\phi_{2x_2}^*)$ are the roots of the quadratic equation:

$$ a \cdot_* \nu^{2*} +_* e^2 \cdot_* b \cdot_* \nu +_* c = 0_*. $$

From here,

$$ (\phi_{1x_1}^*)/_*(\phi_{1x_2}^*) +_* ((\phi_{2x_1}^*)/_*(\phi_{2x_2}^*)) \;=\; -_*((e^2 \cdot_* b)/_*a), $$

$$ ((\phi_{1x_1}^*)/_*(\phi_{1x_2}^*)) \cdot_* ((\phi_{2x_1}^*)/_*(\phi_{2x_2}^*)) \;=\; c/_*a. $$

From the last relations and from the second equation of (3.19), we obtain

$$
\begin{aligned}
0_* \;&=\; a \cdot_* ((\phi_{1x_1}^*)/_*(\phi_{1x_2}^*)) \cdot_* ((\phi_{2x_1}^*)/_*(\phi_{2x_2}^*)) \\[4pt]
&\quad +_* b \cdot_* \left(((\phi_{1x_1}^*)/_*(\phi_{2x_1}^*)) +_* ((\phi_{1x_2}^*)/_*(\phi_{2x_2}^*)) \right) +_* c \\[4pt]
&=\; a \cdot_* (c/_*a) +_* b \left(-_*((e^2 \cdot_* b)/_*a) \right) +_* c \\[4pt]
&=\; e^2 \cdot_* c -_* ((e^2 \cdot_* b^{2*})/_*a) \\[4pt]
&=\; e^2 \cdot_* ((a \cdot_* c -_* b^{2*})/_*a) \\[4pt]
&\neq\; 0_*.
\end{aligned}
$$

Consequently,

$$ a \cdot_* \phi_{1x_1}^{*2*} +_* e^2 \cdot_* b \cdot_* \phi_{1x_1}^* \cdot_* \phi_{1x_2}^* +_* c \cdot_* \phi_{1x_2}^{*2*} \neq 0_*, $$

$$ a \cdot_* \phi_{2x_1}^{*2*} +_* e^2 \cdot_* b \cdot_* \phi_{2x_1}^* \cdot_* \phi_{2x_2}^* +_* c \cdot_* \phi_{2x_2}^{*2*} \neq 0_*, $$

and

$$ \phi_{2x_1}^* \cdot_* \phi_{1x_2}^* -_* \phi_{1x_1}^* \cdot_* \phi_{2x_2}^* \neq 0. \tag{3.23} $$

In other words, the transformation of the differential equation to the canonical form

$$ \alpha \cdot_* \left(u_{\xi_1 \xi_1}^{**} +_* u_{\xi_2 \xi_2}^{**} \right) +_* \cdots = 0_* $$

in a neighborhood of a point is given by any pair of functions satisfying equation (3.21) and having multiplicative nonvanishing multiplicative Jacobian (3.23). Such functions are determined once we have a solution of equation (3.22) with multiplicative nonvanishing multiplicative gradient. If $a, b, c \in C^2_*(U)$, such a solution always exists, at least locally, and hence the system ϕ_1, ϕ_2 may be introduced in a neighborhood of any point.

Definition 3.8 The multiplicative curves

$$\xi_1 = \phi_1(x_1, x_2) = \text{const}$$

and

$$\xi_2 = \phi_2(x_1, x_2) = \text{const}$$

are called the characteristic multiplicative curves of the multiplicative linear multiplicative elliptic differential operator $L(u)$.

Let

$$\phi = \phi_1 +_* i \cdot_* \phi_2.$$

Then,

$$\phi^*_{x_1} = \phi^*_{1x_1} +_* i \cdot_* \phi^*_{2x_1},$$

$$\phi^*_{x_2} = \phi^*_{1x_2} +_* i \cdot_* \phi^*_{2x_2}.$$

Hence, from equation (3.20), we get

$$a \cdot_* \phi^{*2*}_{x_1} +_* e^2 \cdot_* b \cdot_* \phi^*_{x_1} \cdot_* \phi^*_{x_2} +_* c \cdot_* \phi^{*2*}_{x_2} = 0_*,$$

whereupon

$$a \cdot_* \left((\phi^*_{x_1})/_*(\phi^*_{x_2}) \right)^{2*} +_* e^2 \cdot_* b \cdot_* \left((\phi^*_{x_1})/_*(\phi^*_{x_2}) \right) +_* c = 0_*$$

and

$$(\phi^*_{x_1})/_*(\phi^*_{x_2}) = (-_* b \pm_* i \cdot_* (a \cdot_* c -_* b^{2*})^{\frac{1}{2}*})/_* a$$

or

$$a \cdot_* \phi^*_{x_1} +_* \left(b \pm_* i \cdot_* (a \cdot_* c -_* b^{2*})^{\frac{1}{2}*} \right) \cdot_* \phi^*_{x_2} = 0_*.$$

Therefore,

$$(d_* x_1)/_* a = (d_* x_2)/_* (b \pm_* i \cdot_* (a \cdot_* c -_* b^{2*})^{\frac{1}{2}*})$$
$$\text{or} \tag{3.24}$$
$$(d_* x_2)/_* (d_* x_1) = (b \pm_* i \cdot_* (a \cdot_* c -_* b^{2*})^{\frac{1}{2}*})/_* a,$$

and

$$a \cdot_* (d_* x_2)^{2*} -_* e^2 \cdot_* b \cdot_* d_* x_1 \cdot_* d_* x_2 +_* c \cdot_* (d_* x_1)^{2*}$$

$$= a \cdot_* \left((b \pm_* i \cdot_* (a \cdot_* c -_* b^{2*})^{\frac{1}{2}*})/_* a \right)^{2*} \cdot_* (d_* x_1)^{2*}$$

$$-_* e^2 \cdot_* b \cdot_* ((b \pm_* i \cdot_* (a \cdot_* c -_* b^{2*})^{\frac{1}{2}*})/_* a) \cdot_* (d_* x_1)^{2*} +_* c \cdot_* (d_* x_1)^{2*}$$

$$= 0_*.$$

Definition 3.9 The equation

$$a \cdot_* (d_* x_2)^{2*} -_* e^2 \cdot_* b \cdot_* d_* x_1 \cdot_* d_* x_2 +_* c \cdot_* (d_* x_1)^{2*} = 0_*$$

is called the characteristic equation of the multiplicative elliptic operator $L(u)$.

Note that the general multiplicative integrals of equation (3.24) are given by

$$\phi_5(x_1, x_2) \pm_* i \phi_6(x_1, x_2) = \text{const},$$

where ϕ_5 and ϕ_6 are multiplicative real-valued functions. In practice, for convenience, we very often take

$$\xi_1 = \phi_5(x_1, x_2),$$

$$\xi_2 = \phi_6(x_1, x_2).$$

Example 3.10 Consider the equation

$$x_2^{2*} \cdot_* u_{x_1 x_1}^{**} +_* e^2 \cdot_* x_1 \cdot_* x_2 \cdot_* u_{x_1 x_2}^{**} +_* e^2 \cdot_* x_1^{2*} \cdot_* u_{x_2 x_2}^{**}$$

$$+_* x_2 \cdot_* u_{x_2}^{*} = 0_*, \quad x_1, x_2 \neq 0_*.$$

Here

$$a(x_1, x_2) = x_2^{2*},$$

$$b(x_1, x_2) = x_1 \cdot_* x_2,$$

$$c(x_1, x_2) = e^2 \cdot_* x_1^{2*}, \quad x_1, x_2 \neq 0_*.$$

Then,

$$a \cdot_* c -_* b^{2*} = e^2 \cdot_* x_1^{2*} \cdot_* x_2^{2*} -_* x_1^{2*} \cdot_* x_2^{2*}$$

$$= x_1^{2*} \cdot_* x_2^{2*}$$

$$> 0_*.$$

Therefore, the considered equation is a multiplicative elliptic equation. The characteristic equation is

$$x_2^{2*} \cdot_* (d_* x_2)^{2*} -_* e^2 \cdot_* x_1 \cdot_* x_2 \cdot_* d_* x_1 \cdot_* d_* x_2 +_* e^2 \cdot_* x_1^{2*} \cdot_* (d_* x_1)^{2*} = 0_*,$$

whereupon

$$x_2^{2*} \cdot_* ((d_* x_2)/_*(d_* x_1))^{2*} -_* e^2 \cdot_* x_1 \cdot_* x_2 \cdot_* ((d_* x_2)/_*(d_* x_1)) +_* e^2 \cdot_* x_1^{2*} = 0_*.$$

Hence,

$$((d_* x_2)/_*(d_* x_1))_{1,2} = ((x_1 \cdot_* x_2 \pm_* i \cdot_* x_1 \cdot_* x_2)/_*(x_2^{2*})).$$

Consider

$$(d_* x_2)/_*(d_* x_1) = (x_1 +_* i \cdot_* x_1)/_* x_2.$$

Then,

$$x_2 \cdot_* d_* x_2 = (x_1 +_* i \cdot_* x_1) \cdot_* d_* x_1$$

and

$$x_2^{2*} = x_1^{2*} +_* i \cdot_* x_1^{2*} +_* c.$$

We set

$$\xi_1 \;\; = \;\; x_1^{2*} -_* x_2^{2*},$$

$$\xi_2 \;\; = \;\; x_1^{2*}.$$

Then,

$$\xi_{1 x_1}^* \;\; = \;\; e^2 \cdot_* x_1,$$

$$\xi_{1 x_2}^* \;\; = \;\; -_* e^2 \cdot_* x_2,$$

$$\xi_{2 x_1}^* \;\; = \;\; e^2 \cdot_* x_1,$$

$$\xi_{2 x_2}^* \;\; = \;\; 0_*,$$

$$u_{x_1}^* \;\; = \;\; u_{\xi_1}^* \cdot_* \xi_{1 x_1}^* +_* u_{\xi_2}^* \cdot_* \xi_{2 x_1}^*$$

$$= \;\; e^2 \cdot_* x_1 \cdot_* u_{\xi_1}^* +_* e^2 \cdot_* x_1 \cdot_* u_{\xi_2}^*,$$

$$u_{x_1 x_1}^{**} \;\; = \;\; e^2 \cdot_* u_{\xi_1}^* +_* e^2 \cdot_* x_1 \cdot_* \left(u_{\xi_1 \xi_1}^{**} \cdot_* \xi_{1 x_1}^* +_* u_{\xi_1 \xi_2}^{**} \cdot_* \xi_{2 x_1}^* \right)$$

$$+_* e^2 \cdot_* u_{\xi_2}^* +_* e^2 \cdot_* x_1 \cdot_* \left(u_{\xi_1 \xi_2}^{**} \cdot_* \xi_{1 x_1}^* +_* u_{\xi_2 \xi_2}^{**} \cdot_* \xi_{2 x_1}^* \right)$$

$$= \; e^2 \cdot_* u^*_{\xi_1} +_* e^2 \cdot_* u^*_{\xi_2} +_* e^2 \cdot_* x_1 \cdot_*$$

$$\left(e^2 \cdot_* x_1 \cdot_* u^{**}_{\xi_1\xi_1} +_* e^2 \cdot_* x_1 \cdot_* u^{**}_{\xi_1\xi_2} \right)$$

$$+_* e^2 \cdot_* x_1 \cdot_* \left(e^2 \cdot_* x_1 \cdot_* u^{**}_{\xi_1\xi_2} +_* e^2 \cdot_* x_1 \cdot_* u^{**}_{\xi_2\xi_2} \right)$$

$$= \; e^2 \cdot_* u^*_{\xi_1} +_* e^2 \cdot_* u^*_{\xi_2} +_* e^4 \cdot_* x_1^{2*} \cdot_* u^{**}_{\xi_1\xi_1}$$

$$+_* e^8 \cdot_* x_1^{2*} \cdot_* u^{**}_{\xi_1\xi_2} +_* e^4 \cdot_* x_1^{2*} \cdot_* u^{**}_{\xi_2\xi_2},$$

$$u^{**}_{x_1 x_2} = \; e^2 \cdot_* x_1 \cdot_* \left(u^{**}_{\xi_1\xi_1} \cdot_* \xi^*_{1x_2} +_* u^{**}_{\xi_2\xi_2} \cdot_* \xi^*_{2x_2} \right)$$

$$+_* e^2 \cdot_* x_1 \cdot_* \left(u^{**}_{\xi_1\xi_2} \cdot_* \xi^*_{1x_2} +_* u^{**}_{\xi_2\xi_2} \cdot_* \xi^*_{2x_2} \right)$$

$$= \; e^2 \cdot_* x_1 \cdot_* \left(-_* e^2 \cdot_* x_2 \cdot_* u^{**}_{\xi_1\xi_1} \right) +_* e^2 \cdot_* x_1 \cdot_* \left(-_* e^2 \cdot_* x_2 \cdot_* u^{**}_{\xi_1\xi_2} \right)$$

$$= \; -_* e^4 \cdot_* x_1 \cdot_* x_2 \cdot_* u^{**}_{\xi_1\xi_1} -_* e^4 \cdot_* x_1 \cdot_* x_2 \cdot_* u^{**}_{\xi_1\xi_2},$$

$$u^*_{x_2} = \; u^*_{\xi_1} \cdot_* \xi^*_{1x_2} +_* u^*_{\xi_2} \cdot_* \xi^*_{2x_2}$$

$$= \; -_* e^2 \cdot_* x_2 \cdot_* u^*_{\xi_1},$$

$$u^{**}_{x_2 x_2} = \; -_* e^2 \cdot_* u^*_{\xi_1} -_* e^2 \cdot_* x_2 \cdot_* \left(u^{**}_{\xi_1\xi_1} \cdot_* \xi^*_{1x_2} +_* u^{**}_{\xi_1\xi_2} \cdot_* \xi^*_{2x_2} \right)$$

$$= \; -_* e^2 \cdot_* u^*_{\xi_1} +_* e^4 \cdot_* x_2^{2*} \cdot_* u^{**}_{\xi_1\xi_1}.$$

Hence,

$$0_* = \; x_2^{2*} \cdot_* u^{**}_{x_1 x_1} +_* e^2 \cdot_* x_1 \cdot_* x_2 \cdot_* u^{**}_{x_1 x_2} +_* e^2 \cdot_* x_1^{2*} \cdot_* u^{**}_{x_2 x_2} +_* x_2 \cdot_* u^*_{x_2}$$

$$= \; x_2^{2*} \left(e^2 \cdot_* u^*_{\xi_1} +_* e^2 \cdot_* u^*_{\xi_2} +_* e^4 \cdot_* x_1^{2*} \cdot_* u^{**}_{\xi_1\xi_1} \right.$$

$$\left. +_* e^8 \cdot_* x_1^{2*} \cdot_* u^{**}_{\xi_1\xi_2} +_* e^4 \cdot_* x_1^{2*} \cdot_* u^{**}_{\xi_2\xi_2} \right)$$

$$+_* e^2 \cdot_* x_1 \cdot_* x_2 \cdot_* \left(-_* e^4 \cdot_* x_1 \cdot_* x_2 \cdot_* u^{**}_{\xi_1\xi_1} -_* e^4 \cdot_* x_1 \cdot_* x_2 \cdot_* u^{**}_{\xi_1\xi_2} \right)$$

$$+_* e^2 \cdot_* x_1^{2*} \cdot_* \left(-_* e^2 \cdot_* u^*_{\xi_1} +_* e^4 \cdot_* x_2^{2*} \cdot_* u^{**}_{\xi_1\xi_1} \right)$$

$$+_* x_2 \cdot_* \left(-_* e^2 \cdot_* x_2 \cdot_* u^*_{\xi_1} \right)$$

$$= \ -_* e^4 \cdot_* x_1^{2*} \cdot_* u_{\xi_1}^* +_* e^2 \cdot_* x_2^{2*} \cdot_* u_{\xi_2}^*$$

$$+_* e^4 \cdot_* x_1^{2*} \cdot_* x_2^{2*} \cdot_* u_{\xi_1 \xi_1}^{**} +_* e^4 \cdot_* x_1^{2*} \cdot_* x_2^{2*} \cdot_* u_{\xi_2 \xi_2}^{**}$$

$$= \ u_{\xi_1 \xi_1}^{**} +_* u_{\xi_2 \xi_2}^{**} -_* (e/_* (x_2^{2*})) \cdot_* u_{\xi_1}^* +_* (e/_* (e^2 \cdot_* x_1^{2*})) \cdot_* u_{\xi_2}^*$$

$$= \ u_{\xi_1 \xi_1}^{**} +_* u_{\xi_2 \xi_2}^{**} -_* (e/_* (\xi_2 -_* \xi_1)) \cdot_* u_{\xi_1}^* +_* (e/_* (e^2 \cdot_* \xi_2)) \cdot_* u_{\xi_2}^*$$

is the canonical form of the considered equation.

Example 3.11 Consider the equation

$$u_{x_1 x_1}^{**} +_* x_1 \cdot_* u_{x_2 x_2}^{**} = 0_*, \quad x_1 > 0_*.$$

Here

$$a(x_1, x_2) \ = \ e,$$

$$b(x_1, x_2) \ = \ 0_*,$$

$$c(x_1, x_2) \ = \ x_1, \quad x_1 > 0_*.$$

Then,

$$a(x_1, x_2) \cdot_* c(x_1, x_2) -_* (b(x_1, x_2))^{2*} \ = \ x_1$$

$$> \ 0_*,$$

i.e., the considered equation is a multiplicative elliptic equation. The characteristic equation is

$$(d_* x_2)^{2*} +_* x_1 \cdot_* (d_* x_1)^{2*} = 0_*$$

and

$$d_* x_2 \ = \ \pm_* i \cdot_* x_1^{\frac{1}{2}*} \cdot_* d_* x_1,$$

and

$$x_2 = \pm_* e^{\frac{2}{3}} \cdot_* i \cdot_* x_1^{\frac{3}{2}*} +_* c.$$

We set

$$xi_1 \ = \ x_2,$$

$$\xi_2 \ = \ e^{\frac{2}{3}} \cdot_* x_1^{\frac{3}{2}*}.$$

Then,

$$\xi^*_{1x_1} = 0_*,$$

$$\xi^*_{1x_2} = e,$$

$$\xi^*_{2x_1} = x_1^{\frac{1}{2}*},$$

$$\xi^*_{2x_2} = 0_*,$$

$$u^*_{x_1} = u^*_{\xi_1} \cdot_* \xi^*_{1x_1} +_* u^*_{\xi_2} \cdot_* \xi^*_{2x_1}$$

$$= x_1^{\frac{1}{2}*} \cdot_* u^*_{\xi_2},$$

$$u^{**}_{x_1x_1} = e^{\frac{1}{2}} \cdot_* (e/_*(x_1^{\frac{1}{2}*})) \cdot_* u^*_{\xi_2} +_* x_1^{\frac{1}{2}*} \cdot_* \left(u^{**}_{\xi_1\xi_2} \cdot_* \xi^*_{1x_1} +_* u^{**}_{\xi_2\xi_2} \cdot_* \xi^*_{2x_1}\right)$$

$$= (e/_*(e^2 \cdot_* x_1^{\frac{1}{2}*})) \cdot_* u^*_{\xi_2} +_* x_1^{\frac{1}{2}*} \cdot_* \left(x_1^{\frac{1}{2}*} \cdot_* u^{**}_{\xi_2\xi_2}\right)$$

$$= (e/_*(e^2 \cdot_* x_1^{\frac{1}{2}*})) \cdot_* u^*_{\xi_2} +_* x_1 \cdot_* u^{**}_{\xi_2\xi_2},$$

$$u^*_{x_2} = u^*_{\xi_1} \cdot_* \xi^*_{1x_2} +_* u^*_{\xi_2} \cdot_* \xi^*_{2x_2}$$

$$= u^*_{\xi_1},$$

$$u^{**}_{x_2x_2} = u^{**}_{\xi_1\xi_1} \cdot_* \xi^*_{1x_2} +_* u^{**}_{\xi_1\xi_2} \cdot_* \xi^*_{2x_2}$$

$$= u^{**}_{\xi_1\xi_1}.$$

Hence,

$$0_* = u^{**}_{x_1x_1} +_* x_1 \cdot_* u^{**}_{x_2x_2}$$

$$= (e/_*(e^2 \cdot_* x_1^{\frac{1}{2}*}) \cdot_* u^*_{\xi_2} +_* x_1 \cdot_* u^{**}_{\xi_2\xi_2} +_* x_1 \cdot_* u^{**}_{\xi_1\xi_1}$$

$$= u^{**}_{\xi_1\xi_1} +_* u^{**}_{\xi_2\xi_2} +_* \left(e/_*\left(e^2 \cdot_* x_1^{\frac{3}{2}*}\right)\right) \cdot_* u^*_{\xi_2}$$

$$= u^{**}_{\xi_1\xi_1} +_* u^{**}_{\xi_2\xi_2} +_* (e/_*(e^3 \cdot_* \xi_2)) \cdot_* u^*_{\xi_2}$$

is the canonical form of the considered equation.

Example 3.12 Consider the equation

$$(e +_* x_1^{2*}) \cdot_* u_{x_1 x_1}^{**} +_* (e +_* x_2^{2*}) \cdot_* u_{x_2 x_2}^{**} +_* x_1 \cdot_* u_{x_1}^* +_* x_2 \cdot_* u_{x_2}^* = 0_*.$$

Here

$$a(x_1, x_2) \quad = \quad e +_* x_1^{2*},$$

$$b(x_1, x_2) \quad = \quad 0_*,$$

$$c(x_1, x_2) \quad = \quad e +_* x_2^{2*}.$$

Then,

$$a(x_1, x_2) \cdot_* c(x_1, x_2) -_* (b(x_1, x_2))^{2*} \quad = \quad (e +_* x_1^{2*})(e +_* x_2^{2*})$$

$$> \quad 0_*,$$

i.e., the considered equation is a multiplicative elliptic equation. The characteristic equation is

$$(e +_* x_1^{2*}) \cdot_* (d_* x_2)^{2*} +_* (e +_* x_2^{2*}) \cdot_* (d_* x_1)^{2*} = 0_*$$

or

$$(e +_* x_1^{2*})^{\frac{1}{2}*} \cdot_* d_* x_2 = \pm_* i \cdot_* (e +_* x_2^{2*})^{\frac{1}{2}*} \cdot_* d_* x_1,$$

or

$$((d_* x_2)'/_* (e +_* x_2^{2*})^{\frac{1}{2}*}) = \pm_* i \cdot_* ((d_* x_1)/_* (e +_* x_1^{2*})^{\frac{1}{2}*})$$

or

$$\log_* \left(x_2 +_* (e +_* x_2^{2*})^{\frac{1}{2}*} \right) = \pm_* i \cdot_* \log_* (x_1 +_* (e +_* x_1^{2*})^{\frac{1}{2}*}) +_* c.$$

We set

$$\xi_1 \quad = \quad \log_* \left(x_2 +_* (e +_* x_2^{2*})^{\frac{1}{2}*} \right),$$

$$\xi_2 \quad = \quad \log_* (x_1 +_* (e +_* x_1^{2*})^{\frac{1}{2}*}).$$

Then,

$$\xi_{1 x_1}^* \quad = \quad 0_*,$$

$$\xi_{1 x_2}^* \quad = \quad (e/_* (e +_* x_2^{2*})^{\frac{1}{2}*}),$$

$$\xi_{2 x_1}^* \quad = \quad (e/_* (e +_* x_1^{2*})^{\frac{1}{2}*},$$

$$\xi_{2x_2}^* \;=\; 0_*,$$

$$u_{x_1}^* \;=\; u_{\xi_1}^* \cdot_* \xi_{1x_1}^* +_* u_{\xi_2}^* \cdot_* \xi_{2x_1}^*$$

$$=\; (e/_*(e +_* x_1^{2*})^{\frac{1}{2}*}) \cdot_* u_{\xi_2}^*,$$

$$u_{x_1 x_1}^{**} \;=\; -_*(x_1/_*(e +_* x_1^{2*})^{\frac{3}{2}*}) \cdot_* u_{\xi_2}^* +_* (e/_*(e +_* x_1^{2*})^{\frac{1}{2}*})$$

$$\cdot_* \left(u_{\xi_1 \xi_2}^{**} \cdot_* \xi_{1x_1}^* +_* u_{\xi_2 \xi_2}^{**} \cdot_* \xi_{2x_1}^* \right)$$

$$=\; -_*(x_1/_*(e +_* x_1^{2*})^{\frac{3}{2}*}) \cdot_* u_{\xi_2}^* +_* (e/_*(e +_* x_1^{2*})) \cdot_* u_{\xi_2 \xi_2}^{**},$$

$$u_{x_2}^* \;=\; u_{\xi_1}^* \cdot_* \xi_{1x_2}^* +_* u_{\xi_2}^* \cdot_* \xi_{2x_2}^*$$

$$=\; (e/_*(e +_* x_2^{2*})^{\frac{1}{2}*}) \cdot_* u_{\xi_1}^*,$$

$$u_{x_2 x_2}^{**} \;=\; -_*(x_2/_*(e +_* x_2^{2*})^{\frac{3}{2}*}) \cdot_* u_{\xi_1}^* +_* (e/_*(e +_* x_2^{2*})^{\frac{1}{2}*})$$

$$\cdot_* \left(u_{\xi_1 \xi_1}^{**} \cdot_* \xi_{1x_2}^* +_* u_{\xi_1 \xi_2}^{**} \cdot_* \xi_{2x_2}^* \right)$$

$$=\; -_*(x_2/_*(e +_* x_2^{2*})^{\frac{3}{2}*}) \cdot_* u_{\xi_1}^*$$

$$+_*(e/_*(e +_* x_2^{2*})) \cdot_* u_{\xi_1 \xi_1}^{**}.$$

Hence,

$$0_* = (e +_* x_1^{2*}) \cdot_* u_{x_1 x_1}^{**} +_* (e +_* x_2^{2*}) \cdot_* u_{x_2 x_2}^{**} +_* x_1 \cdot_* u_{x_1}^* +_* x_2 \cdot_* u_{x_2}^*$$

$$= (e +_* x_1^{2*}) \cdot_* \left(-_*(x_1/_*(e +_* x_1^{2*})^{\frac{3}{2}*}) \cdot_* u_{\xi_2}^* +_* (e/_*(e +_* x_1^{2*})) \cdot_* u_{\xi_2 \xi_2}^{**} \right)$$

$$+_* (e +_* x_2^{2*}) \cdot_* \left(-_*(x_2/_*(e +_* x_1^{2*})^{\frac{3}{2}*}) \cdot_* u_{\xi_1}^* +_* (e/_*(e +_* x_2^{2*})) \cdot_* u_{\xi_1 \xi_1}^{**} \right)$$

$$+_* (x_1/_*(e +_* x_1^{2*})^{\frac{1}{2}*}) \cdot_* u_{\xi_2}^* +_* (x_2/_*(e +_* x_2^{2*})^{\frac{1}{2}*}) \cdot_* u_{\xi_1}^*$$

$$= u_{\xi_1 \xi_1}^{**} +_* u_{\xi_2 \xi_2}^{**}$$

is the canonical form of the considered equation.

Exercise 3.8 Find the canonical form of the equation

$$x_2 \cdot_* u^{**}_{x_1 x_1} +_* x_1 \cdot_* u^{**}_{x_2 x_2} = 0_*, \quad x_1, x_2 > 0_*.$$

Answer

$$\xi_1 = x_2^{\frac{3}{2}*},$$

$$\xi_2 = x_1^{\frac{3}{2}*},$$

and

$$u^{**}_{\xi_1 \xi_1} +_* u^{**}_{\xi_2 \xi_2} +_* (e/_*(e^3 \cdot_* \xi_1)) \cdot_* u^*_{\xi_1} +_* (e/_*(e^3 \cdot_* \xi_2)) \cdot_* u^*_{\xi_2} = 0_*.$$

3.1.3   The parabolic case

In this case, we have

$$a \cdot_* c -_* b^{2*} = 0_*$$

and

$$\beta = \gamma = 0_*.$$

Then,

$$a \cdot_* \phi^*_{1x_1} \cdot_* \phi^*_{2x_1} +_* b \cdot_* \left(\phi^*_{1x_2} \cdot_* \phi^*_{2x_1} +_* \phi^*_{1x_1} \cdot_* \phi^*_{2x_2} \right) +_* c \cdot_* \phi^*_{1x_2} \cdot_* \phi^*_{2x_2} = 0_*$$

$$a \cdot_* \phi^{*2*}_{2x_1} +_* e^2 \cdot_* b\phi^*_{2x_1} \cdot_* \phi^*_{2x_2} +_* c \cdot_* \phi^{*2*}_{2x_2} = 0_*. \tag{3.25}$$

Let

$$\lambda_1 = ((\phi^*_{2x_1})/_*(\phi^*_{2x_2})).$$

Then, from the second equation of (3.25), we get

$$a \cdot_* \lambda_1^{2*} +_* e^2 \cdot_* b \cdot_* \lambda_1 +_* c = 0_*,$$

whereupon

$$(\lambda_1)_{1,2} = (-_* b \pm_* (b^{2*} -_* a \cdot_* c)^{\frac{1}{2}*})/_* a$$

$$= -_*(b/_* a),$$

or

$$((\phi^*_{2x_1})/_*(\phi^*_{2x_2})) = -_*(b/_* a),$$

or

$$a \cdot_* \phi^*_{2x_1} +_* b \cdot_* \phi^*_{2x_2} = 0_*. \tag{3.26}$$

Hence, from the first equation of (3.25), we find

$$a \cdot_* \phi^*_{1x_1} \cdot_* \left(-_*(b/_* a) \cdot_* \phi^*_{2x_2} \right) +_* b \cdot_* \left(\phi^*_{1x_2} \cdot_* \left(-_*(b/_* a) \cdot_* \phi^*_{2x_2} \right) \right.$$

$$\left. +_* \phi^*_{1x_1} \cdot_* \phi^*_{2x_2} \right) +_* c \cdot_* \phi^*_{1x_2} \cdot_* \phi^*_{2x_2} = 0_*,$$

or

$$-_* b \cdot_* \phi_{1x_1}^* \cdot_* \phi_{2x_2}^* -_* ((b^{2*})/_*a) \cdot_* \phi_{1x_2}^* \cdot_* \phi_{2x_2}^* +_* b \cdot_* \phi_{1x_1}^* \cdot_* \phi_{2x_2}^*$$

$$+_* c \cdot_* \phi_{1x_2}^* \cdot_* \phi_{2x_2}^* = 0_*,$$

or

$$((a \cdot_* c -_* b^{2*})/_*a) \cdot_* \phi_{1x_2}^* \cdot_* \phi_{2x_2}^* = 0_*.$$

Therefore, the function ϕ_2 can be determined by equation (3.26) and the function ϕ_1 is arbitrarily chosen $\mathcal{C}_*^1$-function in U. Equation (3.26) gives a family of solutions of the multiplicative ordinary differential equation:

$$((d_*x_2)/_*(d_*x_1)) = b/_*a, \tag{3.27}$$

where x_2 is considered as a function of x_1 along the multiplicative curves of the family.

Definition 3.10 The curves

$$\xi_1 = \phi_1(x_1, x_2) = \text{const},$$

$$\xi_2 = \phi_2(x_1, x_2) = \text{const}$$

are called the characteristic multiplicative curves of the multiplicative linear multiplicative parabolic operator $L(u)$.

Using equation (3.27), we get

$$a \cdot_* ((d_*x_2)/_*(d_*x_1))^{2*} -_* e^2 \cdot_* b \cdot_* ((d_*x_2)/_*(d_*x_1)) +_* c$$

$$= (b^*)/_*a +_* e^2 \cdot_* ((b^{2*})/_*a) +_* c$$

$$= -_*((b^{2*})/_*a) +_* c$$

$$= 0_*.$$

Definition 3.11 The equation

$$a \cdot_* (d_*x_2)^{2*} -_* e^2 \cdot_* b \cdot_* d_*x_1 \cdot_* d_*x_2 +_* c \cdot_* (d_*x_1)^{2*} = 0_*$$

is called the characteristic equation of the multiplicative parabolic operator $L(u)$.

Remark 3.3 If $\phi(x_1, x_2) = \text{const}$ is the general solution of equation (3.27); in practice, we take

$$\xi_1 = \phi(x_1, x_2),$$

$$\xi_2 = x_1,$$

or

$$\xi_2 \;=\; x_2,$$

or

$$\xi_1 \;=\; x_1 \quad \text{or}$$

$$\xi_1 \;=\; x_2 \quad \text{and}$$

$$\xi_2 \;=\; \phi(x_1, x_2).$$

Example 3.13 Consider the equation

$$x_1^{2*} \cdot_* u_{x_1 x_1}^{**} -_* e^2 \cdot_* x_1 \cdot_* x_2 \cdot_* u_{x_1 x_2}^{**} +_* x_2^{2*} \cdot_* u_{x_2 x_2}^{**} +_* x_1 \cdot_* u_{x_1}^{*}$$
$$+_* x_2 \cdot_* u_{x_2}^{*} = 0_*, \quad x_1, x_2 > 0_*.$$

Here

$$a(x_1, x_2) \;=\; x_1^{2*},$$

$$b(x_1, x_2) \;=\; -_* x_1 \cdot_* x_2,$$

$$c(x_1, x_2) \;=\; x_2^{2*}, \quad x_1, x_2 > 0_*.$$

Then,

$$a(x_1, x_2) \cdot_* c(x_1, x_2) -_* (b(x_1, x_2))^{2*} \;=\; x_1^{2*} \cdot_* x_2^{2*} -_* x_1^{2*} \cdot_* x_2^{2*}$$

$$=\; 0_*.$$

Therefore, the considered equation is multiplicative parabolic. The characteristic equation is

$$x_1^{2*} \cdot_* (d_* x_2)^{2*} +_* e^2 \cdot_* x_1 \cdot_* x_2 \cdot_* d_* x_1 \cdot_* d_* x_2 +_* x_2^{2*} \cdot_* (d_* x_1)^{2*} = 0_*,$$

whereupon

$$(x_1 \cdot_* d_* x_2 +_* x_2 \cdot_* d_* x_1)^{2*} = 0_*$$

and

$$x_1 \cdot_* d_* x_2 = -_* x_2 \cdot_* d_* x_1,$$

and

$$(d_* x_2) /_* x_2 = -_* ((d_* x_1) /_* x_1),$$

and

$$x_1 \cdot_* x_2 = c.$$

We set

$$\xi_1 \;=\; x_1,$$

$$\xi_2 \;=\; x_1 \cdot_* x_2.$$

Then,

$$u^*_{x_1} \;=\; u^*_{\xi_1} \cdot_* \xi^*_{1x_1} +_* u^*_{\xi_2} \cdot_* \xi^*_{2x_1}$$

$$=\; u^*_{\xi_1} +_* x_2 \cdot_* u^*_{\xi_2},$$

$$u^{**}_{x_1 x_1} \;=\; u^{**}_{\xi_1 \xi_1} \cdot_* \xi^*_{1x_1} +_* u^{**}_{\xi_1 \xi_2} \cdot_* \xi^*_{2x_1}$$

$$+_* x_2 \cdot_* \left(u^{**}_{\xi_1 \xi_2} \cdot_* \xi^*_{1x_1} +_* u^{**}_{\xi_2 \xi_2} \cdot_* \xi^*_{2x_1} \right)$$

$$=\; u^{**}_{\xi_1 \xi_1} +_* x_2 \cdot_* u^{**}_{\xi_1 \xi_2} +_* x_2 \cdot_* \left(u^{**}_{\xi_1 \xi_2} +_* x_2 \cdot_* u^{**}_{\xi_2 \xi_2} \right)$$

$$=\; u^{**}_{\xi_1 \xi_1} +_* e^2 \cdot_* x_2 \cdot_* u^{**}_{\xi_1 \xi_2} +_* x_2^{2*} \cdot_* u^{**}_{\xi_2 \xi_2},$$

$$u^{**}_{x_1 x_2} \;=\; u^{**}_{\xi_1 \xi_1} \cdot_* \xi^*_{1x_2} +_* u^{**}_{\xi_1 \xi_2} \cdot_* \xi^*_{2x_2} +_* u^*_{\xi_2}$$

$$+_* x_2 \cdot_* \left(u^{**}_{\xi_1 \xi_2} \cdot_* \xi^*_{1x_2} +_* u^{**}_{\xi_2 \xi_2} \cdot_* \xi^*_{2x_2} \right)$$

$$=\; x_1 \cdot_* u^{**}_{\xi_1 \xi_2} +_* u^*_{\xi_2} +_* x_1 \cdot_* x_2 \cdot_* u^{**}_{\xi_2 \xi_2},$$

$$u^*_{x_2} \;=\; u^*_{\xi_1} \cdot_* \xi^*_{1x_2} +_* u^*_{\xi_2} \cdot_* \xi^*_{2x_2}$$

$$=\; x_1 \cdot_* u^*_{\xi_2},$$

$$u^{**}_{x_2 x_2} \;=\; x_1 \cdot_* \left(u^{**}_{\xi_1 \xi_2} \cdot_* \xi^*_{1x_2} +_* u^{**}_{\xi_2 \xi_2} \cdot_* \xi^*_{2x_2} \right)$$

$$=\; x_1^{2*} \cdot_* u^{**}_{\xi_2 \xi_2}.$$

Hence,

$$0_* \;=\; x_1^{2*} \cdot_* u^{**}_{x_1 x_1} -_* e^2 \cdot_* x_1 \cdot_* x_2 \cdot_* u^{**}_{x_1 x_2} +_* x_2^{2*} \cdot_* u^{**}_{x_2 x_2} +_* x_1 \cdot_* u^*_{x_1}$$

$$+_* x_2 \cdot_* u^*_{x_2}$$

$$= x_1^{2*} \cdot_* \left(u_{\xi_1\xi_1}^{**} +_* e^2 \cdot_* x_2 \cdot_* u_{\xi_1\xi_2}^{**} +_* x_2^{2*} \cdot_* u_{\xi_2\xi_2}^{**} \right)$$

$$-_* e^2 \cdot_* x_1 \cdot_* x_2 \cdot_* \left(x_1 \cdot_* u_{\xi_1\xi_2}^{**} +_* x_1 \cdot_* x_2 \cdot_* u_{\xi_2\xi_2}^{**} +_* u_{\xi_2}^* \right)$$

$$+_* x_1^{2*} \cdot_* x_2^{2*} \cdot_* u_{\xi_2\xi_2}^{**} +_* x_1 \cdot_* \left(u_{\xi_1}^* +_* x_2 \cdot_* u_{\xi_2}^* \right) +_* x_1 \cdot_* x_2 \cdot_* u_{\xi_2}^*$$

$$= \xi_1 \cdot_* u_{\xi_1\xi_1}^{**} +_* u_{\xi_1}^*$$

is the canonical form of the considered equation. Hence,

$$\left(\xi_1 \cdot_* u_{\xi_1}^* \right)_{\xi_1}^* = 0_*$$

and

$$\xi_1 \cdot_* u_{\xi_1}^* = f(\xi_2);$$
$$u_{\xi_1}^* = (e/_* \xi_1) \cdot_* f(\xi_2)$$

and

$$u = \log_*(\xi_1) \cdot_* f(\xi_2) +_* g(\xi_2);$$

and

$$u(x_1, x_2) = \log_*(x_1) \cdot_* f(x_1 \cdot_* x_2) +_* g(x_1 \cdot_* x_2)$$

is the general solution of the considered equation, where f and g are $\mathcal{C}_*^2$-functions.

Example 3.14 Consider the equation

$$x_1^{2*} \cdot_* u_{x_1 x_1}^{**} -_* e^2 \cdot_* x_1 \cdot_* u_{x_1 x_2}^{**} +_* u_{x_2 x_2}^{**} +_* e^2 \cdot_* u_{x_2}^* -_* x_1 \cdot_* u_{x_1}^* = 0_*.$$

Here

$$a(x_1, x_2) = x_1^{2*},$$

$$b(x_1, x_2) = -_* x_1,$$

$$c(x_1, x_2) = e.$$

Then,

$$a(x_1, x_2) \cdot_* c -_* (b(x_1, x_2))^{2*} = x_1^{2*} -_* x_1^{2*}$$

$$= 0_*.$$

Therefore, the considered equation is multiplicative parabolic. The characteristic equation is

$$x_1^{2*} \cdot_* (d_* x_2)^2 +_* e^2 \cdot_* x_1 \cdot_* d_* x_1 \cdot_* d_* x_2 +_* (d_* x_1)^{2*} = 0_*,$$

whereupon

$$x_1 \cdot_* d_* x_2 +_* d_* x_1 = 0_*$$

and

$$x_1 \cdot_* e^{x_2} = c.$$

We set

$$\xi_1 \;=\; x_1 \cdot_* e^{x_2},$$

$$\xi_2 \;=\; x_2.$$

Then,

$$\xi^*_{1x_1} = e^{x_2},$$

$$\xi^*_{1x_2} = x_1 \cdot_* e^{x_2},$$

$$\xi^*_{2x_1} = 0_*,$$

$$\xi^*_{2x_2} = e,$$

$$u^*_{x_1} = u^*_{\xi_1} \cdot_* \xi^*_{1x_1} +_* u^*_{\xi_2} \cdot_* \xi^*_{2x_1}$$

$$= e^{x_2} \cdot_* u^*_{\xi_1},$$

$$u^{**}_{x_1 x_1} = e^{x_2} \cdot_* \left(u^{**}_{\xi_1 \xi_1} \cdot_* \xi^*_{1x_1} +_* u^{**}_{\xi_1 \xi_2} \cdot_* \xi^*_{2x_1} \right)$$

$$= e^{e^2 \cdot_* x_2} \cdot_* u^{**}_{\xi_1 \xi_1},$$

$$u^{**}_{x_1 x_2} = e^{x_2} \cdot_* u^*_{\xi_1} +_* e^{x_2} \cdot_* \left(u^{**}_{\xi_1 \xi_1} \cdot_* \xi^*_{1x_2} +_* u^{**}_{\xi_1 \xi_2} \cdot_* \xi^*_{2x_2} \right)$$

$$= e^{x_2} \cdot_* u^*_{\xi_1} +_* e^{x_2} \cdot_* \left(x_1 \cdot_* e^{x_2} \cdot_* u^{**}_{\xi_1 \xi_1} +_* u^{**}_{\xi_1 \xi_2} \right)$$

$$= e^{x_2} \cdot_* u^*_{\xi_1} +_* x_1 \cdot_* e^{e^2 \cdot_* x_2} \cdot_* u^{**}_{\xi_1 \xi_1} +_* e^{x_2} \cdot_* u^{**}_{\xi_1 \xi_2},$$

$$u^*_{x_2} = u^*_{\xi_1} \cdot_* \xi^*_{1x_2} +_* u^*_{\xi_2} \cdot_* \xi^*_{2x_2}$$

$$= x_1 \cdot_* e^{x_2} \cdot_* u^*_{\xi_1} +_* u^*_{\xi_2},$$

$$u_{x_2 x_2}^{**} = x_1 \cdot_* e^{x_2} \cdot_* u_{\xi_1}^* +_* x_1 \cdot_* e^{x_2} \cdot_* \left(u_{\xi_1 \xi_1}^{**} \cdot_* \xi_{1 x_2}^* +_* u_{\xi_1 \xi_2}^{**} \cdot_* \xi_{2 x_2}^* \right)$$

$$+_* u_{\xi_1 \xi_2}^{**} \cdot_* \xi_{1 x_2}^* +_* u_{\xi_2 \xi_2}^{**} \cdot_* \xi_{2 x_2}^*$$

$$= x_1 \cdot_* e^{x_2} \cdot_* u_{\xi_1}^* +_* x_1 \cdot_* e^{x_2} \cdot_* \left(x_1 \cdot_* e^{x_2} \cdot_* u_{\xi_1 \xi_1}^{**} +_* u_{\xi_1 \xi_2}^{**} \right)$$

$$+_* x_1 \cdot_* e^{x_2} \cdot_* u_{\xi_1 \xi_2}^{**} +_* u_{\xi_2 \xi_2}^{**}$$

$$= x_1 \cdot_* e^{x_2} \cdot_* u_{\xi_1}^* +_* x_1^{2*} \cdot_* e^{e^2 \cdot_* x_2} \cdot_* u_{\xi_1 \xi_1}^{**} +_* e^2 \cdot_* x_1 \cdot_* e^{x_2} \cdot_* u_{\xi_1 \xi_2}^{**}$$

$$+_* u_{\xi_2 \xi_2}^{**}.$$

Hence,

$$0_* = x_1^{2*} \cdot_* u_{x_1 x_1}^{**} -_* e^2 \cdot_* x_1 \cdot_* u_{x_1 x_2}^{**} +_* u_{x_2 x_2}^{**} +_* e^2 \cdot_* u_{x_2}^* -_* x_1 \cdot_* u_{x_1}^*$$

$$= x_1^{2*} \cdot_* e^{e^2 \cdot_* x_2} \cdot_* u_{\xi_1 \xi_1}^{**} -_* e^2 \cdot_* x_1 \cdot_*$$

$$\left(e^{x_2} \cdot_* u_{\xi_1}^* +_* x_1 \cdot_* e^{e^2 \cdot_* x_2} \cdot_* u_{\xi_1 \xi_1}^{**} +_* e^{x_2} \cdot_* u_{\xi_1 \xi_2}^{**} \right)$$

$$+_* x_1 \cdot_* e^{x_2} \cdot_* u_{\xi_1}^* +_* x_1^{2*} \cdot_* e^{e^2 \cdot_* x_2} \cdot_* u_{\xi_1 \xi_1}^{**}$$

$$+_* e^2 \cdot_* x_1 \cdot_* e^{x_2} \cdot_* u_{\xi_1 \xi_2}^{**} +_* u_{\xi_2 \xi_2}^{**}$$

$$+_* e^2 \cdot_* x_1 \cdot_* e^{x_2} \cdot_* u_{\xi_1}^* +_* e^2 \cdot_* u_{\xi_2}^* -_* x_1 \cdot_* e^{x_2} \cdot_* u_{\xi_1}^*$$

$$= u_{\xi_2 \xi_2}^{**} +_* e^2 \cdot_* u_{\xi_2}^*$$

is the canonical form of the considered equation. Let

$$v = u_{\xi_2}^*.$$

Then,

$$v_{\xi_2}^* = -_* e^2 \cdot_* v$$

and

$$v = e^{-_* e^2 \cdot_* \xi_2} \cdot_* f(\xi_1),$$

and

$$u_{\xi_2}^* = e^{-_* e^2 \cdot_* \xi_2} \cdot_* f_1(\xi_1)$$

and

$$u = f(\xi_1) \cdot_* e^{-_* e^2 \cdot_* \xi_2} +_* g(\xi_1),$$

i.e.,

$$u(x_1, x_2) = e^{-_* e^2 \cdot_* x_2} \cdot_* f\left(x_1 \cdot_* e^{x_2}\right) +_* g\left(x_1 \cdot_* e^{x_2}\right)$$

is the general solution of the considered equation, where f and g are C_*^2-functions.

Example 3.15 Consider the equation

$$u^{**}_{x_1 x_1} -_* e^2 \cdot_* \sin_* x_1 u^{**}_{x_1 x_2} +_* (\sin_* x_1)^{2*} \cdot_* u^{**}_{x_2 x_2} = 0_*.$$

Here

$$a(x_1, x_2) \quad = \quad e,$$

$$b(x_1, x_2) \quad = \quad -_* \sin_* x_1,$$

$$c(x_1, x_2) \quad = \quad (\sin_* x_1)^{2*}.$$

Then,

$$a(x_1, x_2) \cdot_* c(x_1, x_2) -_* (b(x_1, x_2))^{2*} \quad = \quad (\sin_* x_1)^{2*} -_* (\sin_* x_1)^{2*}$$

$$= \quad 0_*.$$

Therefore, the considered equation is multiplicative parabolic. The characteristic equation is

$$(d_* x_2)^{2*} +_* 2 \sin_* x_1 \cdot_* d_* x_1 \cdot_* d_* x_2 +_* (\sin_* x_1)^{2*} \cdot_* (d_* x_1)^{2*} = 0_*,$$

whereupon

$$(d_* x_2 +_* \sin_* x_1 \cdot_* d_* x_1)^2 = 0_*$$

and

$$d_* x_2 +_* \sin_* x_1 \cdot_* d_* x_1 = 0_*,$$

and

$$x_2 -_* \cos_* x_1 = c.$$

We set

$$\xi_1 \quad = \quad x_2 -_* \cos_* x_1,$$

$$\xi_2 \quad = \quad x_1.$$

Then,

$$\xi^*_{1x_1} = \sin_* x_1,$$

$$\xi^*_{1x_2} = e,$$

$$\xi^*_{2x_1} = e,$$

$$\xi^*_{2x_2} = 0_*,$$

$$u^*_{x_1} = u^*_{\xi_1} \cdot_* \xi^*_{1x_1} +_* u^*_{\xi_2} \cdot_* \xi^*_{2x_1}$$

$$= \sin_* x_1 \cdot_* u^*_{\xi_1} +_* u^*_{\xi_2},$$

$$u^*_{x_1 x_1} = \cos_* x_1 \cdot_* u^*_{\xi_1} +_* \sin_* x_1 \cdot_* \left(u^{**}_{\xi_1 \xi_1} \cdot_* \xi^*_{1x_1} +_* u^{**}_{\xi_1 \xi_2} \cdot_* \xi^*_{2x_1} \right)$$

$$+_* u^{**}_{\xi_1 \xi_2} \cdot_* \xi^*_{1x_1} +_* u^{**}_{\xi_2 \xi_2} \cdot_* \xi^*_{2x_1}$$

$$= \cos_* x_1 \cdot_* u^*_{\xi_1} +_* \sin_* x_1 \cdot_* \left(\sin_* x_1 \cdot_* u^{**}_{\xi_1 \xi_1} +_* u^{**}_{\xi_1 \xi_2} \right)$$

$$+_* \sin_* x_1 \cdot_* u^{**}_{\xi_1 \xi_2} +_* u^{**}_{\xi_2 \xi_2}$$

$$= \cos_* x_1 \cdot_* u^*_{\xi_1} +_* (\sin_* x_1)^{2*} \cdot_* u^{**}_{\xi_1 \xi_1} +_* e^2 \cdot_* \sin_* x_1 \cdot_* u^{**}_{\xi_1 \xi_2}$$

$$+_* u^{**}_{\xi_2 \xi_2},$$

$$u^{**}_{x_1 x_2} = \sin_* x_1 \cdot_* \left(u^{**}_{\xi_1 \xi_1} \cdot_* \xi^*_{1x_2} +_* u^{**}_{\xi_1 \xi_2} \cdot_* \xi^*_{2x_2} \right)$$

$$+_* u^{**}_{\xi_1 \xi_2} \cdot_* \xi^*_{1x_2} +_* u^{**}_{\xi_2 \xi_2} \cdot_* \xi^*_{2x_2}$$

$$= \sin_* x_1 \cdot_* u^{**}_{\xi_1 \xi_1} +_* u^{**}_{\xi_1 \xi_2},$$

$$u^*_{x_2} = u^*_{\xi_1} \cdot_* \xi^*_{1x_2} +_* u^*_{\xi_2} \cdot_* \xi^*_{2x_2}$$

$$= u^*_{\xi_1},$$

$$u^{**}_{x_2 x_2} = u^{**}_{\xi_1 \xi_1} \cdot_* \xi^*_{1x_2} +_* u^{**}_{\xi_1 \xi_2} \cdot_* \xi^*_{2x_2}$$

$$= u^{**}_{\xi_1 \xi_1}.$$

Hence,

$$
\begin{aligned}
0_* &= u^{**}_{x_1 x_1} -_* e^2 \cdot_* \sin_* x_1 \cdot_* u^{**}_{x_1 x_2} +_* (\sin_* x_1)^{2*} \cdot_* u^{**}_{x_2 x_2} \\
&= \cos_* x_1 u^*_{\xi_1} +_* (\sin_* x_1)^{2*} \cdot_* u^{**}_{\xi_1 \xi_1} +_* e^2 \cdot_* \sin_* x_1 \cdot_* u^{**}_{\xi_1 \xi_2} +_* u^{**}_{\xi_2 \xi_2} \\
&\quad -_* e^2 \cdot_* \sin_* x_1 \cdot_* \left(\sin_* x_1 \cdot_* u^{**}_{\xi_1 \xi_1} +_* u^{**}_{\xi_1 \xi_2} \right) +_* (\sin_* x_1)^{2*} \cdot_* u^{**}_{\xi_1 \xi_1} \\
&= \cos_* x_1 \cdot_* u^*_{\xi_1} +_* u^{**}_{\xi_2 \xi_2} \\
&= \cos_* (\xi_2) \cdot_* u^*_{\xi_1} +_* u^{**}_{\xi_2 \xi_2}
\end{aligned}
$$

is the canonical form of the considered equation.

Exercise 3.9 Find the canonical form of the equation

$$
u^*_{x_1 x_1} +_* e^6 \cdot_* u^{**}_{x_1 x_2} +_* e^9 \cdot_* u^{**}_{x_2 x_2} +_* u^*_{x_1} +_* u^*_{x_2} = 0_*.
$$

Answer

$$
\xi_1 = x_1,
$$

$$
\xi_2 = x_2 -_* e^3 \cdot_* x_1,
$$

and

$$
u^{**}_{\xi_1 \xi_1} +_* u^*_{\xi_1} -_* e^2 \cdot_* u^*_{\xi_2} = 0_*.
$$

Exercise 3.10 Consider the equation

$$
e^4 \cdot_* x_2^{2*} \cdot_* u^{**}_{x_1 x_1} +_* e^4 \cdot_* x_2 \cdot_* u^{**}_{x_1 x_2} +_* u^{**}_{x_2 x_2} +_* e^2 \cdot_* u^*_{x_1} = 0_*.
$$

1. Find the canonical form.
2. Find the general solution.
3. Find the solution $u(x_1, x_2)$ for which

$$
u(x_1, 1) = e^9 \cdot_* \sin_* x_1,
$$

$$
u^*_{x_2}(x_1, 1) = e^{x_1}.
$$

Solution   1. Here

$$
a(x_1, x_2) = e^4 \cdot_* x_2^{2*},
$$

$$
b(x_1, x_2) = e^2 \cdot_* x_2,
$$

$$
c(x_1, x_2) = e.
$$

Then,

$$(b(x_1, x_2))^{2*} -_* a(x_1, x_2) \cdot_* c(x_1, x_2) \quad = \quad e^4 \cdot_* x_2^{2*} -_* e^4 \cdot_* x_2^{2*}$$

$$- \quad 0_*.$$

Therefore, the considered equation is multiplicative parabolic. The characteristic equation is

$$e^4 \cdot_* x_2^{2*} \cdot_* (d_* x_2)^{2*} -_* e^4 \cdot_* x_2 \cdot_* d_* x_1 \cdot_* d_* x_2 +_* (d_* x_1)^{2*} = 0_*,$$

whereupon

$$(e^2 \cdot_* x_2 \cdot_* d_* x_2 -_* d_* x_1)^{2*} = 0_*$$

and

$$x_2 \cdot_* d_* x_2 -_* d_* x_1 = 0_*$$

and

$$x_2^{2*} -_* x_1 = c.$$

We set

$$\xi_1 \quad = \quad x_2^{2*} -_* x_1,$$

$$\xi_2 \quad = \quad x_2.$$

Then,

$$\xi_{1x_1}^* \quad = \quad -_* e,$$

$$\xi_{1x_2}^* \quad = \quad e^2 \cdot_* x_2,$$

$$\xi_{2x_1}^* \quad = \quad 1,$$

$$\xi_{2x_2}^* \quad = \quad e,$$

$$u_{x_1}^* \quad = \quad u_{\xi_1}^* \cdot_* \xi_{1x_1}^* +_* u_{\xi_2}^* \cdot_* \xi_{2x_1}^*$$

$$= \quad -_* u_{\xi_1}^*,$$

$$u_{x_1 x_1}^* \quad = \quad -_* \left(u_{\xi_1 \xi_1}^{**} \cdot_* \xi_{1x_1}^* +_* u_{\xi_1 \xi_2}^{**} \cdot_* \xi_{2x_1}^* \right)$$

$$= \quad u_{\xi_1 \xi_1}^{**},$$

$$u^{**}_{x_1 x_2} = -_*\left(u^{**}_{\xi_1 \xi_1} \cdot_* \xi^*_{1 x_2} +_* u^{**}_{\xi_1 \xi_2} \cdot_* \xi^*_{2 x_2}\right)$$

$$= -_* e^2 \cdot_* x_2 \cdot_* u^{**}_{\xi_1 \xi_1} -_* u^{**}_{\xi_1 \xi_2},$$

$$u^*_{x_2} = u^*_{\xi_1} \cdot_* \xi^*_{1 x_2} +_* u^*_{\xi_2} \cdot_* \xi^*_{2 x_2}$$

$$= e^2 \cdot_* x_2 \cdot_* u^*_{\xi_1} +_* u^*_{\xi_2},$$

$$u^{**}_{x_2 x_2} = e^2 \cdot_* u^*_{\xi_1} +_* e^2 \cdot_* x_2 \cdot_* \left(u^{**}_{\xi_1 \xi_1} \cdot_* \xi^*_{1 x_2} +_* u^{**}_{\xi_1 \xi_2} \cdot_* \xi^*_{2 x_2}\right)$$

$$+_* u^{**}_{\xi_1 \xi_2} \cdot_* \xi^*_{1 x_2} +_* u^{**}_{\xi_2 \xi_2} \cdot_* \xi^*_{2 x_2}$$

$$= e^2 \cdot_* u^*_{\xi_1} +_* e^2 \cdot_* x_2 \cdot_* \left(e^2 \cdot_* x_2 \cdot_* u^{**}_{\xi_1 \xi_1} +_* u^{**}_{\xi_1 \xi_2}\right)$$

$$+_* e^2 \cdot_* x_2 \cdot_* u^{**}_{\xi_1 \xi_2} +_* u^{**}_{\xi_2 \xi_2}$$

$$= e^2 \cdot_* u^*_{\xi_1} +_* e^4 \cdot_* x_2^{2*} \cdot_* u^{**}_{\xi_1 \xi_1} +_* e^4 \cdot_* x_2 \cdot_* u^{**}_{\xi_1 \xi_2} +_* u^{**}_{\xi_2 \xi_2}.$$

Hence,

$$0_* = e^4 \cdot_* x_2^{2*} \cdot_* u^{**}_{x_1 x_1} +_* e^4 \cdot_* x_2 \cdot_* u^{**}_{x_1 x_2} +_* u^{**}_{x_2 x_2} +_* e^2 \cdot_* u^*_{x_1}$$

$$= e^4 \cdot_* x_2^{2*} \cdot_* u^{**}_{\xi_1 \xi_1} +_* e^4 \cdot_* x_2 \cdot_* \left(-_* e^2 \cdot_* x_2 \cdot_* u^{**}_{\xi_1 \xi_1} -_* u^{**}_{\xi_1 \xi_2}\right)$$

$$+_* e^2 \cdot_* u^*_{\xi_1} +_* e^4 \cdot_* x_2^{2*} \cdot_* u^{**}_{\xi_1 \xi_1}$$

$$+_* e^4 \cdot_* x_2 \cdot_* u^{**}_{\xi_1 \xi_2} +_* u^{**}_{\xi_2 \xi_2} -_* e^2 \cdot_* u^*_{\xi_1}$$

$$= u^*_{\xi_2 \xi_2}$$

is the canonical form of the considered equation.

2. Let
$$v = u^*_{\xi_2}.$$

Then,
$$v_{\xi_2} = 0_*$$

and
$$v = f(\xi_1);$$

$$u^*_{\xi_2} = f(\xi_1),$$

and
$$u = \xi_2 \cdot_* f(\xi_1) +_* g(\xi_1);$$

and
$$u(x_1, x_2) = x_2 \cdot_* f(x_2^{2*} -_* x_1) +_* g(x_2^{2*} -_* x_1)$$

is the general solution of the considered equation, where f and g are C^2_*-functions.

3. We have

$$u(x_1, 1) \;=\; g(-_* x_1),$$

$$u^*_{x_2}(x_1, x_2) \;=\; f(x_2^{2*} -_* x_1) +_* e^2 \cdot_* x_2^{2*} \cdot_* f^*(x_2 -_* x_1)$$

$$+_* e^2 \cdot_* x_2 \cdot_* g^* \cdot_* (x_2^{2*} -_* x_1),$$

$$u^*_{x_2}(x_1, 1) \;=\; f(-_* x_1).$$

In this way we obtain the system

$$f(-_* x_1) \;=\; e^{x_1}$$

$$g(-_* x_1) \;=\; e^9 \cdot_* \sin_* x_1,$$

whereupon

$$f(x_1) \;=\; e^{-_* x_1}$$

$$g(x_1) \;=\; -_* e^9 \cdot_* \sin_* x_1.$$

Consequently,

$$u(x_1, x_2) = x_2 \cdot_* e^{x_1 -_* x_2^{2*}} -_* e^9 \cdot_* \sin_* \left(x_2^{2*} -_* x_1\right).$$

3.2 Classification and Canonical Form of Multiplicative Quasilinear Second-Order Multiplicative Partial Differential Equations in Two Independent Variables

Here we consider the equation

$$L(u) = a \cdot_* u^{**}_{x_1 x_1} +_* e^2 \cdot_* b \cdot_* u^{**}_{x_1 x_2} +_* c \cdot_* u^{**}_{x_2 x_2} +_* d = 0_* \quad \text{in} \quad U, \quad (3.28)$$

where

$$a \;=\; a(x_1, x_2, u, u^*_{x_1}, u^*_{x_2}),$$

$$b \;=\; b(x_1, x_2, u, u^*_{x_1}, u^*_{x_2}),$$

$$c \;=\; c(x_1, x_2, u, u^*_{x_1}, u^*_{x_2}),$$

$$d \;=\; d(x_1, x_2, u, u^*_{x_1}, u^*_{x_2})$$

are given functions and u is unknown.

Definition 3.12 The operator $L(u)$ will be called as follows:

1. Multiplicative hyperbolic at the point $(x_1^0, x_2^0) \in U$, if

$$a \cdot_* c -_* b^{2_*} < 0_*.$$

2. Multiplicative elliptic at the point $(x_1^0, x_2^0) \in U$, if

$$a \cdot_* c -_* b^{2_*} > 0_*.$$

3. Multiplicative parabolic at the point $(x_1^0, x_2^0) \in U$, if

$$a \cdot_* c -_* b^{2_*} = 0_*.$$

Definition 3.13 The operator $L(u)$ will be called as follows:

1. Multiplicative hyperbolic in U, if it is multiplicative hyperbolic at any point of U.
2. Multiplicative elliptic in U, if it is multiplicative elliptic at any point of U.
3. Multiplicative parabolic in U, if it is multiplicative parabolic at any point of U.

However, since a, b, c and d depend on the unknown u and its derivatives $u_{x_1}^*$, $u_{x_2}^*$, then the character of the operator $L(u)$ at a point (x_1, x_2) also depends on u, $u_{x_1}^*$, $u_{x_2}^*$. We introduce the characteristic parameters:

$$\xi_1 = \phi_1(x_1, x_2),$$

$$\xi_2 = \phi_2(x_1, x_2).$$

We consider u, x_1 and x_2 as functions of ξ_1 and ξ_2:

$$u = u(\xi_1, \xi_2),$$

$$x_1 = x_1(\xi_1, \xi_2),$$

$$x_2 = x_2(\xi_1, \xi_2).$$

Now we multiplicative differentiate with respect to x_1 and x_2 the second and the third equations of the last system and we get

$$e = x_{1\xi_1}^* \cdot_* \xi_{1x_1}^* +_* x_{1\xi_2}^* \cdot_* \xi_{2x_1}^*,$$

$$1 = x_{1\xi_1}^* \cdot_* \xi_{1x_2}^* +_* x_{1\xi_2}^* \cdot_* \xi_{2x_2}^*,$$

$$1 = x_{2\xi_1}^* \cdot_* \xi_{1x_1}^* +_* x_{2\xi_2}^* \cdot_* \xi_{2x_1}^*,$$

$$e = x_{2\xi_1}^* \cdot_* \xi_{1x_2}^* +_* x_{2\xi_2}^* \cdot_* \xi_{2x_2}^*.$$

Hence, setting

$$D = e/_* (x_{1\xi_1}^* \cdot_* x_{2\xi_2}^* -_* x_{1\xi_2}^* \cdot_* x_{2\xi_1}^*),$$

we go to the system

$$\xi_{1x_1}^* = x_{2\xi_2}^* \cdot_* D,$$

$$\xi_{1x_2}^* = -_* x_{1\xi_2}^* \cdot_* D,$$

$$\xi_{2x_1}^* = -_* x_{2\xi_1}^* \cdot_* D,$$

$$\xi_{2x_2}^* = x_{1\xi_1}^* \cdot_* D. \tag{3.29}$$

Next, we multiplicative differentiate u with respect to ξ_1 and ξ_2 and we get

$$u_{\xi_1}^* = u_{x_1}^* \cdot_* x_{1\xi_1}^* +_* u_{x_2}^* \cdot_* x_{2\xi_1}^*,$$

$$u_{\xi_2}^* = u_{x_1}^* \cdot_* x_{1\xi_2}^* +_* u_{x_2}^* \cdot_* x_{2\xi_2}^*, \tag{3.30}$$

whereupon

$$u_{x_1}^* = u_{\xi_1}^* \cdot_* x_{2\xi_2}^* \cdot_* D -_* u_{\xi_2}^* \cdot_* x_{2\xi_1}^* \cdot_* D,$$

$$u_{x_2}^* = -_* u_{\xi_1}^* \cdot_* x_{1\xi_2}^* \cdot_* D +_* u_{\xi_2}^* \cdot_* x_{1\xi_1}^* \cdot_* D. \tag{3.31}$$

Now we multiplicative differentiate equation (3.30) with respect to ξ_1 and ξ_2 and we get

$$u_{x_1 x_1}^{**} \cdot_* x_{1\xi_1}^{*2*} +_* e^2 \cdot_* u_{x_1 x_2}^{**} \cdot_* x_{1\xi_1}^* \cdot_* x_{2\xi_1}^* +_* u_{x_2 x_2}^{**} \cdot_* x_{2\xi_1}^{*2*}$$

$$= u_{\xi_1 \xi_1}^{**} -_* u_{x_1}^* \cdot_* x_{1\xi_1 \xi_1}^{**} -_* u_{x_2}^* \cdot_* x_{2\xi_1 \xi_1}^{**},$$

$$u_{x_1 x_1}^{**} \cdot_* x_{1\xi_1}^* \cdot_* x_{1\xi_2}^* +_* u_{x_1 x_2}^{**} \cdot_* \left(x_{1\xi_1}^* \cdot_* x_{2\xi_2}^* +_* x_{1\xi_2}^* \cdot_* x_{2\xi_1}^* \right)$$

$$+_* u_{x_2 x_2}^{**} \cdot_* x_{2\xi_1}^* \cdot_* x_{2\xi_2}^*$$

$$= u_{\xi_1 \xi_2}^{**} -_* u_{x_1}^* \cdot_* x_{1\xi_1 \xi_2}^{**} -_* u_{x_2}^* \cdot_* x_{2\xi_1 \xi_2}^{**},$$

$$u_{x_1 x_1}^{**} \cdot_* x_{1\xi_2}^{*2*} +_* e^2 \cdot_* u_{x_1 x_2}^{**} \cdot_* x_{1\xi_2}^* \cdot_* x_{2\xi_2}^* +_* u_{x_2 x_2}^{**} \cdot_* x_{2\xi_2}^{*2*}$$

$$= u_{\xi_2 \xi_2}^{**} -_* u_{x_1}^* \cdot_* x_{1\xi_2 \xi_2}^{**} -_* u_{x_2}^* \cdot_* x_{2\xi_2 \xi_2}^{**}.$$

Let

$$
A = \begin{pmatrix}
x^{*2*}_{1\xi_1} & e^2 \cdot_* x^*_{1\xi_1} \cdot_* x^*_{2\xi_1} & x^{*2*}_{2\xi_1} \\[6pt]
x^*_{1\xi_1} \cdot_* x^*_{1\xi_2} & x^*_{1\xi_1} \cdot_* x^*_{2\xi_2} +_* x^*_{1\xi_2} \cdot_* x^*_{2\xi_1} & x^*_{2\xi_1} \cdot_* x^*_{2\xi_2} \\[6pt]
x^{*2*}_{1\xi_2} & e^2 \cdot_* x^*_{1\xi_2} \cdot_* x^*_{2\xi_2} & x^{*2*}_{2\xi_2}
\end{pmatrix},
$$

$$
B = \begin{pmatrix}
u^{**}_{\xi_1\xi_1} -_* u^*_{x_1} \cdot_* x^{**}_{1\xi_1\xi_1} -_* u^*_{x_2} \cdot_* x^{**}_{2\xi_1\xi_1} \\[6pt]
u^{**}_{\xi_1\xi_2} -_* u^*_{x_1} \cdot_* x^{**}_{1\xi_1\xi_2} -_* u^*_{x_2} \cdot_* x^{**}_{2\xi_1\xi_2} \\[6pt]
u^{**}_{\xi_2\xi_2} -_* u^*_{x_1} \cdot_* x^{**}_{1\xi_2\xi_2} -_* u^*_{x_2} \cdot_* x^{**}_{2\xi_2\xi_2}
\end{pmatrix}.
$$

We have that

$$
\det A = e/_*(D^{3*}).
$$

The multiplicative cofactors of the matrix A are as follows:

$$
a_{11} = \begin{vmatrix}
x^*_{1\xi_1} \cdot_* x^*_{2\xi_2} +_* x^*_{1\xi_2} \cdot_* x^*_{2\xi_1} & x^*_{2\xi_1} \cdot_* x^*_{2\xi_2} \\[6pt]
e^2 \cdot_* x^*_{1\xi_2} \cdot_* x^*_{2\xi_2} & x^{*2*}_{2\xi_2}
\end{vmatrix}_*
$$

$$
= (e/_*D) \cdot_* x^{*2*}_{2\xi_2},
$$

$$
a_{12} = -_* \begin{vmatrix}
x^*_{1\xi_1} \cdot_* x^*_{1\xi_2} & x^*_{2\xi_1} \cdot_* x^*_{2\xi_2} \\[6pt]
x^{*2*}_{1\xi_2} & x^{*2*}_{2\xi_2}
\end{vmatrix}_*
$$

$$
= -_*(e/_*D) \cdot_* x^*_{1\xi_2} \cdot_* x^*_{2\xi_2},
$$

$$
a_{13} = \begin{vmatrix}
x^*_{1\xi_1} \cdot_* x^*_{1\xi_2} & x^*_{1\xi_1} \cdot_* x^*_{2\xi_2} +_* x^*_{1\xi_2} \cdot_* x^*_{2\xi_1} \\[6pt]
x^{*2*}_{1\xi_2} & e^2 \cdot_* x^*_{1\xi_2} \cdot_* x^*_{2\xi_2}
\end{vmatrix}_*
$$

$$
= (e/_*D) \cdot_* x^{*2*}_{1\xi_2},
$$

$$
a_{21} = -_* \begin{vmatrix}
e^2 \cdot_* x^*_{1\xi_1} \cdot_* x^*_{2\xi_1} & x^{*2*}_{2\xi_1} \\[6pt]
e^2 \cdot_* x^*_{1\xi_2} \cdot_* x^*_{2\xi_2} & x^{*2*}_{2\xi_2}
\end{vmatrix}_*
$$

$$
= -_*((e^2)/_*D) \cdot_* x^*_{2\xi_1} \cdot_* x^*_{2\xi_2},
$$

$$a_{22} = \begin{vmatrix} x^{*2*}_{1\xi_1} & x^{*2*}_{2\xi_1} \\ x^{*2*}_{1\xi_2} & x^{*2*}_{2\xi_2} \end{vmatrix}_*$$

$$= (c/_*D) *_\left(x^*_{1\xi_1} *_* x^*_{2\xi_2} \;|\; *_* x^*_{1\xi_2} \cdot_* x^*_{2\xi_1} \right),$$

$$a_{23} = -_* \begin{vmatrix} x^{*2*}_{1\xi_1} & e^2 \cdot_* x^*_{1\xi_1} \cdot_* x^*_{2\xi_1} \\ x^{*2*}_{1\xi_2} & e^2 \cdot_* x^*_{1\xi_2} \cdot_* x^*_{2\xi_2} \end{vmatrix}_*$$

$$= -_*((e^2)/_*D) \cdot_* x^*_{1\xi_1} \cdot_* x^*_{1\xi_2},$$

$$a_{31} = \begin{vmatrix} e^2 \cdot_* x^*_{1\xi_1} \cdot_* x^*_{2\xi_1} & x^{*2*}_{2\xi_1} \\ x^*_{1\xi_1} \cdot_* x^*_{2\xi_2} +_* x^*_{1\xi_2} \cdot_* x^*_{2\xi_1} & x^*_{2\xi_1} \cdot_* x^*_{2\xi_2} \end{vmatrix}_*$$

$$= (e/_*D) \cdot_* x^{*2*}_{2\xi_1},$$

$$a_{32} = -_* \begin{vmatrix} x^{*2*}_{1\xi_1} \cdot_* & x^{*2*}_{2\xi_1} \\ x^*_{1\xi_1} \cdot_* x^*_{1\xi_2} & x^*_{2\xi_1} \cdot_* x^*_{2\xi_2} \end{vmatrix}_*$$

$$= -_*(e/_*D) \cdot_* x^*_{1\xi_1} \cdot_* x^*_{2\xi_1},$$

$$a_{33} = \begin{vmatrix} x^{*2*}_{1\xi_1} & e^2 \cdot_* x^*_{1\xi_1} \cdot_* x^*_{2\xi_1} \\ x^*_{1\xi_1} \cdot_* x^*_{1\xi_2} & x^*_{1\xi_1} \cdot_* x^*_{2\xi_2} +_* x^*_{1\xi_2} \cdot_* x^*_{2\xi_1} \end{vmatrix}_*$$

$$= (e/_*D) \cdot_* x^{*2*}_{1\xi_1}.$$

Hence,

$$A^{-*1} = D^{2*} \cdot_* \begin{pmatrix} x^{*2*}_{2\xi_2} & -_*e^2 \cdot_* x^*_{2\xi_1} \cdot_* x^*_{2\xi_2} & x^{*2*}_{2\xi_1} \\ -_*x^*_{1\xi_2} \cdot_* x^*_{2\xi_2} & x^*_{1\xi_1} \cdot_* x^*_{2\xi_2} +_* x^*_{1\xi_2} \cdot_* x^*_{2\xi_1} & -_*x^*_{1\xi_1} \cdot_* x^*_{2\xi_1} \\ x^{*2*}_{1\xi_2} & -_*e^2 \cdot_* x^*_{1\xi_1} \cdot_* x^*_{1\xi_2} & x^{*2*}_{1\xi_1} \end{pmatrix}.$$

Therefore,

$$\begin{pmatrix} u^{**}_{x_1 x_1} \\ u^{**}_{x_1 x_2} \\ u^{**}_{x_2 x_2} \end{pmatrix} = A^{-*1} \cdot_* B$$

and the canonical form of the operator $L(u)$ is as follows:

$$
L(u) = D^{2*} \cdot_* \left(\left(\left(a \cdot_* x_{2\xi_2}^{*2*} -_* e^2 \cdot_* b \cdot_* x_{1\xi_2}^{*} \cdot_* x_{2\xi_2}^{*} \right. \right. \right.
$$

$$
\left. +_* c \cdot_* x_{1\xi_2}^{*2*} \right) \cdot_* u_{\xi_1\xi_1}^{**}
$$

$$
+_* \left(-_* e^2 \cdot_* a \cdot_* x_{2\xi_1}^{*} \cdot_* x_{2\xi_2}^{*} +_* e^2 \cdot_* b \left(x_{1\xi_1}^{*} \cdot_* x_{2\xi_2}^{*} +_* x_{1\xi_2}^{*} \cdot_* x_{2\xi_1}^{*} \right) \right.
$$

$$
\left. -_* e^2 \cdot_* c \cdot_* x_{1\xi_1}^{*} \cdot_* x_{1\xi_2}^{*} \right) \cdot_* u_{\xi_1\xi_2}^{**}
$$

$$
+_* \left(a \cdot_* x_{2\xi_1}^{*2*} -_* e^2 \cdot_* b \cdot_* x_{1\xi_1}^{*} \cdot_* x_{2\xi_1}^{*} +_* c \cdot_* x_{1\xi_1}^{*2*} \right) \cdot_* u_{\xi_2\xi_2}^{**}
$$

$$
-_* u_{x_1}^{*} \cdot_* \left(a \cdot_* \left(x_{2\xi_2}^{*2*} \cdot_* x_{1\xi_1\xi_1}^{**} -_* e^2 \cdot_* x_{2\xi_1}^{*} \cdot_* x_{2\xi_2}^{*} \cdot_* x_{1\xi_1\xi_2}^{**} \right. \right.
$$

$$
\left. +_* x_{2\xi_1}^{*2*} \cdot_* x_{1\xi_2\xi_2}^{**} \right)
$$

$$
+_* e^2 \cdot_* b \cdot_* \left(-_* x_{1\xi_2}^{*} \cdot_* x_{2\xi_2}^{*} \cdot_* x_{1\xi_1\xi_1}^{**} \right.
$$

$$
+_* x_{1\xi_1\xi_2}^{**} \cdot_* \left(x_{1\xi_1}^{*} \cdot_* x_{2\xi_2}^{*} +_* x_{1\xi_2}^{*} \cdot_* x_{2\xi_1}^{*} \right)
$$

$$
\left. -_* x_{1\xi_1}^{*} \cdot_* x_{2\xi_1}^{*} \cdot_* x_{1\xi_2\xi_2}^{**} \right)
$$

$$
+_* c \cdot_* \left(\left(x_{1\xi_2}^{*2*} \cdot_* x_{1\xi_1\xi_1}^{**} -_* e^2 \cdot_* x_{1\xi_1}^{*} \cdot_* x_{1\xi_2}^{*} \cdot_* x_{1\xi_1\xi_2}^{**} \right. \right.
$$

$$
\left. \left. +_* x_{1\xi_1}^{*2*} \cdot_* x_{1\xi_2\xi_2}^{**} \right) \right)
$$

$$
-_* u_{x_2}^{*} \cdot_* \left(a \cdot_* \left(x_{2\xi_1\xi_1}^{**} \cdot_* x_{2\xi_2}^{*2*} -_* e^2 \cdot_* x_{2\xi_1}^{*} \cdot_* x_{2\xi_2}^{*} \cdot_* x_{2\xi_1\xi_2}^{**} \right. \right.
$$

$$
\left. +_* x_{2\xi_1}^{*2*} \cdot_* x_{2\xi_2\xi_2}^{**} \right)
$$

$$
+_* e^2 \cdot_* b \cdot_* \left(-_* x_{1\xi_2}^{*} \cdot_* x_{2\xi_2}^{*} \cdot_* x_{2\xi_1\xi_1}^{**} \right.
$$

$$
+_* \left(x_{1\xi_1}^{*} \cdot \right)_* x_{2\xi_2}^{*} +_* x_{1\xi_2}^{*} \cdot_* x_{2\xi_1}^{*} \right) \cdot_* x_{2\xi_1\xi_2}^{**}
$$

$$
\left. -_* x_{1\xi_1}^{*} \cdot_* x_{2\xi_1}^{*} \cdot_* x_{2\xi_2\xi_2}^{**} \right)
$$

$$
+_* c \cdot_* \left(x_{1\xi_2}^{*2*} \cdot_* x_{2\xi_1\xi_1}^{**} -_* e^2 \cdot_* x_{1\xi_1}^{*} \cdot_* x_{1\xi_2}^{*} \cdot_* x_{2\xi_1\xi_2}^{**} \right.
$$

$$
\left. \left. \left. \left. +_* x_{1\xi_1}^{*2*} \cdot_* \cdot_* x_{2\xi_2\xi_2}^{**} \right) \right) \right) \right) +_* d
$$

$$
= 0_*.
$$

$$
(3.32)
$$

3.2.1 The multiplicative hyperbolic case

In this case, we have $a \cdot_* c -_* b^2 < 0_*$. The characteristic parameters

$$\xi_1 \;=\; \phi_1(x_1, x_2),$$

$$\xi_2 \;=\; \phi_2(x_1, x_2),$$

satisfy, as in the multiplicative linear case, the equations

$$\phi^*_{1x_1} -_* \lambda_1 \cdot_* \phi^*_{1x_2} = 0_*,$$
$$\phi^*_{2x_1} -_* \lambda_2 \cdot_* \phi^*_{2x_2} = 0_*,$$
(3.33)

where λ_1 and λ_2 are the roots of the multiplicative quadratic equation (3.5). By equation (3.33), we find

$$\xi^*_{1x_1} = \lambda_1 \cdot_* \xi^*_{1x_2},$$

$$\xi^*_{2x_1} = \lambda_2 \cdot_* \xi^*_{2x_2}.$$

Hence, from equation (3.29), we get

$$x^*_{2\xi_2} = -_*\lambda_1 \cdot_* x^*_{1\xi_2},$$
$$x^*_{2\xi_1} = -_*\lambda_2 \cdot_* x^*_{1\xi_1}.$$
(3.34)

In the multiplicative hyperbolic case, we have

$$\alpha = \gamma = 0_*$$

or

$$a \cdot_* \xi^{*2*}_{1x_1} +_* e^2 \cdot_* b \cdot_* \xi^*_{1x_1} \cdot_* \xi^*_{1x_2} +_* c \cdot_* \xi^{*2*}_{1x_2} = 0_*$$

$$a \cdot_* \xi^{*2*}_{2x_1} +_* e^2 \cdot_* b \cdot_* \xi^*_{2x_1} \cdot_* \xi^*_{2x_2} +_* c \cdot_* \xi^{*2*}_{2x_2} = 0_*.$$

Hence, applying equation (3.29), we get

$$a \cdot_* x^{*2*}_{2\xi_2} -_* e^2 \cdot_* b \cdot_* x^*_{2\xi_2} \cdot_* x^*_{1\xi_2} +_* c \cdot_* x^{*2*}_{1\xi_2} = 0_*,$$

$$a \cdot_* x^{*2*}_{2\xi_1} -_* e^2 \cdot_* b \cdot_* x^*_{2\xi_1} \cdot_* x^*_{1\xi_1} +_* c \cdot_* x^{*2*}_{1\xi_1} = 0_*.$$

Then the canonical form of the operator $L(u)$ is

$$L(u) = D^{2*} \cdot_* \left\{ \left(-_* e^2 \cdot_* a \cdot_* x^*_{2\xi_1} \cdot_* x^*_{2\xi_2} +_* e^2 \cdot_* b \cdot_* \right. \right.$$

$$\left(x^*_{1\xi_1} \cdot_* x^*_{2\xi_2} +_* x^*_{1\xi_2} \cdot_* x^*_{2\xi_1} \right) -_* e^2 \cdot_* c \cdot_* x^*_{1\xi_1} \cdot_* x^*_{1\xi_2} \right) \cdot_* u^{**}_{\xi_1\xi_2}$$

$$-_* u^*_{x_1} \cdot_* \left(a \cdot_* \left(x^{*2*}_{2\xi_2} \cdot_* x^{**}_{1\xi_1\xi_1} -_* e^2 \cdot_* x^*_{2\xi_1} \cdot_* x^*_{2\xi_2} \cdot_* x^{**}_{1\xi_1\xi_2} \right. \right.$$

$$\left. +_* x^{*2*}_{2\xi_1} \cdot_* x^{**}_{1\xi_2\xi_2} \right) +_* e^2 \cdot_* b \cdot_* \left(-_* x^*_{1\xi_2} \cdot_* x^*_{2\xi_2} \cdot_* x^{**}_{1\xi_1\xi_1} \right.$$

$$\left. +_* x^{**}_{1\xi_1\xi_2} \cdot_* \left(x^*_{1\xi_1} \cdot_* x^*_{2\xi_2} +_* x^*_{1\xi_2} \cdot_* x^*_{2\xi_1} \right) \right.$$

$$\left. -_* x^*_{1\xi_1} \cdot_* x^*_{2\xi_1} \cdot_* x^{**}_{1\xi_2\xi_2} \right)$$

$$\left. +_* c \cdot_* \left(x^{*2*}_{1\xi_2} \cdot_* x^{**}_{1\xi_1\xi_1} -_* e^2 \cdot_* x^*_{1\xi_1} \cdot_* x^*_{1\xi_2} \cdot_* x^{**}_{1\xi_1\xi_2} +_* x^{*2*}_{1\xi_1} \cdot_* x^{**}_{1\xi_2\xi_2} \right) \right)$$

$$-_* u^*_{x_2} \cdot_* \left(a \cdot_* \left(x^{**}_{2\xi_1\xi_1} \cdot_* x^{*2*}_{2\xi_2} -_* e^2 \cdot_* x^*_{2\xi_1} \cdot_* x^*_{2\xi_2} \cdot_* x^{**}_{2\xi_1\xi_2} \right. \right.$$

$$\left. +_* x^{*2*}_{2\xi_1} \cdot_* x^{**}_{2\xi_2\xi_2} \right) +_* e^2 \cdot_* b \cdot_* \left(-_* x^*_{1\xi_2} \cdot_* x^*_{2\xi_2} \cdot_* x^{**}_{2\xi_1\xi_1} \right.$$

$$\left. +_* \left(x^*_{1\xi_1} \cdot_* x^*_{2\xi_2} +_* x^*_{1\xi_2} \cdot_* x^*_{2\xi_1} \right) \cdot_* x^{**}_{2\xi_1\xi_2} \right.$$

$$\left. -_* x^*_{1\xi_1} \cdot_* x^*_{2\xi_1} \cdot_* x^{**}_{2\xi_2\xi_2} \right)$$

$$\left. \left. +_* c \cdot_* \left(x^{*2*}_{1\xi_2} \cdot_* x^{**}_{2\xi_1\xi_1} -_* e^2 \cdot_* x^*_{1\xi_1} \cdot_* x^*_{1\xi_2} \cdot_* x^{**}_{2\xi_1\xi_2} +_* x^{*2*}_{1\xi_1} \cdot_* x^{**}_{2\xi_2\xi_2} \right) \right) \right\}$$

$$+_* d.$$

Now, using equation (3.34), we get

$$L(u) = D^{2*} \cdot_* \left\{ \left(-_* e^2 \cdot_* a \cdot_* \lambda_1 \cdot_* \lambda_2 -_* e^2 \cdot_* b \cdot_* (\lambda_1 +_* \lambda_2) -_* e^2 \cdot_* c \right) \right.$$

$$\cdot_* x^*_{1\xi_1} \cdot_* x^*_{1\xi_2} \cdot_* u^{**}_{\xi_1\xi_2}$$

$$-_* u^*_{x_1} \cdot_* \left((a \cdot_* \lambda_1^{2*} +_* e^2 \cdot_* b \cdot_* \lambda_1 +_* c) \cdot_* x^{*2*}_{1\xi_2} \cdot_* x^{**}_{1\xi_1\xi_1} \right.$$

$$+_* \left(-_* e^2 \cdot_* a \cdot_* \lambda_1 \cdot_* \lambda_2 -_* e^2 \cdot_* b \cdot_* (\lambda_1 +_* \lambda_2) -_* e^2 \cdot_* c \right)$$

$$\cdot_* x^*_{1\xi_1} \cdot_* x^*_{1\xi_2} \cdot_* x^{**}_{1\xi_1\xi_2}$$

$$+_* \left(a \cdot_* \lambda_2^{2*} +_* e^2 \cdot_* b \cdot_* \lambda_2 +_* c \right) \cdot_* x_{1\xi_1}^{*2*} \cdot_* x_{1\xi_1\xi_2}^{**} \right)$$

$$-_* u_{x_2}^* \cdot_* \left(\left(a \cdot_* \lambda_1^{2*} +_* e^2 \cdot_* b \cdot_* \lambda_1 +_* c \right) \cdot_* x_{1\xi_2}^{*2*} \cdot_* x_{2\xi_1\xi_1}^{**}$$

$$+_* \left(-_* e^2 \cdot_* a \cdot_* \lambda_1 \cdot_* \lambda_2 -_* e^2 \cdot_* b \cdot_* (\lambda_1 +_* \lambda_2) -_* e^2 \cdot_* c \right)$$

$$\cdot_* x_{1\xi_1}^* \cdot_* x_{1\xi_2}^* \cdot_* x_{2\xi_1\xi_2}^{**}$$

$$+_* \left(a \cdot_* \lambda_2^{2*} +_* e^2 \cdot_* b \cdot_* \lambda_2 +_* c \right) \cdot_* x_{1\xi_1}^{*2*} \cdot_* x_{2\xi_2\xi_2}^{**} \right) \right\} +_* d$$

$$= 0_*.$$

Next, using

$$a \cdot_* \lambda_1^{2*} +_* e^2 \cdot_* b \cdot_* \lambda_1 +_* c = 0_*,$$

$$a \cdot_* \lambda_2^{2*} +_* e^2 \cdot_* b \cdot_* \lambda_2 +_* c = 0_*,$$

and

$$\lambda_1 +_* \lambda_2 = -_*((e^2 \cdot_* b)/_* a),$$

$$\lambda_1 \cdot_* \lambda_2 = c/_* a,$$

we get

$$D^{2*} \cdot_* \left(-_* e^4 \cdot_* ((a \cdot_* c -_* b^{2*})/_* a) \cdot_* x_{1\xi_1}^* \cdot_* x_{1\xi_2}^* \cdot_* u_{\xi_1\xi_2}^{**} \right.$$

$$+_* e^4 ((a \cdot_* c -_* b^{2*})/_* a) \cdot_* x_{1\xi_1}^* \cdot_* x_{1\xi_2}^* \cdot_* x_{1\xi_1\xi_2}^{**} \cdot_* u_{x_1}^*$$

$$\left. +_* e^4 \cdot_* ((a \cdot_* c -_* b^{2*})/_* a) \cdot_* x_{1\xi_1}^* \cdot_* x_{1\xi_2}^* \cdot_* x_{1\xi_1\xi_2}^{**} \cdot_* u_{x_2}^* \right) +_* d$$

$$= 0_*$$

or

$$x_{1\xi_1}^* \cdot_* x_{1\xi_2}^* \cdot_* \left(u_{\xi_1\xi_2}^{**} -_* x_{1\xi_1\xi_2}^{**} \cdot_* D \cdot_* \left(-_* \lambda_1 \cdot_* u_{\xi_1}^* \cdot_* x_{1\xi_2}^* \right. \right.$$

$$+_* \lambda_2 \cdot_* u_{\xi_2}^* \cdot_* x_{1\xi_1}^*)$$

$$-_* x_{1\xi_1\xi_2}^{**} \cdot_* D \cdot_* \left(-_* u_{\xi_1}^* \cdot_* x_{1\xi_2}^* +_* u_{\xi_2}^* \cdot_* x_{1\xi_1}^* \right) \right)$$

$$= ((a \cdot_* d)/_* (e^4 \cdot_* (a \cdot_* c -_* b^{2*}))) \cdot_* D^{2*}$$

is the canonical form of $L(u)$.

3.2.2   The multiplicative elliptic case

In this case, we have

$$a \cdot_* c -_* b^{2*} > 0_*$$

and

$$\alpha = \gamma, \beta = 0_*.$$

Then,

$$a \cdot_* \xi_{1x_1}^{*2*} +_* e^2 \cdot_* b \cdot_* \xi_{1x_1}^* \cdot_* \xi_{1x_2}^* +_* c \cdot_* \xi_{1x_2}^{*2*}$$

$$= a \cdot_* \xi_{2x_1}^{*2*} +_* e^2 \cdot_* b \cdot_* \xi_{2x_1}^* \cdot_* \xi_{2x_2}^* +_* c \cdot_* \xi_{2x_2}^{*2*}$$

$$a \cdot_* \xi_{1x_1}^* \cdot_* \xi_{2x_1}^* +_* b \cdot_* \left(\xi_{1x_2}^* \cdot_* \xi_{2x_1}^* +_* \xi_{1x_1}^* \cdot_* \xi_{2x_2}^* \right)$$

$$+_* c \cdot_* \xi_{1x_2}^* \cdot_* \xi_{2x_2}^*$$

$$= 0_*.$$

Hence, from equation (3.29), we obtain

$$a \cdot_* x_{2\xi_2}^{*2*} -_* e^2 \cdot_* b \cdot_* x_{2\xi_2}^* \cdot_* x_{1\xi_2}^* +_* c \cdot_* x_{1\xi_2}^{*2*}$$

$$= a \cdot_* x_{2\xi_1}^{*2*} -_* e^2 \cdot_* bx_{2\xi_1}^* \cdot_* x_{1\xi_1}^* +_* c \cdot_* x_{1\xi_1}^{*2*}$$

$$-_* a \cdot_* x_{2\xi_1}^* \cdot_* x_{2\xi_2}^* +_* b \cdot_* \left(x_{1\xi_2}^* \cdot_* x_{2\xi_1}^* +_* x_{1\xi_1}^* \cdot_* x_{2\xi_2}^* \right) -_* c \cdot_* x_{1\xi_1}^* \cdot_* x_{1\xi_2}^*$$

$$= 0_*.$$

Consequently, the canonical form of the operator $L(u)$ is as follows:

$$L(u) = D^{2*} \cdot_* \left(\left(a \cdot_* x_{2\xi_2}^{*2*} -_* e^2 \cdot_* b \cdot_* x_{1\xi_2}^* \cdot_* x_{2\xi_2}^* +_* c \cdot_* x_{1\xi_2}^{*2*} \right) \right.$$

$$\cdot_* \left(u_{\xi_1\xi_1}^{**} +_* u_{\xi_2\xi_2}^{**} \right)$$

$$-_* u_{x_1}^* \cdot_* \left(a \cdot_* x_{2\xi_2}^{*2*} -_* e^2 \cdot_* b \cdot_* x_{2\xi_2}^* \cdot_* x_{1\xi_2}^* +_* c \cdot_* x_{1\xi_2}^{*2*} \right)$$

$$\cdot_* \left(x_{1\xi_1\xi_1}^{**} +_* x_{1\xi_2\xi_2}^{**} \right)$$

$$-_* u_{x_2}^* \cdot_* \left(a \cdot_* x_{2\xi_2}^{*2*} -_* e^2 \cdot_* b \cdot_* x_{2\xi_2}^* \cdot_* x_{1\xi_2}^* +_* c \cdot_* x_{1\xi_2}^{*2*} \right)$$

$$\left. \cdot_* \left(x_{2\xi_1\xi_1}^{**} +_* x_{2\xi_2\xi_2}^{**} \right) \right) +_* d$$

$$= 0_*$$

or

$$D^{2*} \cdot_* \left(a \cdot_* x_{2\xi_2}^{*2*} -_* e^2 \cdot_* b \cdot_* x_{1\xi_2}^* \cdot_* x_{2\xi_2}^* +_* c \cdot_* x_{1\xi_2}^{*2*} \right)$$

$$\cdot_* \left(\Delta_* u -_* u_{x_1}^* \cdot_* \Delta_* x_1 -_* u_{x_2}^* \cdot_* \Delta_* x_2 \right) +_* d$$

$$= 0_*,$$

where $u_{x_1}^*$ and $u_{x_2}^*$ are determined with equation (3.31).

3.2.3   The multiplicative parabolic case

In this case, we have

$$a \cdot_* c -_* b^{2*} = 0_*$$

and

$$\beta = \gamma = 0_*.$$

Then,

$$a \cdot_* \xi_{1x_1}^* \cdot_* \xi_{2x_1}^* +_* b \cdot_* \left(\xi_{1x_2}^* \cdot_* \xi_{2x_1}^* +_* \xi_{1x_1}^* \cdot_* \xi_{2x_2}^* \right)$$

$$+_* c \cdot_* \xi_{1x_2}^* \cdot_* \xi_{2x_2}^*$$

$$= 0_*$$

$$a \cdot_* \xi_{2x_1}^{*2*} +_* e^2 \cdot_* b \cdot_* \xi_{2x_1}^* \cdot_* \xi_{2x_2}^* +_* c \cdot_* \xi_{2x_2}^{*2*} = 0_*.$$

Hence, from equation (3.29), we obtain

$$a \cdot_* x_{2\xi_1}^* \cdot_* x_{2\xi_2}^* -_* b \cdot_* \left(x_{1\xi_2}^* \cdot_* x_{2\xi_1}^* +_* x_{1\xi_1}^* \cdot_* x_{2\xi_2}^* \right)$$

$$+_* c \cdot_* x_{1\xi_1}^* \cdot_* x_{1\xi_2}^*$$

$$= 0_*$$

$$a \cdot_* x_{2\xi_1}^{*2*} -_* e^2 \cdot_* b \cdot_* x_{2\xi_1}^* \cdot_* x_{1\xi_1}^* +_* c \cdot_* x_{1\xi_1}^{*2*} = 0_*.$$

Therefore, the canonical form of the operator $L(u)$ is as follows:

$$L(u) = D^{2*} \cdot_* \left(\left(a \cdot_* x_{2\xi_2}^{*2*} -_* e^2 \cdot_* b \cdot_* x_{1\xi_2}^* \cdot_* x_{2\xi_2}^* +_* c \cdot_* x_{1\xi_2}^{*2*} \right) \cdot_* u_{\xi_1\xi_1}^{**} \right.$$

$$-_* u_{x_1}^* \cdot_* \left(a \cdot_* x_{2\xi_2}^{*2*} -_* e^2 \cdot_* b \cdot_* x_{1\xi_2}^* \cdot_* x_{2\xi_2}^* +_* c \cdot_* x_{1\xi_2}^{*2*} \right) \cdot_* x_{1\xi_1\xi_1}^{**}$$

$$\left. -_* u_{x_2}^* \cdot_* \left(a \cdot_* x_{2\xi_2}^{*2*} -_* e^2 \cdot_* b \cdot_* x_{1\xi_2}^* \cdot_* x_{2\xi_2}^* +_* c \cdot_* x_{1\xi_2}^{*2*} \right) \cdot_* x_{2\xi_1\xi_1}^{**} \right)$$

$$+_* d$$

$$= 0_*,$$

where $u_{x_1}^*$ and $u_{x_2}^*$ are determined with equation (3.31).

3.3   Classification and Canonical Form of Second-Order Linear Partial Differential Equations in n Multiplicative Independent Variables

Consider a general second-order multiplicative linear multiplicative partial differential equation in n multiplicative independent variables:

$$\sum_{*i,j=1}^{n} a_{ij} \cdot_* u_{x_i x_j}^{**} +_* \sum_{*i=1}^{n} b_i \cdot_* u_{x_i}^{*} +_* c \cdot_* u +_* d = 0_*, \qquad (3.35)$$

where the coefficients $a_{ij}, b_i, 1 \le i, j \le n, c, d$ and the unknown u are functions of $x = (x_1, \ldots, x_n)$. Let

$$A \;=\; \begin{pmatrix} a_{11} & a_{12} & \cdots & a_{1n} \\ a_{21} & a_{22} & \cdots & a_{2n} \\ \vdots & & & \\ a_{n1} & a_{n2} & \cdots & a_{nn} \end{pmatrix},$$

$$b \;=\; (b_1, \cdots, b_n).$$

Assume that the matrix A is symmetric. Otherwise, we can always find a symmetric matrix:

$$\bar{a}_{ij} \;=\; e^{\frac{1}{2}} \cdot_* (a_{ij} +_* a_{ji})$$

$$=\; e^{\frac{1}{2}} \cdot_* (a_{ij} a_{ji})$$

$$=\; e^{\log\left(e^{\frac{1}{2}}\right) \log(a_{ij} a_{ji})}$$

$$=\; e^{\frac{\log(a_{ij}) + \log(a_{ji})}{2}}$$

such that equation (3.35) can be rewritten in the form

$$\sum_{*i,j=1}^{n} \bar{a}_{ij} \cdot_* u_{x_i x_j}^{**} +_* \sum_{*i=1}^{n} b_i \cdot_* u_{x_i}^{*} +_* c \cdot_* u +_* d = 0_*.$$

Now we consider the transformation

$$\xi = Q \cdot_* x,$$

where $\xi = (\xi_1, \ldots, \xi_n)$ and $Q = (q_{ij})$ is an $n \times n$ matrix. We have that

$$\xi_i^* = \sum_{*j=1}^{n} q_{ij} \cdot_* x_j.$$

Using the chain rule, we have

$$u_{x_i}^* = \sum_{*k=1}^{n} u_{\xi_k^*}^* \cdot_* \xi_{kx_i}^*$$

$$= \sum_{*k=1}^{n} u_{\xi_k^*}^* \cdot_* q_{ki},$$

$$u_{x_i x_j}^{**} = \sum_{*k,l=1}^{n} u_{\xi_k^* \xi_l^*}^{**} \cdot_* \xi_{kx_i}^* \cdot_* \xi_{lx_j}^*$$

$$= \sum_{*k,l=1}^{n} u_{\xi_k^* \xi_l^*}^{**} \cdot_* q_{ki} \cdot_* q_{lj}.$$

This allows equation (3.35) to be expressed as

$$\sum_{*k,l=1}^{n} \left(\sum_{*i,j=1}^{n} q_{ki} \cdot_* a_{ij} \cdot_* q_{lj} \right) \cdot_* u_{\xi_k^* \xi_l^*}^{**} +_* \text{lower} \quad \text{order} \quad \text{terms} = 0_*.$$

The coefficient matrix of the terms $u_{\xi_k^* \xi_l^*}^{**}$ in this transformed expression is equal to

$$(q_{ki} \cdot_* a_{ij} \cdot_* q_{lj}) = Q^T \cdot_* A \cdot_* Q.$$

We know that for any real symmetric matrix A, there is an associative multiplicative orthogonal matrix Q_1 such that $Q_1^T \cdot_* A \cdot_* Q_1 = \Lambda$. Here Q_1 is called diagonalizing matrix of the matrix A and Λ is a diagonal matrix whose elements are the multiplicative eigenvalues λ_i of the matrix A and the columns of Q_1 are multiplicative linearly independent eigenvectors of A, $q_i = (q_{1i}, \ldots, q_{ni})$,

$$|q_i|_* = (q_{1i}^{2*} +_* \cdots +_* q_{ni}^{2*})^{\frac{1}{2}*}_* = e.$$

So, we have

$$Q = (q_{ij}),$$

$$\Lambda = (\lambda_i \cdot_* \delta_{ij}), \quad i, j = 1, \ldots, n,$$

where δ_{ij} are the multiplicative Kronecker delta. Now, if Q is taken to be a diagonalizing matrix of A, it follows that

$$Q^T \cdot_* A \cdot_* Q = \Lambda$$

$$= \begin{pmatrix} \lambda_1 & & & \\ & \lambda_2 & & \\ & & \ddots & \\ & & & \lambda_n \end{pmatrix}.$$

The members $\lambda_1, \lambda_2, \ldots, \lambda_n$ are real numbers because the matrix A is symmetric.

Definition 3.14 Equation (3.35) is called as follows:

1. Multiplicative elliptic, if all multiplicative eigenvalues λ_i of A are multiplicative nonzero and have the same sign.

2. Multiplicative hyperbolic, if all multiplicative eigenvalues λ_i of A are multiplicative nonzero and have the same sign except for one of the multiplicative eigenvalues.

3. Multiplicative parabolic, if the multiplicative eigenvalues λ_i of A are all multiplicative positive or all multiplicative negative, save one that is multiplicative zero.

4. Multiplicative ultrahyperbolic, if there is more than one multiplicative positive multiplicative eigenvalue and more than one multiplicative negative multiplicative eigenvalue, and there are multiplicative nonzero multiplicative eigenvalues.

The multiplicative nonsingular transformation

$$x = Q^T \cdot_* \xi$$

transforms equation (3.35) in the canonical form:

$$\sum_{*i=1}^{n} \lambda_i \cdot_* u^{**}_{\xi_i \xi_i} +_* \text{lower} \quad \text{order} \quad \text{terms} = 1.$$

Example 3.16 Consider the equation

$$u^{**}_{x_1 x_1} -_* u^{**}_{x_1 x_2} +_* u^{**}_{x_2 x_2} = f(x_1, x_2).$$

We can rewrite this equation in the following form:

$$u^{**}_{x_1 x_1} -_* e^{\frac{1}{2}} \cdot_* u^{**}_{x_1 x_2} -_* e^{\frac{1}{2}} \cdot_* u^{**}_{x_2 x_1} +_* u^{**}_{x_2 x_2} = f(x_1, x_2).$$

Then,

$$A = \begin{pmatrix} e & e^{-\frac{1}{2}} \\ e^{-\frac{1}{2}} & e \end{pmatrix}.$$

We will find the multiplicative eigenvalues of the matrix A. We have

$$\det_*(A -_* \lambda \cdot_* I_*) = \begin{vmatrix} e -_* \lambda & e^{-\frac{1}{2}} \\ e^{-\frac{1}{2}} & e -_* \lambda \end{vmatrix}_*$$

$$= 0_*$$

if and only if

$$(\lambda -_* e)^{2*} -_* e^{\frac{1}{4}} = 1,$$

if and only if

$$\lambda^{2*} -_* e^2 \cdot_* \lambda +_* e^{\frac{3}{4}} = 1,$$

if and only if

$$\lambda_1 = e^{\frac{3}{2}},$$

$$\lambda_2 = e^{\frac{1}{2}}.$$

Since $\lambda_1, \lambda_2 > 0_*$, the considered equation is a multiplicative elliptic equation. Now we will find the matrix Q. Let $A_1 = (e^a, e^b)$ be an eigenvector of the matrix A corresponding to the eigenvalue $\lambda_1 = e^{\frac{3}{2}}$. Then,

$$\begin{pmatrix} -_*e^{\frac{1}{2}} & -_*e^{\frac{1}{2}} \\ -_*e^{\frac{1}{2}} & e^{\frac{1}{2}} \end{pmatrix} \cdot_* \begin{pmatrix} e^a \\ e^b \end{pmatrix} = \begin{pmatrix} 1 \\ 1 \end{pmatrix},$$

i.e.,

$$e^a +_* e^b = 1.$$

We take

$$q_1 = e^{\frac{1}{\sqrt{2}}} \cdot_* (e, -_*e).$$

Let $A_2 = (e^a, e^b)$ be an eigenvector of the matrix A corresponding to the eigenvalue $\lambda_2 = e^{\frac{1}{2}}$. Then,

$$\begin{pmatrix} e^{\frac{1}{2}} & -_*e^{\frac{1}{2}} \\ -_*e^{\frac{1}{2}} & e^{\frac{1}{2}} \end{pmatrix} \cdot_* \begin{pmatrix} e^a \\ e^b \end{pmatrix} = \begin{pmatrix} 1 \\ 1 \end{pmatrix},$$

i.e.,

$$e^a = e^b.$$

We take

$$q_2 = e^{\frac{1}{\sqrt{2}}} \cdot_* (e, e).$$

Therefore,

$$Q = (q_1^T, q_2^T)$$

$$= e^{\frac{1}{\sqrt{2}}} \cdot_* \begin{pmatrix} e & e \\ -_*e & e \end{pmatrix}.$$

Let

$$\xi = Q \cdot_* x$$

$$= \begin{pmatrix} e^{\frac{1}{\sqrt{2}}} & e^{\frac{1}{\sqrt{2}}} \\ -_*e^{\frac{1}{\sqrt{2}}} & e^{\frac{1}{\sqrt{2}}} \end{pmatrix} \cdot_* \begin{pmatrix} x_1 \\ x_2 \end{pmatrix}$$

$$= \begin{pmatrix} e^{\frac{1}{\sqrt{2}}} \cdot_* (x_1 +_* x_2) \\ e^{\frac{1}{\sqrt{2}}} \cdot_* (-_*x_1 +_* x_2) \end{pmatrix}.$$

Then,

$$\xi^*_{1x_1} = e^{\frac{1}{\sqrt{2}}},$$

$$\xi^*_{1x_2} = e^{\frac{1}{\sqrt{2}}},$$

$$\xi^*_{2x_1} = -_*e^{\frac{1}{\sqrt{2}}},$$

$$\xi^*_{2x_2} = e^{\frac{1}{\sqrt{2}}},$$

$$u^*_{x_1} = u^*_{\xi_1} \cdot_* \xi^*_{1x_1} +_* u^*_{\xi_2} \cdot_* \xi^*_{2x_1}$$

$$= e^{\frac{1}{\sqrt{2}}} \cdot_* u^*_{\xi_1} -_* e^{\frac{1}{\sqrt{2}}} \cdot_* u^*_{\xi_2},$$

$$u^{**}_{x_1x_1} = e^{\frac{1}{\sqrt{2}}} \cdot_* \left(u^{**}_{\xi_1\xi_1} \cdot_* \xi^*_{1x_1} +_* u^{**}_{\xi_1\xi_2} \cdot_* \xi^*_{2x_1} \right)$$

$$-_*e^{\frac{1}{\sqrt{2}}} \cdot_* \left(u^{**}_{\xi_1\xi_2} \cdot_* \xi^*_{1x_1} +_* u^{**}_{\xi_2\xi_2} \cdot_* \xi^*_{2x_1} \right)$$

$$= e^{\frac{1}{\sqrt{2}}} \cdot_* \left(e^{\frac{1}{\sqrt{2}}} \cdot_* u^{**}_{\xi_1\xi_1} -_* e^{\frac{1}{\sqrt{2}}} \cdot_* u^{**}_{\xi_1\xi_2} \right)$$

$$-_*e^{\frac{1}{\sqrt{2}}} \cdot_* \left(e^{\frac{1}{\sqrt{2}}} \cdot_* u^{**}_{\xi_1\xi_2} -_* e^{\frac{1}{\sqrt{2}}} \cdot_* u^{**}_{\xi_2\xi_2} \right)$$

$$= e^{\frac{1}{2}} \cdot_* u^{**}_{\xi_1\xi_1} -_* u^{**}_{\xi_1\xi_2} +_* e^{\frac{1}{2}} \cdot_* u^{**}_{\xi_2\xi_2},$$

$$u^{**}_{x_1x_2} = e^{\frac{1}{\sqrt{2}}} \cdot_* \left(u^{**}_{\xi_1\xi_1} \cdot_* \xi^*_{1x_2} +_* u^{**}_{\xi_1\xi_2} \cdot_* \xi^*_{2x_2} \right)$$

$$-_*e^{\frac{1}{\sqrt{2}}} \cdot_* \left(u^{**}_{\xi_1\xi_1} \cdot_* \xi^*_{1x_2} +_* u^{**}_{\xi_2\xi_2} \cdot_* \xi^*_{2x_2} \right)$$

$$= e^{\frac{1}{\sqrt{2}}} \cdot_* \left(e^{\frac{1}{\sqrt{2}}} \cdot_* u^{**}_{\xi_1\xi_1} +_* e^{\frac{1}{\sqrt{2}}} \cdot_* u^{**}_{\xi_1\xi_2} \right)$$

$$-_*e^{\frac{1}{\sqrt{2}}} \cdot_* \left(e^{\frac{1}{\sqrt{2}}} \cdot_* u^{**}_{\xi_1\xi_2} +_* e^{\frac{1}{\sqrt{2}}} \cdot_* u^{**}_{\xi_2\xi_2} \right)$$

$$= e^{\frac{1}{2}} \cdot_* u^*_{\xi_1\xi_1} -_* e^{\frac{1}{2}} \cdot_* u^*_{\xi_2\xi_2},$$

$$u^*_{x_2} = u^*_{\xi_1} \cdot_* \xi^*_{1x_2} +_* u^*_{\xi_2} \cdot_* \xi^*_{2x_2}$$

$$= e^{\frac{1}{\sqrt{2}}} \cdot_* u^*_{\xi_1} +_* e^{\frac{1}{\sqrt{2}}} \cdot_* u^*_{\xi_2},$$

$$u^{**}_{x_2 x_2} = e^{\frac{1}{\sqrt{2}}} \cdot_* \left(u^{**}_{\xi_1 \xi_1} \cdot_* \xi^*_{1x_2} +_* u^{**}_{\xi_1 \xi_2} \cdot_* \xi^*_{2x_2} \right)$$

$$+_* e^{\frac{1}{\sqrt{2}}} \cdot_* \left(u^{**}_{\xi_1 \xi_2} \cdot_* \xi^*_{1x_2} +_* u^{**}_{\xi_2 \xi_2} \cdot_* \xi^*_{2x_2} \right)$$

$$= e^{\frac{1}{\sqrt{2}}} \cdot_* \left(e^{\frac{1}{\sqrt{2}}} \cdot_* u^{**}_{\xi_1 \xi_1} +_* e^{\frac{1}{\sqrt{2}}} \cdot_* u^{**}_{\xi_1 \xi_2} \right)$$

$$+_* e^{\frac{1}{\sqrt{2}}} \cdot_* \left(e^{\frac{1}{\sqrt{2}}} \cdot_* u^{**}_{\xi_1 \xi_2} +_* e^{\frac{1}{\sqrt{2}}} \cdot_* u^{**}_{\xi_2 \xi_2} \right)$$

$$= e^{\frac{1}{2}} \cdot_* u^{**}_{\xi_1 \xi_1} +_* u^{**}_{\xi_1 \xi_2} +_* e^{\frac{1}{2}} \cdot_* u^{**}_{\xi_2 \xi_2}.$$

Therefore,

$$u^{**}_{x_1 x_1} -_* u^{**}_{x_1 x_2} +_* u^{**}_{x_2 x_2} = e^{\frac{1}{2}} \cdot_* u^{**}_{\xi_1 \xi_1} -_* u^{**}_{\xi_1 \xi_2} +_* e^{\frac{1}{2}} \cdot_* u^{**}_{\xi_2 \xi_2}$$

$$-_* e^{\frac{1}{2}} \cdot_* u^{**}_{\xi_1 \xi_1} +_* e^{\frac{1}{2}} \cdot_* u^{**}_{\xi_2 \xi_2}$$

$$+_* e^{\frac{1}{2}} \cdot_* u^{**}_{\xi_1 \xi_1} +_* u^{**}_{\xi_1 \xi_2} +_* e^{\frac{1}{2}} \cdot_* u^{**}_{\xi_2 \xi_2}$$

$$= e^{\frac{1}{2}} \cdot_* u^{**}_{\xi_1 \xi_1} +_* e^{\frac{3}{2}} \cdot_* u^{**}_{\xi_2 \xi_2}.$$

Hence, the canonical form of the considered equation is

$$e^{\frac{1}{2}} \cdot_* u^{**}_{\xi_1 \xi_1} +_* e^{\frac{3}{2}} \cdot_* u^{**}_{\xi_2 \xi_2} = f\left(e^{\frac{1}{\sqrt{2}}} \cdot_* (\xi_2 -_* \xi_2), e^{\frac{1}{\sqrt{2}}} \cdot_* (\xi_1 +_* \xi_2) \right).$$

Example 3.17 Consider the equation

$$u^{**}_{x_1 x_1} +_* e^2 \cdot_* u^{**}_{x_1 x_3} +_* u^{**}_{x_2 x_2} +_* u^{**}_{x_3 x_3} = 0_*.$$

We can rewrite it in the following form:

$$u^{**}_{x_1 x_1} +_* u^{**}_{x_1 x_3} +_* u^{**}_{x_2 x_2} +_* u^{**}_{x_3 x_1} +_* u^{**}_{x_3 x_3} = 0_*.$$

Then,

$$A = \begin{pmatrix} e & 1 & e \\ 1 & e & 1 \\ e & 1 & e \end{pmatrix}.$$

We will find the multiplicative eigenvalues of the matrix A. We have

$$\det_*(A -_* \lambda \cdot_* I_*) = \begin{vmatrix} e -_* \lambda & 1 & e \\ 1 & e -_* \lambda & 1 \\ e & 1 & e -_* \lambda \end{vmatrix}_*$$

$$= 1$$

if and only if
$$-_*(\lambda -_* e)^{3*} +_* \lambda -_* e = 1,$$

if and only if
$$-_*(\lambda -_* e) \cdot_* ((\lambda -_* e)^{2*} -_* e) = 1,$$

if and only if
$$(\lambda -_* e) \cdot_* (\lambda -_* e^2) \cdot_* \lambda = 1,$$

if and only if
$$\lambda_1 = 1,$$

$$\lambda_2 = e,$$

$$\lambda_3 = e^2.$$

Since $\lambda_1 = 0_*$ and $\lambda_2, \lambda_3 > 0_*$, the considered equation is a multiplicative parabolic equation. Let $A_1 = (a, b, c)$ be a multiplicative eigenvector of the matrix A corresponding to the multiplicative eigenvalue $\lambda_1 = 0_*$. Then,

$$\begin{pmatrix} e & 1 & e \\ 1 & e & 1 \\ e & 1 & e \end{pmatrix} \cdot_* \begin{pmatrix} a \\ b \\ c \end{pmatrix} = \begin{pmatrix} 1 \\ 1 \\ 1 \end{pmatrix},$$

i.e.,

$$a +_* c = 1,$$

$$b = 1.$$

We take
$$q_1 = e^{\frac{1}{\sqrt{2}}} \cdot_* (e, 1, -_*e).$$

Let $A_2 = (a, b, c)$ be a multiplicative eigenvector of the matrix A corresponding to the multiplicative eigenvalue $\lambda_2 = e$. Then,

$$\begin{pmatrix} 1 & 1 & e \\ 1 & 1 & 1 \\ e & 1 & 1 \end{pmatrix} \cdot_* \begin{pmatrix} a \\ b \\ c \end{pmatrix} = \begin{pmatrix} 1 \\ 1 \\ 1 \end{pmatrix},$$

i.e.,

$$a = 1,$$

$$c = 1.$$

We take
$$q_2 = (1, e, 1).$$

Let $A_3 = (a, b, c)$ be a multiplicative eigenvector of the matrix A corresponding to the multiplicative eigenvalue $\lambda_3 = e^2$. Then,

$$
\begin{pmatrix} -_*e & 1 & e \\ 1 & -_*e & 1 \\ e & 1 & -_*e \end{pmatrix} \cdot_* \begin{pmatrix} a \\ b \\ c \end{pmatrix} = \begin{pmatrix} 1 \\ 1 \\ 1 \end{pmatrix},
$$

i.e.,

$$
a = c,
$$

$$
b = 1.
$$

We take

$$
q_3 = e^{\frac{1}{\sqrt{2}}} \cdot_* (e, 1, e).
$$

Hence,

$$
Q = (q_1^T, q_2^T, q_3^T)
$$

$$
= \begin{pmatrix} e^{\frac{1}{\sqrt{2}}} & 1 & e^{\frac{1}{\sqrt{2}}} \\ 1 & e & 1 \\ -_*e^{\frac{1}{\sqrt{2}}} & 1 & e^{\frac{1}{\sqrt{2}}} \end{pmatrix}
$$

and

$$
\begin{pmatrix} \xi_1 \\ \xi_2 \\ \xi_3 \end{pmatrix} = \begin{pmatrix} e^{\frac{1}{\sqrt{2}}} & 1 & e^{\frac{1}{\sqrt{2}}} \\ 1 & e & 1 \\ -_*e^{\frac{1}{\sqrt{2}}} & 1 & e^{\frac{1}{\sqrt{2}}} \end{pmatrix} \cdot_* \begin{pmatrix} x_1 \\ x_2 \\ x_3 \end{pmatrix}
$$

$$
= \begin{pmatrix} e^{\frac{1}{\sqrt{2}}} \cdot_* x_1 +_* e^{\frac{1}{\sqrt{2}}} \cdot_* x_3 \\ x_2 \\ -_*e^{\frac{1}{\sqrt{2}}} \cdot_* x_1 +_* e^{\frac{1}{\sqrt{2}}} \cdot_* x_3 \end{pmatrix}.
$$

Then,

$$
\xi^*_{1x_1} = e^{\frac{1}{\sqrt{2}}},
$$

$$
\xi^*_{1x_2} = 1,
$$

$$
\xi^*_{1x_3} = -_*e^{\frac{1}{\sqrt{2}}},
$$

$$
\xi^*_{2x_1} = 1,
$$

$$\xi^*_{2x_2} \;=\; e,$$

$$\xi^*_{2x_3} \;=\; 1,$$

$$\xi^*_{3x_1} \;=\; e^{\frac{1}{\sqrt{2}}},$$

$$\xi^*_{3x_2} \;=\; 1,$$

$$\xi^*_{3x_3} \;=\; e^{\frac{1}{\sqrt{2}}},$$

$$u^*_{x_1} \;=\; u^*_{\xi_1} \cdot_* \xi^*_{1x_1} +_* u^*_{\xi_2} \cdot_* \xi^*_{2x_1} +_* u^*_{\xi_3} \cdot_* \xi^*_{3x_1}$$

$$\;=\; e^{\frac{1}{\sqrt{2}}} \cdot_* u^*_{\xi_1} -_* e^{\frac{1}{\sqrt{2}}} \cdot_* u^*_{\xi_3},$$

$$u^{**}_{x_1 x_1} \;=\; e^{\frac{1}{\sqrt{2}}} \cdot_* \left(u^{**}_{\xi_1\xi_1} \cdot_* \xi^*_{1x_1} +_* u^{**}_{\xi_1\xi_2} \cdot_* \xi^*_{2x_1} +_* u^{**}_{\xi_1\xi_3} \cdot_* \xi^*_{3x_1} \right)$$

$$-_* e^{\frac{1}{\sqrt{2}}} \cdot_* \left(u^{**}_{\xi_1\xi_3} \cdot_* \xi^*_{1x_1} +_* u^{**}_{\xi_2\xi_3} \cdot_* \xi^*_{2x_1} \right.$$

$$\left. +_* u^{**}_{\xi_3\xi_3} \cdot_* \xi^*_{3x_1} \right)$$

$$\;=\; e^{\frac{1}{\sqrt{2}}} \cdot_* \left(e^{\frac{1}{\sqrt{2}}} \cdot_* u^{**}_{\xi_1\xi_1} -_* e^{\frac{1}{\sqrt{2}}} \cdot_* u^{**}_{\xi_1\xi_3} \right)$$

$$-_* e^{\frac{1}{\sqrt{2}}} \cdot_* \left(e^{\frac{1}{\sqrt{2}}} \cdot_* u^{**}_{\xi_1\xi_3} -_* e^{\frac{1}{\sqrt{2}}} \cdot_* u^{**}_{\xi_3\xi_3} \right)$$

$$\;=\; e^{\frac{1}{2}} \cdot_* u^{**}_{\xi_1\xi_1} -_* u^{**}_{\xi_1\xi_3} +_* e^{\frac{1}{2}} \cdot_* u^{**}_{\xi_3\xi_3},$$

$$u^{**}_{x_1 x_3} \;=\; e^{\frac{1}{\sqrt{2}}} \cdot_* \left(u^{**}_{\xi_1\xi_1} \cdot_* \xi^*_{1x_3} +_* u^{**}_{\xi_1\xi_2} \cdot_* \xi^*_{2x_3} +_* u^{**}_{\xi_1\xi_3} \cdot_* \xi^*_{3x_3} \right)$$

$$-_* e^{\frac{1}{\sqrt{2}}} \cdot_* \left(u^{**}_{\xi_1\xi_3} \cdot_* \xi^*_{1x_3} +_* u^{**}_{\xi_2\xi_3} \cdot_* \xi^*_{2x_3} +_* u^{**}_{\xi_3\xi_3} \cdot_* \xi^*_{3x_3} \right)$$

$$\;=\; e^{\frac{1}{\sqrt{2}}} \cdot_* \left(e^{\frac{1}{\sqrt{2}}} \cdot_* u^{**}_{\xi_1\xi_1} +_* e^{\frac{1}{\sqrt{2}}} \cdot_* u^{**}_{\xi_1\xi_3} \right) -_* e^{\frac{1}{\sqrt{2}}} \cdot_*$$

$$\left(e^{\frac{1}{\sqrt{2}}} \cdot_* u^{**}_{\xi_1\xi_3} +_* e^{\frac{1}{\sqrt{2}}} \cdot_* u^{**}_{\xi_3\xi_3} \right)$$

$$\;=\; e^{\frac{1}{2}} \cdot_* u^{**}_{\xi_1\xi_1} -_* e^{\frac{1}{2}} \cdot_* u^{**}_{\xi_3\xi_3},$$

$$u^*_{x_2} \;=\; u^*_{\xi_1} \cdot_* \xi^*_{1x_2} +_* u^*_{\xi_2} \cdot_* \xi^*_{2x_2} +_* u^{**}_{\xi_3} \xi^*_{3x_2}$$

$$=\; u^*_{\xi_2},$$

$$u^{**}_{x_2 x_2} \;=\; u^{**}_{\xi_1 \xi_2} \cdot_* \xi^*_{1x_2} +_* u^{**}_{\xi_2 \xi_2} \cdot_* \xi^*_{2x_2} +_* u^{**}_{\xi_2 \xi_3} \cdot_* \xi^*_{3x_2}$$

$$=\; u^{**}_{\xi_2 \xi_2},$$

$$u^*_{x_3} \;=\; u^*_{\xi_1} \cdot_* \xi^*_{1x_3} +_* u^*_{\xi_2} \cdot_* \xi^*_{2x_3} +_* u^*_{\xi_3} \cdot_* \xi^*_{3x_3}$$

$$=\; e^{\frac{1}{\sqrt{2}}} \cdot_* u^*_{\xi_1} +_* e^{\frac{1}{\sqrt{2}}} \cdot_* u^*_{\xi_3},$$

$$u^{**}_{x_3 x_3} \;=\; e^{\frac{1}{\sqrt{2}}} \cdot_* \left(u^{**}_{\xi_1 \xi_1} \cdot_* \xi^*_{1x_3} +_* u^{**}_{\xi_1 \xi_2} \cdot_* \xi^*_{2x_3} +_* u^{**}_{\xi_1 \xi_3} \cdot_* \xi^*_{3x_1} \right)$$

$$+_* e^{\frac{1}{\sqrt{2}}} \cdot_* \left(u^{**}_{\xi_1 \xi_3} \cdot_* \xi^*_{1x_3} +_* u^{**}_{\xi_2 \xi_3} \cdot_* \xi^*_{2x_3} +_* u^{**}_{\xi_3 \xi_3} \cdot_* \xi^*_{3x_3} \right)$$

$$=\; e^{\frac{1}{\sqrt{2}}} \cdot_* \left(e^{\frac{1}{\sqrt{2}}} \cdot_* u^{**}_{\xi_1 \xi_1} +_* e^{\frac{1}{\sqrt{2}}} \cdot_* u^{**}_{\xi_1 \xi_3} \right)$$

$$+_* e^{\frac{1}{\sqrt{2}}} \cdot_* \left(u^{**}_{\xi_1 \xi_3} +_* e^{\frac{1}{\sqrt{2}}} \cdot_* u^{**}_{\xi_3 \xi_3} \right)$$

$$=\; e^{\frac{1}{2}} \cdot_* u^{**}_{\xi_1 \xi_1} +_* u^{**}_{\xi_1 \xi_3} +_* e^{\frac{1}{2}} \cdot_* u^{**}_{\xi_3 \xi_3}.$$

From here,

$$u^{**}_{x_1 x_1} +_* e^2 \cdot_* u^{**}_{x_1 x_3} +_* u^{**}_{x_2 x_2} +_* u^{**}_{x_3 x_3} = e^{\frac{1}{2}} \cdot_* u^{**}_{\xi_1 \xi_1} -_* u^{**}_{\xi_1 \xi_3} +_* e^{\frac{1}{2}} \cdot_* u^{**}_{\xi_3 \xi_3}$$

$$+_* e^2 \cdot_* \left(e^{\frac{1}{2}} \cdot_* u^{**}_{\xi_1 \xi_1} -_* e^{\frac{1}{2}} \cdot_* u^{**}_{\xi_1 \xi_3} \right)$$

$$+_* u^{**}_{\xi_2 \xi_2} +_* e^{\frac{1}{2}} \cdot_* u^{**}_{\xi_1 \xi_1} +_* u^{**}_{\xi_1 \xi_3}$$

$$+_* e^{\frac{1}{2}} \cdot_* u^{**}_{\xi_3 \xi_3}$$

$$=\; e^2 \cdot_* u^{**}_{\xi_1 \xi_1} +_* u^{**}_{\xi_2 \xi_2}.$$

Therefore, the canonical form of the considered equation is

$$e^2 \cdot_* u^{**}_{\xi_1 \xi_1} +_* u^{**}_{\xi_2 \xi_2} = 0_*.$$

Exercise 3.11 Classify the three-dimensional multiplicative wave equation:

$$u_{x_1 x_1}^{**} -_* e^{c^2} \cdot_* \left(u_{x_2 x_2}^{**} +_* u_{x_3 x_3}^{**} +_* u_{x_4 x_4}^{**}\right) = 1,$$

where c is a constant, $c \neq 0$.

Answer *Multiplicative hyperbolic.*

3.4　Classification of First-Order Systems with Two Multiplicative Independent Variables

Consider the system

$$A \cdot_* u_{x_1}^* +_* B \cdot_* u_{x_2}^* = F, \tag{3.36}$$

where

$$A = \begin{pmatrix} a_{11} & a_{12} & \cdots & a_{1n} \\ a_{21} & a_{22} & \cdots & a_{2n} \\ \vdots & & & \\ a_{n1} & a_{n2} & \cdots & a_{nn} \end{pmatrix},$$

$$B = \begin{pmatrix} b_{11} & b_{12} & \cdots & b_{1n} \\ b_{21} & b_{22} & \cdots & b_{2n} \\ \vdots & & & \\ b_{n1} & b_{n2} & \cdots & b_{nn} \end{pmatrix},$$

$$F = \begin{pmatrix} f_1 \\ f_2 \\ \vdots \\ f_n \end{pmatrix},$$

$$u = \begin{pmatrix} u_1^* \\ u_2^* \\ \vdots \\ u_n^* \end{pmatrix}, \quad u_{x_i}^* = \begin{pmatrix} u_{1x_i}^* \\ u_{2x_i}^* \\ \vdots \\ u_{nx_i}^* \end{pmatrix}, \quad i = 1, 2,$$

$a_{ij}, b_{ij}, f_i, i, j \in \{1, 2, \ldots, n\}$, are given functions, u is unknown. Now we consider the transformation

$$v = P^{-1} *_* u, \quad \text{i.e.,} \quad u = P \cdot_* v,$$

where P is a multiplicative nonsingular $n \times n$-matrix. Then,

$$u_{x_i}^* = (P \cdot_* v)_{x_i}$$

$$= P_{x_i}^* \cdot_* v +_* P \cdot_* v_{x_i}^*, \quad i = 1, 2,$$

and equation (3.36) takes the following form:

$$A \cdot_* \left(P_{x_1}^* \cdot_* v +_* P \cdot_* v_{x_1}^* \right) +_* B \cdot_* \left(P_{x_2}^* \cdot_* v +_* P \cdot_* v_{x_2}^* \right) = F$$

or

$$A \cdot_* P \cdot_* v_{x_1}^* +_* B \cdot_* P \cdot_* v_{x_2}^* = F -_* \left(A \cdot_* P_{x_1}^* +_* B \cdot_* P_{x_2}^* \right) \cdot_* v = G.$$

Assuming that A is multiplicative nonsingular, we multiplicative multiply the above equation by $(A \cdot_* P)^{-1_*}$ to obtain

$$(A \cdot_* P)^{-1_*} \cdot_* A \cdot_* P \cdot_* v_{x_1}^* +_* (A \cdot_* P)^{-1_*} \cdot_* B \cdot_* P \cdot_* v_{x_2}^* = (A \cdot_* P)^{-1_*} \cdot_* G$$

or

$$v_{x_1}^* +_* P^{-1_*} \cdot_* A^{-1_*} \cdot_* B \cdot_* P \cdot_* v_{x_2}^* = (A \cdot_* P)^{-1_*} \cdot_* G. \tag{3.37}$$

We set

$$D = A^{-1_*} \cdot_* B,$$

$$H = (A \cdot_* P)^{-1_*} \cdot_* G.$$

Then,

$$v_{x_1}^* +_* P^{-1_*} \cdot_* D \cdot_* P v_{x_2}^* = H.$$

If P is taken to be the diagonalizing matrix of D and Λ is a diagonal matrix whose elements are the multiplicative eigenvalues λ_i of D and the columns of P are multiplicative linearly independent multiplicative eigenvectors of D, $p_i = (p_{1i}, p_{2i}, \ldots, p_{ni})$, $|p_i| = 1$, $i \in \{1, 2, \ldots, n\}$. So,

$$P = (p_{ij}),$$

$$\Lambda = (\lambda_j \delta_{ij}), \quad i, j \in \{1, 2, \ldots, n\},$$

where δ_{ij} is the multiplicative Kronecker delta. It follows that

$$P^{-1_*} \cdot_* D \cdot_* P = \Lambda.$$

Thus, we can write equation (3.37) in the following form:

$$v_{x_1}^* +_* \Lambda \cdot_* v_{x_2}^* = H \tag{3.38}$$

or

$$v_{x_1}^* +_* \lambda_i \cdot_* v_{x_2}^* = h_i, \quad i \in \{1, 2, \ldots, n\},$$

with n characteristics given by

$$(d_* x_2)/_* (d_* x_1) = \lambda_i,$$

$i \in \{1, 2, \ldots, n\}$.

Definition 3.15 Equation (3.38) is said to be the canonical form of equation (3.36).

The classification of equation (3.36) is done basing on the nature of the multiplicative eigenvalues λ_i of D.

Definition 3.16   1. If all the n multiplicative eigenvalues of D are real and distinct, then equation (3.36) is said to be multiplicative hyperbolic.

2. If all the n multiplicative eigenvalues of D are complex, then equation (3.36) is said to be multiplicative elliptic.

3. If some of the n multiplicative eigenvalues of D are real and other complex, equation (3.36) is considered as hybrid of multiplicative elliptic-hyperbolic type.

4. If all the n multiplicative eigenvalues of D are real and some of them are repeated, then equation (3.36) is said to be multiplicative parabolic.

Example 3.18 Consider the system

$$u^*_{1x_1} +_* \alpha \cdot_* u^*_{1x_2} +_* \beta \cdot_* u^*_{2x_2} = 0_*$$

$$u^*_{2x_1} +_* \gamma \cdot_* u^*_{1x_2} +_* \delta \cdot_* u^*_{2x_2} = 0_*.$$

Then,

$$A = \begin{pmatrix} e & 1 \\ 1 & e \end{pmatrix},$$

$$B = \begin{pmatrix} \alpha & \beta \\ \gamma & \delta \end{pmatrix}.$$

The matrix A is multiplicative nonsingular and $D = B$. We will find the multiplicative eigenvalues of the matrix D. We have

$$\det_*(D -_* \lambda \cdot_* I_*) = \begin{vmatrix} \alpha -_* \lambda & \beta \\ \gamma & \delta -_* \lambda \end{vmatrix}_*$$

$$= 0_*$$

if and only if

$$(\lambda -_* \alpha) \cdot_* (\lambda -_* \delta) -_* \beta \cdot_* \gamma = 0_*,$$

if and only if

$$\lambda^{2_*} -_* (\alpha +_* \delta) \cdot_* \lambda +_* \alpha \cdot_* \delta -_* \beta \cdot_* \gamma = 0_*.$$

Therefore,

1. If
$$(\alpha -_* \delta)^{2*} +_* e^4 \cdot_* \beta \cdot_* \gamma > 0_*,$$
then the considered system is multiplicative hyperbolic.

2. If
$$(\alpha -_* \delta)^{2*} +_* e^4 \cdot_* \beta \cdot_* \gamma = 0_*,$$
then the considered system is multiplicative parabolic.

3. If
$$(\alpha -_* \delta)^{2*} +_* e^4 \cdot_* \beta \cdot_* \gamma < 0_*,$$
then the considered system is multiplicative elliptic.

Example 3.19 Consider the system

$$u^*_{1x_1} -_* u^*_{2x_2} = 0_*,$$

$$u^*_{1x_2} -_* u^*_{2x_1} = 0_*.$$

Here

$$A = \begin{pmatrix} e & 1 \\ 1 & -_*e \end{pmatrix},$$
$$B = \begin{pmatrix} 1 & -_*e \\ e & 1 \end{pmatrix}.$$

We have
$$\det A = -_*e \neq 0_*,$$
i.e., the matrix A is multiplicative nonsingular. Also,

$$A^{-1*} = \begin{pmatrix} e & 1 \\ 1 & -_*e \end{pmatrix},$$
$$D = A^{-1*} \cdot_* B$$
$$= \begin{pmatrix} 1 & -_*e \\ -_*e & 1 \end{pmatrix}.$$

We will find the multiplicative eigenvalues of the matrix D. We have

$$\det(D -_* \lambda \cdot_* I_*) = \begin{vmatrix} -_*\lambda & -_*e \\ -_*e & -_*\lambda \end{vmatrix}_*$$
$$= 0_*$$

if and only if
$$\lambda^{2*} -_* e = 1,$$

if and only if

$$\lambda_1 = 1,$$

$$\lambda_2 = -_*e.$$

Therefore, the considered system is multiplicative hyperbolic.

Example 3.20 Consider the system

$$u^*_{1x_2} -_* u^*_{2x_1} = 1$$

$$(\rho \cdot_* u_1)^*_{x_1} +_* (\rho \cdot_* u_2)^*_{x_2} = 0_*,$$

where

$$\rho = \left(e +_* u_1^{2*} +_* u_2^{2*}\right)^{\sigma_*}$$

and σ is a multiplicative constant. We have

$$\rho^*_{x_i} = \sigma \cdot_* \left(e +_* u_1^{2*} +_* u_2^{2*}\right)^{\sigma_* -_* e} \cdot_* \left(e^2 \cdot_* u_1 u^*_{1x_i} +_* e^2 \cdot_* u_2 u^*_{2x_i}\right),$$

$$(\rho \cdot_* u_j)^*_{x_i} = \rho^*_{x_i} \cdot_* u_j +_* \rho \cdot_* u^*_{jx_i}, \quad i,j = 1,2.$$

Therefore, we can rewrite the considered system we can rewrite in this form:

$$u^*_{1x_2} -_* u^*_{2x_1} = 0_*$$

$$\rho \cdot_* u^*_{1x_1} +_* \rho \cdot_* u^*_{2x_2} = -_* \rho^*_{x_1} \cdot_* u_1 -_* \rho^(_{x_2} \cdot_* u_2.$$

Here

$$A = \begin{pmatrix} 1 & -_*e \\ \rho & 1 \end{pmatrix},$$

$$B = \begin{pmatrix} e & 1 \\ 1 & \rho \end{pmatrix}.$$

Since

$$\det_* A = \rho \neq 0_*,$$

the matrix A is multiplicative nonsingular and

$$A^{-1*} = \begin{pmatrix} 1 & e/_*\rho \\ -_*e & 1 \end{pmatrix},$$

$$D = A^{-1*} \cdot_* B$$

$$= \begin{pmatrix} 1 & e \\ -_*e & 1 \end{pmatrix}.$$

We will find the multiplicative eigenvalues of the matrix D. We have

$$\det_*(D -_* \lambda \cdot_* I_*) = \begin{vmatrix} -_*\lambda & e \\ -_*e & -_*\lambda \end{vmatrix}_*$$

$$= 0_*.$$

if and only if

$$\lambda^{2*} +_* e = 1,$$

if and only if

$$\lambda_{1,2} = \pm_* i.$$

Therefore, the considered system is multiplicative elliptic.

Exercise 3.12 Prove that the system

$$u^*_{1x_2} -_* u^*_{2x_1} = 0_*,$$

$$u^*_{1x_1} -_* e^9 \cdot_* u^*_{2x_2} = 0_*$$

is a multiplicative hyperbolic system.

3.5 Advanced Practical Exercises

Problem 3.1 Determine the type of the operator L, where

1.
$$L(u) = e^{\sqrt{3}} \cdot_* u^{**}_{x_1 x_1} -_* e^{4\sqrt{3}} \cdot_* u^{**}_{x_1 x_2} +_* u^{**}_{x_2 x_2} +_* u.$$

2.
$$L(u) = -_* u^{**}_{x_1 x_1} -_* e^{2\sqrt{2}} \cdot_* u^{**}_{x_1 x_2} +_* u^{**}_{x_2 x_2} +_* u^*_{x_2}.$$

3.
$$L(u) = u^{**}_{x_1 x_1} +_* u^{**}_{x_2 x_2} +_* u.$$

4.
$$L(u) = u^{**}_{x_1 x_1} -_* u^{**}_{x_2 x_2} +_* u^*_{x_1}.$$

5.
$$L(u) = e^2 \cdot_* u^{**}_{x_1 x_2} -_* u^*_{x_1} -_* u^*_{x_2}.$$

6.
$$L(u) = u^{**}_{x_1 x_1} +_* e^{2\sqrt{3}} \cdot_* u^{**}_{x_1 x_2} +_* e^3 \cdot_* u^{**}_{x_2 x_2}.$$

7.
$$L(u) = u^*_{x_1} -_* u^{**}_{x_2 x_2} +_* u^*_{x_2}.$$

Answer   *1. Multiplicative hyperbolic.*

2. *Multiplicative hyperbolic.*

3. *Multiplicative elliptic.*

4. *Multiplicative hyperbolic.*

5. *Multiplicative hyperbolic.*

6. *Multiplicative parabolic.*

7. *Multiplicative parabolic.*

Problem 3.2 Find the canonical form of the following equations:

1.

$$u^{**}_{x_1 x_1} -_* e^3 \cdot_* u^{**}_{x_1 x_2} +_* e^2 \cdot_* u^{**}_{x_2 x_2} = 0_*.$$

2.

$$u^{**}_{x_1 x_1} -_* u^{**}_{x_1 x_2} -_* e^2 \cdot_* u^{**}_{x_2 x_2} = 0_*.$$

3.

$$u^{**}_{x_2 x_2} -_* (x_1/_* x_2) \cdot_* u^{**}_{x_1 x_2} +_* (e^2/_*(e^3 \cdot_* x_2)) \cdot_* u^{*}_{x_2} = 0_*.$$

Answer   1.

$$u^{**}_{\xi_1 \xi_2} = 0_*,$$

$$\xi_1 = x_1 +_* x_2,$$

$$\xi_2 = x_2 +_* e^2 \cdot_* x_1.$$

2.

$$u^{**}_{\xi_1 \xi_2} = 0_*,$$

$$\xi_1 = x_2 +_* e^2 \cdot_* x_1,$$

$$\xi_2 = x_2 -_* x_1.$$

3.

$$u^{**}_{\xi_1 \xi_2} +_* (e/_*(e^3 \cdot_* \xi_2)) \cdot_* u^{*}_{\xi_1} = 0,$$

$$\xi_1 = x_1 \cdot_* x_2,$$

$$\xi_2 = x_1.$$

Problem 3.3 Consider the equation

$$u^{**}_{x_1 x_1} -_* u^{**}_{x_2 x_2} = e.$$

1. Find the canonical form.

2. Find the general solution.

3. Find the solution $u(x_1, x_2)$ for which

$$u(x_1, 1) = \phi_0(x_1),$$

$$u^*_{x_2}(x_1, 1) = \phi_1(x_1),$$

where $\phi_0 \in C^2_*(\mathbb{R}^1_*)$, $\phi_1 \in C_*^{\,1}(\mathbb{R}^1_*)$.

Answer *1.*

$$e^4 \cdot_* u^{**}_{\xi_1 \xi_2} = e,$$

$$\xi_1 = x_1 +_* x_2,$$

$$\xi_2 = x_1 -_* x_2.$$

2.

$$u(x_1, x_2) = (x_1^{2*} -_* x_2^{2*})/_* e^4 +_* f(x_1 +_* x_2) +_* g(x_1 -_* x_2),$$

where f and g are C^2_*-functions.

3.

$$u(x_1, x_2) = -_* (x_2^{2*}/_* e^2) +_* (\phi_0(x_1 +_* x_2) +_* \phi_0(x_1 -_* x_2))/_* e^2$$

$$+_* e^{\frac{1}{2}} \cdot_* \int_{*x_1 -_* x_2}^{x_1 +_* x_2} \phi_1(x) \cdot_* d_* x.$$

Problem 3.4 Consider the equation

$$u^{**}_{x_1 x_1} +_* e^4 \cdot_* u^{**}_{x_1 x_2} -_* e^4 \cdot_* u^*_{x_1} = 0_*.$$

1. Find the canonical form.

2. Find the general solution.

3. Find a solution $u(x_1, x_2)$ such that

$$u(x_1, e^8 \cdot_* x_1) = 1,$$

$$u^*_{x_1}(x_1, e^8 \cdot_* x_1) = -_* e^{32} \cdot_* x_1.$$

Answer   *1.*

$$u^{**}_{\xi_1\xi_2} -_* u^*_{\xi_2} = 0_*,$$

$$\xi_1 = x_2,$$

$$\xi_2 = x_2 -_* e^4 \cdot_* x_1.$$

2.

$$u(x_1, x_2) = e^{x_2} \cdot_* f(x_2 -_* e^4 \cdot_* x_1) +_* g(x_2),$$

where f is a C^2_*-function and g is a C^1_*-function,

3.

$$u(x_1, x_2) = -_* \left(x_2 -_* e^4 \cdot_* x_1 +_* e^{\frac{1}{2}}\right) \cdot_* e^{-_* x_2 +_* e^8 \cdot_* x_1}$$

$$+_* (x_2 /_* e^2) +_* e^{\frac{1}{2}}.$$

Problem 3.5 Find the canonical form of the equation

$$u^{**}_{x_1x_1} +_* x_2 \cdot_* u^{**}_{x_2x_2} = 0_*, \quad x_2 > 0_*.$$

Answer

$$\xi_1 = e^2 \cdot_* (x_2)^{\frac{1}{2}}_*,$$

$$\xi_2 = x_1$$

and

$$u^{**}_{\xi_1\xi_1} +_* u^{**}_{\xi_2\xi_2} -_* (e/_*\xi_1) \cdot_* u^*_{\xi_1} = 0_*.$$

Problem 3.6 Find the canonical form of the equation

$$u^{**}_{x_1x_1} +_* e^4 \cdot_* u^{**}_{x_1x_2} +_* e^4 \cdot_* u^{**}_{x_2x_2} -_* e^2 \cdot_* u^*_{x_1} +_* u^*_{x_2} = 0_*.$$

Answer

$$\xi_1 = x_2,$$

$$\xi_2 = x_2 -_* e^2 \cdot_* x_1$$

and

$$e^4 \cdot_* u^{**}_{\xi_1\xi_1} +_* u^*_{\xi_1} +_* e^5 \cdot_* u^*_{\xi_2} = 0_*.$$

Problem 3.7 Consider the equation

$$x_1^{2_*} \cdot_* u^{**}_{x_1x_1} -_* e^2 \cdot_* x_1 \cdot_* u^{**}_{x_1x_2} +_* u^{**}_{x_2x_2} +_* x_1 \cdot_* u^*_{x_1} = 0_*, \quad x_1 > 0_*.$$

1. Find the canonical form.
2. Find the general solution.
3. Find the solution $u(x_1, x_2)$ for which

$$u(e, x_2) = e^{e^2 \cdot_* x_2},$$

$$u^*_{x_1}(e, x_2) = 1.$$

Answer   *1.*

$$\xi_1 = x_1 \cdot_* e^{x_2},$$

$$\xi_2 = x_2,$$

$$u^{**}_{\xi_2 \xi_2} = 0_*.$$

2.

$$u(x_1, x_2) = x_2 \cdot_* g\left(x_1 \cdot_* e^{x_2}\right) +_* f\left(x_1 \cdot_* e^{x_2}\right),$$

where f and g are C^2_*-functions.

3.

$$u(x_1, x_2) = \left(e -_* e^2 \cdot_* \log_* x_1\right) \cdot_* x_1^{2_*} \cdot_* e^{e^2 \cdot_* x_2}.$$

Problem 3.8 Classify the three-dimensional multiplicative heat equation:

$$u^*_{x_1} -_* \alpha^{2_*} \cdot_* \left(u^{**}_{x_2 x_2} +_* u^{**}_{x_3 x_3} +_* u^{**}_{x_4 x_4}\right) = 0_*,$$

where α is a multiplicative constant, $\alpha \neq 0_*$.

Answer *Multiplicative parabolic.*

Problem 3.9 Prove that the system

$$u^*_{1x_2} -_* u^*_{2x_1} = 0_*,$$

$$u^*_{1x_1} +_* e^2 \cdot_* u^*_{1x_2} +_* e^4 \cdot_* u^*_{2x_2} = 0_*$$

is a multiplicative elliptic system.

4

The Multiplicative Wave Equation

4.1 The One-Dimensional Multiplicative Wave Equation

4.1.1 The Cauchy problem and D'Alambert formula

The multiplicative homogeneous wave equation in one (spatial) dimension has the form

$$u_{tt}^{**} -_* c^{2*} \cdot_* u_{xx}^{**} = 0_*, \quad -_*\infty \le a < x < b \le \infty, \quad b > 0_*, \qquad (4.1)$$

where $c \in \mathbb{R}_*$ is called the multiplicative wave speed. Introducing the new multiplicative variables

$$\xi_1 = x +_* c \cdot_* t,$$

$$\xi_2 = x -_* c \cdot_* t,$$

we get the canonical form of equation (4.1):

$$u_{tt}^{**} -_* c^{2*} \cdot_* u_{xx}^{**} = -_* e^4 \cdot_* c^{2*} \cdot_* u_{\xi_1 \xi_2}^{**}$$

$$= 0_*.$$

Therefore, its general solution is given by

$$u(x,t) = f(x +_* c \cdot_* t) +_* g(x -_* c \cdot_* t), \qquad (4.2)$$

where $f, g \in \mathcal{C}_*^2(\mathbb{R}_*)$ are two arbitrary functions. Conversely, any two functions $f, g \in \mathcal{C}_*^2(\mathbb{R}_*)$ define a solution of the multiplicative wave equation (4.1) via equation (4.2). The function $g(x -_* c \cdot_* t)$ represents a multiplicative wave moving to the right with multiplicative velocity c and it is called a multiplicative forward wave. The function $f(x +_* c \cdot_* t)$ is a multiplicative wave traveling to the left with the same multiplicative speed, and it is called a multiplicative backward wave. Equation (4.2) shows the fact that any solution of the multiplicative wave equation is the multiplicative sum of two such multiplicative traveling waves. Since for any two piecewise continuous functions f and g, equation (4.2) defines a piecewise continuous function u that is a

DOI: 10.1201/9781003440116-4

multiplicative superposition of a multiplicative backward and a multiplicative forward wave traveling in multiplicative opposite directions with multiplicative speed c. Let $\{f_n(s)\}_{n \in \mathbb{N}}$ and $\{g_n(s)\}_{n \in \mathbb{N}}$ be sequences of multiplicative smooth functions converging at any point t to f and g, respectively, which converge multiplicative uniformly to these functions in any bounded and closed interval that does not contain points of discontinuity. The function

$$u_n(x,t) = f_n(x +_* c \cdot_* t) +_* g_n(x -_* c \cdot_* t)$$

is a proper solution of the wave equation, but the limiting function

$$u(x,t) = f(x +_* c \cdot_* t) +_* g(x -_* c \cdot_* t)$$

is not necessarily twice multiplicative differentiable, therefore it might be not a solution of (4.1).

Definition 4.1 We call a function u that satisfies (4.2) with piecewise continuous functions f and g a generalized solution of the multiplicative wave equation (4.1).

The Cauchy problem for the one-dimensional multiplicative homogeneous wave equation is given by

$$u_{tt}^{**} -_* c^{2*} \cdot_* u_{xx}^{**} = 0_*, \quad -_*\infty < x < \infty, \quad t > 0_*, \tag{4.3}$$

$$u(x,1) = \phi(x),$$
$$\tag{4.4}$$
$$u_t^*(x,1) = \psi(x), \quad -_*\infty < x < \infty,$$

where $\phi \in \mathcal{C}_*^2(\mathbb{R}_*)$ and $\psi \in \mathcal{C}_*^1(\mathbb{R}_*)$.

Definition 4.2 A classical (proper) solution of the Cauchy problem (4.3), (4.4) is a function u that is

1. continuously twice multiplicative differentiable for all $t > 0_*$,

2. u and u_t^* are continuous in the half-space $t \geq 0_*$ and such that (4.3), (4.4) are satisfied.

Recall that the general solution of the multiplicative wave equation is of the form (4.2). We will find the functions f and g using the initial condition (4.4). Substituting $t = 0_*$ in (4.2), we obtain

$$u(x,1) = f(x) +_* g(x)$$
$$\tag{4.5}$$
$$= \phi(x).$$

Multiplicative differentiating (4.2) with respect to t and substituting $t = 0_*$, we get

$$u_t^*(x,1) = c \cdot_* f^*(x) -_* c \cdot_* g^*(x)$$

$$= \psi(x).$$

Multiplicative integrating the last equation over $[1, x]$, we get

$$f(x) -_* g(x) = (e/_*c) \cdot_* \int_{*1}^{x} \psi(s) \cdot_* d_*s +_* C, \qquad (4.6)$$

where $C = f(1) -_* g(1)$. Equations (4.5) and (4.6) are two multiplicative linear algebraic equations for f and g. The solution of this system of equations is given by

$$f(x) \;=\; e^{\frac{1}{2}} \cdot_* \phi(x) +_* e^{\frac{1}{2c}} \cdot_* \int_{*1}^{x} \psi(s) \cdot_* d_*s +_* e^{\frac{C}{2}},$$

$$g(x) \;=\; e^{\frac{1}{2}} \cdot_* \phi(x) -_* e^{\frac{1}{2c}} \cdot_* \int_{*1}^{x} \psi(s) \cdot_* d_*s -_* e^{\frac{C}{2}}.$$

Substituting these expressions for f and g into the general solution (4.2), we obtain

$$u(x,t) = (\phi(x +_* c \cdot_* t) +_* \phi(x -_* c \cdot_* t))/_*e^2 +_* e^{\frac{1}{2c}} \cdot_* \int_{*x -_* c \cdot_* t}^{x +_* c \cdot_* t} \psi(s) \cdot_* d_*s,$$

$$(4.7)$$

which is called multiplicative d'Alambert's formula, shortly d'Alambert's formula.

Example 4.1 Consider the following Cauchy problem:

$$u_{tt}^{**} -_* e^9 \cdot_* u_{xx}^{**} \;=\; 0_*, \quad 0_* < x < \infty, \quad t > 0_*,$$

$$u(x,1) \;=\; u_t^*(x,1) = x^{2*}, \quad 0_* < x < \infty.$$

Here

$$c \;=\; 3,$$

$$\phi(x) \;=\; x^{2*},$$

$$\psi(x) \;=\; x^2.$$

Then, using d'Alambert's formula, we get

$$u(x,t) \;=\; (((x +_* e^3 \cdot_* t)^{2*} +_* (x -_* e^3 \cdot_* t)^{2*})/_*e^2)$$

$$+_* e^{\frac{1}{6}} \cdot_* \int_{*x -_* e^3 \cdot_* t}^{x +_* e^3 \cdot_* t} s^{2*} \cdot_* d_*s$$

$$= \quad x^{2*} +_* e^{9} \cdot_* t^{2*} +_* e^{\frac{1}{18}} \cdot_* s^{3*} \Big|_{s=x-_*e^{3}\cdot_*t}^{s=x+_*e^{3}\cdot_*t}$$

$$= \quad x^{2*} +_* e^{9} \cdot_* t^{2*} +_* e^{\frac{1}{18}} \cdot_* \left((x +_* e^{3} \cdot_* t)^{3*} -_* (x -_* e^{3} \cdot_* t)^{3*} \right)$$

$$= \quad x^{2*} +_* e^{9} \cdot_* t^{2*} +_* x^{2*} \cdot_* t +_* e^{3} \cdot_* t^{3*}.$$

Example 4.2 Consider the following Cauchy problem:

$$u_{tt}^{**} -_* e^{4} \cdot_* u_{xx}^{**} \quad = \quad 0_*, \quad -_*\infty < x < \infty, \quad t > 0,$$

$$u(x,1) \quad = \quad \phi(x)$$

$$= \quad \begin{cases} e -_* x^{2*} & |x|_* \le e \\ 1 & \text{otherwise,} \end{cases}$$

$$u_t^*(x,1) \quad = \quad \psi(x)$$

$$= \quad \begin{cases} (x -_* e) \cdot_* (x -_* e^{2}) & e \le x \le e^{2} \\ 1 & \text{otherwise.} \end{cases}$$

We will find $u(e,e)$. Using d'Alambert's formula, we have

$$u(e,e) \quad = \quad ((\phi(e^{3}) +_* \phi(-_*e))/_*e^{2}) +_* e^{\frac{1}{4}} \cdot_* \int_{*-_*e}^{e^{3}} \psi(s) \cdot_* d_*s$$

$$= \quad e^{\frac{1}{4}} \cdot_* \int_{*e}^{e^{2}} (s -_* e) \cdot_* (s -_* e^{2}) \cdot_* d_*s$$

$$= \quad e^{\frac{1}{4}} \cdot_* \int_{*e}^{e^{2}} (s^{2*} -_* e^{3} \cdot_* s +_* e^{2}) \cdot_* d_*s$$

$$= \quad e^{\frac{1}{4}} \cdot_* \left(e^{\frac{1}{3}} \cdot_* s^{3*} \Big|_{s=e}^{s=e^{2}} -_* e^{\frac{3}{2}} \cdot_* s^{2*} \Big|_{s=e}^{s=e^{2}} +_* e^{2} \cdot_* s \Big|_{s=e}^{s=e^{2}} \right)$$

$$= \quad -_* e^{\frac{1}{24}}.$$

Exercise 4.1 Solve the Cauchy problem

$$u_{tt}^{**} -_* u_{xx}^{**} \quad = \quad 0_*, \quad -_*\infty < x < \infty, \quad t > 0_*,$$

$$u(x,1) \quad = \quad x,$$

$$u_t^*(x,1) \quad = \quad \cos_* x, \quad -_*\infty < x < \infty.$$

Answer

$$u(x,t) = x +_* \sin_* t \cdot_* \cos_* x.$$

Theorem 4.1 *Fix $T > 0_*$. The Cauchy problem (4.3), (4.4) in the domain $0 < x < \infty$, $1 \leq t \leq T$, is well-posed for $\phi \in C_*^2(\mathbb{R}_*)$, $\psi \in C_*^1(\mathbb{R}_*)$.*

Proof D'Alambert's formula provides us with a solution, and we have shown that any solution of the Cauchy problem (4.3), (4.4) is necessarily equal to d'Alambert solution. Since $\phi \in C_*^2(\mathbb{R}_*)$, $\psi \in C_*^1(\mathbb{R}_*)$, we have that

$$u \in C_*^2(\mathbb{R}_* \times (1, \infty)) \cap C_*^1(\mathbb{R}_* \times [1, \infty)).$$

Therefore, the d'Alambert solution is a classical solution. Now we will prove the stability of the Cauchy problem. Let $\epsilon > 0_*$ be arbitrarily chosen. We take $1 < \delta < (\epsilon/_*(e +_* T))$. Let also u_1 and u_2 be solutions of the Cauchy problem with initial conditions ϕ_1, ψ_1 and ϕ_2, ψ_2, respectively, such that

$$|\phi_1(x) -_* \phi_2(x)|_* \quad < \quad \delta,$$

$$|\psi_1(x) -_* \psi_2(x)|_* \quad < \quad \delta,$$

for all $x \in \mathbb{R}_*$. Then for all $x \in \mathbb{R}_*$ and $1 \leq t \leq T$, we have

$$
\begin{aligned}
|u_1(x,t) -_* u_2(x,t)|_* \;=\; & \Big|((\phi_1(x +_* c \cdot_* t) -_* \phi_2(x +_* c \cdot_* t))/_* e^2) \\[2mm]
& +_* ((\phi_1(x -_* c \cdot_* t) -_* \phi_2(x -_* c \cdot_* t))/_* e^2) \\[2mm]
& +_* e^{\frac{1}{2c}} \cdot_* \int_{*x -_* ct}^{x +_* ct} (\psi_1(s) -_* \psi_2(s)) \cdot_* d_* s \Big|_* \\[2mm]
\leq\; & ((|\phi_1(x +_* c \cdot_* t) -_* \phi_2(x +_* c \cdot_* t)|_*)/_* e^2) \\[2mm]
& +_* ((|\phi_1(x -_* c \cdot_* t) -_* \phi_2(x -_* c \cdot_* t)|_*)/_* e^2) \\[2mm]
& +_* e^{\frac{1}{2c}} \cdot_* \int_{*x -_* c \cdot_* t}^{x +_* c \cdot_* t} |\psi_1(s) -_* \psi_2(s)|_* \cdot_* d_* s \\[2mm]
\leq\; & (\delta/_* e^2) +_* (\delta/_* e^2) +_* \delta \cdot_* T \\[2mm]
=\; & \delta \cdot_* (e +_* T) \\[2mm]
<\; & \epsilon,
\end{aligned}
$$

which completes the proof.

Remark 4.1 The d'Alambert formula is also valid for $1 < x < \infty$, $T < t \leq 1$, and the Cauchy problem is well-posed in this domain.

Remark 4.2 The Cauchy problem is ill-posed on the domain $1 < x < \infty$, $t \geq 1$. Indeed, consider the following Cauchy problem:

$$u_{tt}^{**} -_* u_{xx}^{**} = 0_*, \quad 1 < x < \infty, \quad t > 1,$$

$$u^n(x,1) = (e/_* e^{n^2}) \cdot_* \sin_*(n \cdot_* x),$$

$$u_t^n(x,1) = 1, \quad 1 < x < \infty.$$

We have that

$$u^n(x,t) = e^{\frac{1}{n^2}} \cdot_* \cosh_*(n \cdot_* t) \cdot_* \sin_*(n \cdot_* x)$$

is its solution. When n is large enough, the initial conditions describe an arbitrary small perturbation of the trivial solution $u = 1$. On the other hand, $\sup_{x \in \mathbb{R}_*} |u^n(x,t)|$ grows fast as $n \to \infty$ for any $t > 1$.

4.1.2 The Cauchy problem for the nonhomogeneous wave equation

Consider the following Cauchy problem:

$$u_{tt}^{**} -_* c^{2*} \cdot_* u_{xx}^{**} = f(x,t), \quad 0_* < x < \infty, \quad t > 1, \tag{4.8}$$

$$u(x,1) = \phi(x),$$

$$u_t^*(x,1) = \psi(x), \quad 0_* < x < \infty, \tag{4.9}$$

where $f \in \mathcal{C}_*(\mathbb{R}_* \times (1,\infty))$, $\phi \in \mathcal{C}_*^2(\mathbb{R}_*)$, and $\psi \in \mathcal{C}_*^1(\mathbb{R}_*)$ are given functions.

Theorem 4.2 *The Cauchy problem* (4.8), (4.9) *admit at most one solution.*

Proof Suppose that u_1 and u_2 are solutions to the problem (4.8), (4.9). Then the function $u = u_1 -_* u_2$ is a solution to the homogeneous problem

$$u_{tt}^{**} -_* c^{2*} \cdot_* u_{xx}^{**} = 0_*, \quad 1 < x < \infty, \quad t > 1, \tag{4.10}$$

$$u(x,1) = 1,$$

$$u_t^*(x,1) = 1, \quad 1 < x < \infty. \tag{4.11}$$

Note that $v = 1$ is also a solution to the homogeneous problem (4.10), (4.11). Hence from Theorem 4.1, we conclude that

$$u = v = 1,$$

i.e.,

$$u_1 = u_2.$$

This completes the proof.

Theorem 4.3 *Let* $f, f_x^* \in \mathcal{C}_*(\mathbb{R}_* \times [1, \infty))$, $\phi \in \mathcal{C}_*^2(\mathbb{R}_*)$, $\psi \in \mathcal{C}_*^1(\mathbb{R}_*)$. *Then the Cauchy problem* (4.8), (4.9) *has a solution given by*

$$u(x,t) = ((\phi(x +_* c \cdot_* t) +_* \phi(x -_* c \cdot_* t))/_* e^2)$$

$$+_* e^{\frac{1}{2c}} \cdot_* \int\limits_{*x -_* c \cdot_* t}^{x +_* c \cdot_* t} \psi(s) \cdot_* d_* s \tag{4.12}$$

$$+_* e^{\frac{1}{2c}} \cdot_* \int\limits_{*1}^{t} \int\limits_{*x -_* c \cdot_* (t -_* \tau)}^{x +_* c \cdot_* (t -_* \tau)} f(\xi, \tau) \cdot_* d_* \xi \cdot_* d_* \tau.$$

Definition 4.3 Equation (4.12) is also called d'Alambert's formula.

Remark 4.3 Note that for $f = 1$ both d'Alambert's formulas (4.12) and (4.7) coincide.

Proof Let $(x_0, t_0) \in \mathbb{R}_* \times (1, \infty)$ be arbitrarily chosen. Let $\triangle$ be the multiplicative triangle with edges the points (x_0, t_0), $(x_0 -_* c \cdot_* t_0, 1)$ and $(x_0 +_* c \cdot_* t_0, 1)$. Multiplicative integrating both sides of equation (4.8) over the multiplicative triangle $\triangle$, we get

$$\iint\limits_{*\triangle} \left(e^{c^2} \cdot_* u_{xx}^{**}(x, t) -_* u_{tt}^{**}(x, t) \right) \cdot_* d_* x \cdot_* d_* t$$

$$= -_* \iint\limits_{*\triangle} f(x, t) \cdot_* d_* x \cdot_* d_* t.$$

Using the multiplicative Green formula, we obtain

$$-_* \iint\limits_{*\triangle} f(x, t) \cdot_* d_* x \cdot_* d_* t = \oint\limits_{*\partial\triangle} \left(u_t^*(x, t) \cdot_* d_* x +_* c^{2*} \cdot_* u_x^*(x, t) \cdot_* d_* t \right) \tag{4.13}$$

$$= \int\limits_{(x_0, t_0)}^{*(x_0 -_* c \cdot_* t_0, 1)} \left(u_t^*(x, t) \cdot_* d_* x +_* c^{2*} \cdot_* u_x^*(x, t) \cdot_* d_* t \right)$$

$$+_* \int_{*(x_0 -_* c \cdot_* t_0, 1)}^{(x_0 +_* c \cdot_* t_0, 1)} \left(u_t^*(x,t) \cdot_* \cdot_* d_* x +_* c^{2*} \cdot_* u_x^*(x,t) \cdot_* d_* t \right)$$

$$+_* \int_{*(x_0 +_* c \cdot_* t_0, 1)}^{(x_0, t_0)} \left(u_t^*(x,t) \cdot_* d_* x +_* c^{2*} \cdot_* u_x^*(x,t) \cdot_* d_* t \right).$$

Note that

$$\int_{*(x_0, t_0)}^{(x_0 -_* c \cdot_* t_0, 0)} \left(u_t^*(x,t) \cdot_* d_* x +_* c^{2*} \cdot_* u_x^*(x,t) \cdot_* d_* t \right)$$

$$= \int_{*(x_0, t_0)}^{(x_0 -_* c \cdot_* t_0, 1)} \left(c \cdot_* u_t^*(x,t) \cdot_* d_* t \right.$$

$$\left. +_* c \cdot_* u_x^*(x,t) \cdot_* d_* x \right) \tag{4.14}$$

$$= c \cdot_* \int_{*(x_0, t_0)}^{(x_0 -_* c \cdot_* t_0, 1)} d_* u$$

$$= c \cdot_* \left(u(x_0 -_* c \cdot_* t_0, 1) -_* u(x_0, t_0) \right)$$

$$= c \cdot_* \left(\phi(x_0 -_* c \cdot_* t_0) -_* u(x_0, t_0) \right),$$

$$\int_{*(x_0 -_* c \cdot_* t_0, 1)}^{(x_0 +_* c \cdot_* t_0, 1)} \left(u_t^*(x,t) \cdot_* d_* x +_* c^{2*} \cdot_* u_x^*(x,t) \cdot_* d_* t \right)$$

$$= \int_{*x_0 -_* c \cdot_* t_0}^{x_0 +_* c \cdot_* t_0} u_t^*(x, 0_*) \cdot_* d_* x \tag{4.15}$$

$$= \int_{*x_0 -_* c \cdot_* t_0}^{x_0 +_* c \cdot_* t_0} \psi(x) \cdot_* d_* x,$$

and

$$\int\limits_{*(x_0+_*c\cdot_*t_0,1)}^{(x_0,t_0)} \left(u_t^*(x,t)\cdot_* d_*x +_* c^{2*}\cdot_* u_x^*(x,t)\cdot_* d_*t\right)$$

$$= \int\limits_{*(x_0+_*c\cdot_*t_0,1)}^{(x_0,t_0)} \left(-_*c\cdot_* u_t^*(x,t)\cdot_* d_*t\right.$$

$$\left. -_*c\cdot_* u_x^*(x,t)\cdot_* d_*x\right) \tag{4.16}$$

$$= -_*c\cdot_* \int\limits_{*(x_0+_*c\cdot_*t_0,1)}^{(x_0,t_0)} \cdot_*d_*u$$

$$= -_*c\cdot_* \left(u(x_0,t_0) -_* u(x_0 +_* c\cdot_* t_0,1)\right)$$

$$= -_*c\cdot_* \left(u(x_0,t_0) -_* \phi(x_0 +_* c\cdot_* t_0)\right).$$

We substitute (4.14), (4.15) and (4.16) into (4.13) and we find

$$-_*\int\int\limits_{*\triangle} f(x,t)\cdot_* d_*x \cdot_* d_*t$$

$$= c\cdot_* \left(\phi(x_0 -_* c\cdot_* t_0) -_* u(x_0,t_0)\right) +_* \int\limits_{*x_0-_*c\cdot_*t_0}^{x_0+_*c\cdot_*t_0} \psi(x)\cdot_* d_*x$$

$$-_*c\cdot_* \left(u(x_0,t_0) -_* \phi(x_0 +_* c\cdot_* t_0)\right),$$

or

$$u(x_0,t_0) \quad = \quad \left((\phi(x_0 +_* c\cdot_* t_0) +_* \phi(x_0 -_* c\cdot_* t_0))/_*e^2\right)$$

$$+_*e^{\frac{1}{2c}}\cdot_* \int\limits_{*x_0-_*c\cdot_*t_0}^{x_0+_*c\cdot_*t_0} \psi(x)\cdot_* d_*x$$

$$+_*e^{\frac{1}{2c}}\cdot_* \int\int\limits_{*\triangle} f(x,t)\cdot_* d_*x \cdot_* d_*t.$$

Because $(x_0,t_0) \in \mathbb{R}_* \times (1,\infty)$ was arbitrarily chosen, we finally obtain an explicit formula for the solutions of the Cauchy problem (4.8), (4.9) at an arbitrary point (x,t) given by (4.12). Now we will prove that the function

$u(x,t)$ given by the formula (4.12) is indeed a solution to the Cauchy problem (4.8), (4.9). We have

$$u(x,0) \;=\; \phi(x),$$

$$
\begin{aligned}
u_t^*(x,t) \;=\;& (c \cdot_* \phi^*(x +_* c \cdot_* t) -_* c \cdot_* \phi^*(x -_* c \cdot_* t))/_* e^2 \\[4pt]
&+_* (\psi(x +_* c \cdot_* t) +_* \psi(x -_* c \cdot_* t))/_* e^2 \\[4pt]
&+_* e^{\frac{1}{2}} \cdot_* \int_{*1}^{t} (f(x +_* c \cdot_* (t -_* \tau), \tau) \\[4pt]
&+_* f(x -_* c \cdot_* (t -_* \tau), \tau)) \cdot_* d_* \tau,
\end{aligned}
$$

$$u_t^*(x,1) \;=\; \psi(x),$$

$$
\begin{aligned}
u_{tt}^{**}(x,t) \;=\;& c^{2*} \cdot_* (\phi^{**}(x +_* c \cdot_* t) +_* \phi^{**}(x -_* c \cdot_* t))/_* e^2 \\[4pt]
&+_* (c \cdot_* \psi^*(x +_* c \cdot_* t) -_* c \cdot_* \psi^*(x -_* c \cdot_* t))/_* e^2 +_* f(x,t) \\[4pt]
&+_* e^{\frac{1}{2}} \cdot_* \int_{*1}^{t} (c \cdot_* f_x^*(x +_* c \cdot_* (t -_* \tau), \tau) \\[4pt]
&-_* c \cdot_* f_x(x -_* c \cdot_* (t -_* \tau), \tau)) \cdot_* d_* \tau
\end{aligned}
$$

$$
\begin{aligned}
u_x^*(x,t) \;=\;& (\phi^*(x +_* c \cdot_* t) +_* \phi^*(x -_* c \cdot_* t))/_* e^2 \\[4pt]
&+_* (\psi(x +_* c \cdot_* t) -_* \psi(x -_* c \cdot_* t))/_* e^{e^2 \cdot_* c} \\[4pt]
&+_* e^{\frac{1}{2c}} \cdot_* \int_{*1}^{t} (f(x +_* c \cdot_* (t -_* \tau), \tau) \\[4pt]
&-_* f(x -_* c \cdot_* (t -_* \tau), \tau)) \cdot_* d_* \tau,
\end{aligned}
$$

$$
\begin{aligned}
u_{xx}^*(x,t) \;=\;& (\phi^{**}(x +_* c \cdot_* t) +_* \phi^{**}(x -_* c \cdot_* t))/_* e^2 \\[4pt]
&+_* (\psi^*(x +_* c \cdot_* t) -_* \psi^*(x -_* c \cdot_* t))/_* e^{e^2 \cdot_* c} \\[4pt]
&+_* e^{\frac{1}{2c}} \cdot_* \int_{*1}^{t} (f_x^*(x +_* c \cdot_* (t -_* \tau), \tau) \\[4pt]
&-_* f_x^*(x -_* c \cdot_* (t -_* \tau), \tau)) \cdot_* d_* \tau,
\end{aligned}
$$

whereupon

$$u_{tt}^{**}(x,t) -_* c^{2*} \cdot_* u_{xx}^{**}(x,t) = f(x,t).$$

Therefore, u is a solution to the Cauchy problem (4.8), (4.9). This completes the proof.

Example 4.3 Consider the following Cauchy problem:

$$u_{tt}^{**} -_* e^4 \cdot_* u_{xx}^{**} = e^x +_* \sin_* t, \quad 1 < x < \infty, \quad t > 1,$$

$$u(x,1) = 1,$$

$$u_t^*(x,1) = e/_*(= e +_* x^{2*}), \quad 1 < x < \infty.$$

Here

$$c = e^2,$$

$$f(x,t) = e^x +_* \sin_* t,$$

$$\phi(x) = 1, \quad \psi(x) = e/_*(e +_* x^{2*}).$$

Then, using d'Alanbert's formula (4.12), we get

$$u(x,t) = e^{\frac{1}{4}} \cdot_* \int_{*x-_* e^2 \cdot_* t}^{x+_* e^2 \cdot_* t} (e/_*(e +_* s^{2*})) \cdot_* d_* s$$

$$+_* e^{\frac{1}{4}} \cdot_* \int_{*1}^{t} \int_{*x-_* e^2 \cdot_*(t-_* \tau)}^{x+_* e^2 \cdot_*(t-_* \tau)} (e^\xi +_* \sin_* \tau) \cdot_* d_* \tau$$

$$= e^{\frac{1}{4}} \cdot_* \arctan_* s \Big|_{s=x-_* e^2 \cdot_* t}^{s=x+_* e^2 \cdot_* t} +_* e^{\frac{1}{4}} \cdot_*$$

$$\int_{*1}^{t} (e^\xi +_* \xi \cdot_* \sin_* \tau) \Big|_{\xi=x-_* e^2 \cdot_*(t-_* \tau)}^{\xi=x+_* e^2 \cdot_*(t-_* \tau)} \cdot_* d_* \tau$$

$$= (\arctan_*(x +_* e^2 \cdot_* t) -_* \arctan(x -_* e^2 \cdot_* t))/_* e^4$$

$$+_* e^{\frac{1}{4}} \cdot_* \int_{*1}^{t} \left(e^{x+_* e^2 \cdot_* t -_* e^2 \cdot_* \tau} -_* e^{x-_* e^2 \cdot_* t +_* e^2 \cdot_* \tau} \right.$$

$$+_* e^4 \cdot_* (t -_* \tau) \cdot_* \sin_* \tau) \cdot_* d_* \tau$$

$$= \quad e^{\frac{1}{4}} \cdot_* \left(\arctan_* (x +_* e^2 \cdot_* t) -_* \arctan_* (x -_* e^2 \cdot_* t) -_* e^x \right.$$

$$+_* e^x \cdot_* \cosh_* (e^2 \cdot_* t) \bigg) +_* t -_* \sin_* t.$$

Example 4.4 Consider the following Cauchy problem.

$$u_{tt}^{**} -_* u_{xx}^{**} \quad = \quad \cos_* (x +_* t), \quad 1 < x < \infty, \quad t > 1,$$

$$u(x, 1) \quad = \quad x,$$

$$u_t^*(x, 1) \quad = \quad \sin_* x, \quad 1 < x < \infty.$$

Here

$$c \quad = \quad e,$$

$$f(x, t) \quad = \quad \cos_* (x +_* t),$$

$$\phi(x) \quad = \quad x,$$

$$\psi(x) \quad = \quad \sin_* x.$$

Then, using d'Alambert's formula (4.12), we have

$$u(x, t) \quad = \quad (x +_* t +_* x -_* t)/_* e^2 +_* e^{\frac{1}{2}} \cdot_* \int_{*x -_* t}^{x +_* t} \sin_* s \cdot_* d_* s$$

$$+_* e^{\frac{1}{2}} \cdot_* \int_{*1}^{t} \int_{x -_* (t -_* \tau)}^{x +_* (t -_* \tau)} \cos_* (\xi +_* \tau) \cdot_* d_* \xi \cdot_* d_* \tau$$

$$= \quad x -_* e^{\frac{1}{2}} \cdot_* \cos_* s \Big|_{s = x -_* t}^{s = x +_* t} +_* e^{\frac{1}{2}} \cdot_* \int_{*1}^{t} \sin_* (\xi +_* \tau) \Big|_{\xi = x -_* (t -_* \tau)}^{\xi = x +_* (t -_* \tau)} \cdot_* d_* \tau$$

$$= \quad x -_* e^{\frac{1}{2}} \cdot_* (\cos_* (x +_* t) -_* \cos_* (x -_* t))$$

$$+_* e^{\frac{1}{2}} \cdot_* \int_{*1}^{t} (\sin_* (x +_* t) -_* \sin_* (x -_* t +_* e^2 \cdot_* \tau)) \cdot_* d_* \tau$$

$$= \quad x +_* e^{\frac{1}{2}} \cdot_* \sin_* x \cdot_* \sin_* t +_* e^{\frac{1}{2}} \cdot_* t \cdot_* \sin_* (x +_* t).$$

Exercise 4.2 Solve the Cauchy problem

$$u_{tt}^{**} -_* u_{xx}^{**} = x \cdot_* t, \quad 1 < x < \infty, \quad t > 1,$$

$$u(x, 1) = 1,$$

$$u_t^*(x, 1) = e^x, \quad 1 < x < \infty.$$

Answer

$$u(x, t) = e^x \cdot_* \sinh_* t +_* e^{\frac{1}{6}} \cdot_* x \cdot_* t^{3*}.$$

Theorem 4.4 *Let* $T > 1$ *be fixed and* $f, f_x^* \in C_*(\mathbb{R}_* \times [1, \infty))$, $\phi \in C_*^2(\mathbb{R}_*)$, $\psi \in C_*^1(\mathbb{R}_*)$. *Then the Cauchy problem* (4.8), (4.9) *is well-posed in the domain* $1 < x < \infty$, $1 \leq t \leq T$.

Proof Let $\epsilon > 1$ be arbitrarily chosen and $1 < \delta < \epsilon/_*(e +_* T +_* ((T^{2*})/_* e^2))$. Let also f_i, ϕ_i and ψ_i, $i = 1, 2$, be such that $f_i, f_{ix}^* \in C_*(\mathbb{R}_* \times [1, \infty))$, $\phi_i \in C_*^2(\mathbb{R}_*)$, $\psi_i \in C_*^1(\mathbb{R}_*)$, and

$$|f_1(t, x) -_* f_2(t, x)|_* < \delta,$$

$$|\phi_1(x) -_* \phi_2(x)|_* < \delta,$$

$$|\psi_1(x) -_* \psi_2(x)|_* < \delta,$$

for all $x \in \mathbb{R}_*$ and $1 \leq t \leq T$. Let $u_1(x, t)$ and $u_2(x, t)$ be the solutions of the Cauchy problems:

$$u_{tt}^{**} -_* c^{2*} \cdot_* u_{xx}^{**} = f_1(x, t), \quad 1 < x < \infty, \quad t > 1,$$

$$u(x, 1) = \phi_1(x),$$

$$u_t^*(x, 1) = \psi_1(x), \quad 1 < x < \infty,$$

and

$$u_{tt}^{**} -_* c^{2*} \cdot_* u_{xx}^{**} = f_2(x, t), \quad 1 < x < \infty, \quad t > 1,$$

$$u(x, 1) = \phi_2(x),$$

$$u_t^*(x, 1) = \psi_2(x), \quad 1 < x < \infty,$$

respectively. Then,

$$|u_1(x,t) -_* u_2(x,t)|_* = \left|(\phi_1(x +_* c \cdot_* t) -_* \phi_2(x +_* c \cdot_* t))/_* e^2\right.$$

$$+_* (\phi_1(x -_* c \cdot_* t) -_* \phi_2(x -_* c \cdot_* t))/_* e^2$$

$$+_* e^{\frac{1}{2c}} \cdot_* \int_{x-_* c \cdot_* t}^{x+_* c \cdot_* t} (\psi_1(s) -_* \psi_2(s)) \cdot_* d_* s$$

$$+_* e^{\frac{1}{2c}} \cdot_* \int_{*1}^{t} \int_{x-_* c \cdot_* (t-_* \tau)}^{x+_* c \cdot_* (t-_* \tau)} (f_1(\xi,\tau) -_* f_2(\xi,\tau))$$

$$\left. \cdot_* d_* \xi \cdot_* d_* \tau\right|_*$$

$$\leq \left|(\phi_1(x +_* c \cdot_* t) -_* \phi_2(x +_* c \cdot_* t))/_* e^2\right|_*$$

$$+_* \left|(\phi_1(x -_* c \cdot_* t) -_* \phi_2(x -_* c \cdot_* t))/_* e^2\right|_*$$

$$+_* e^{\frac{1}{2c}} \cdot_* \int_{x-_* c \cdot_* t}^{x+_* c \cdot_* t} |\psi_1(s) -_* \psi_2(s)|_* \cdot_* d_* s$$

$$+_* e^{\frac{1}{2c}} \cdot_* \int_{*1}^{t} \int_{*x-_* c \cdot_* (t-_* \tau)}^{x+_* c \cdot_* (t-_* \tau)} |f_1(\xi,\tau) -_* f_2(\xi,\tau)|_*$$

$$\cdot_* d_* \xi \cdot_* d_* \tau$$

$$< \delta/_* e^2 +_* \delta/_* e^2 +_* \delta \cdot_* T +_* \delta \cdot_* (T^{2*})/_* e^2$$

$$= \delta \cdot_* \left(1 +_* T +_* (T^{2*})/_* e^2\right)$$

$$< \epsilon,$$

which completes the proof.

Theorem 4.5 *Suppose that $f(\cdot,t)$, $\phi(\cdot)$, $\psi(\cdot)$ are multiplicative even functions for all $t \geq 1$. Then the solution $u(\cdot,t)$ of the Cauchy problem (4.8), (4.9) is also a multiplicative even function for every $t \geq 1$.*

Proof We set

$$v(x,t) = u(-_*x,t).$$

Then,

$$
\begin{aligned}
v_t^*(x,t) &= u_t^*(-_*x,t), \\[2mm]
v_{tt}^{**}(x,t) &= u_{tt}^*(-_*x,t), \\[2mm]
v_x^*(x,t) &= -_*u_x^*(-_*x,t), \\[2mm]
v_{xx}^{**}(x,t) &= u_{xx}^*(-_*x,t).
\end{aligned}
$$

Hence,

$$
\begin{aligned}
v_{tt}^{**}(x,t) -_* c^{2*} \cdot_* v_{xx}^{**}(x,t) &= u_{tt}^{**}(-_*x,t) -_* c^{2*} \cdot_* u_{xx}^{**}(-_*x,t) \\[2mm]
&= f(-_*x,t) \\[2mm]
&= f(x,t)
\end{aligned}
$$

and

$$
\begin{aligned}
v(x,1) &= u(-_*x,1) \\[2mm]
&= \phi(-_*x) \\[2mm]
&= \phi(x), \\[2mm]
v_t(x,1) &= u_t^*(-_*x,1) \\[2mm]
&= \psi(-_*x) \\[2mm]
&= \psi(x).
\end{aligned}
$$

Therefore, v is a solution to the Cauchy problem (4.8), (4.9). Hence from Theorem 4.2, it follows that

$$u(x,t) = u(-_*x,t).$$

Exercise 4.3 Suppose that $f(\cdot,t)$, $\phi(\cdot)$, $\psi(\cdot)$ are multiplicative odd functions for all $t \geq 1$. Prove that the solution $u(\cdot,t)$ of the Cauchy problem (4.8), (4.9) is also a multiplicative odd function for every $t \geq 1$.

Exercise 4.4 Let $\omega > 1$. Suppose that $f(\cdot, t)$, $\phi(\cdot)$, $\psi(\cdot)$ are multiplicative ω-periodic functions for all $t \geq 1$. Prove that the solution $u(\cdot, t)$ of the Cauchy problem (4.8), (4.9) is also a multiplicative ω-periodic function for every $t \geq 1$.

4.1.3 Separation of multiplicative variables

We consider the problem

$$u_{tt}^{**} -_* e^{c^2} \cdot_* u_{xx}^{**} = 0_*, \quad 1 < x < e^L, \quad t > 1, \tag{4.17}$$

$$u_x^*(1, t) = u_x^*(e^L, t) = 1, \quad t \geq 1, \tag{4.18}$$

$$u(x, 1) = \phi(x), \quad u_t^*(x, 1) = \psi(x), \quad 1 \leq x \leq e^L, \tag{4.19}$$

where $\phi \in C_*^2(\mathbb{R}_*)$, $\psi \in C_*^1(\mathbb{R}_*)$ are given functions. The compatibility conditions are given as follows:

$$\phi^*(1) = \phi^*(e^L) = \psi^*(1) = \psi^*(e^L) = 1.$$

Equations (4.17)–(4.19) is a multiplicative linear homogeneous initial boundary value problem.

Definition 4.4 The condition (4.18) is called Neumann boundary condition.

We will find nontrivial separated solutions of equation (4.17), i.e., solutions of the form

$$u(x, t) = X(x) \cdot_* T(t), \tag{4.20}$$

that satisfy the boundary condition (4.18). Substituting (4.20) into (4.17), we find

$$X(x) \cdot_* T^{**}(t) = e^{c^2} \cdot_* X^{**}(x) \cdot_* T(t).$$

By separating the multiplicative variables, we see

$$((T^{**}(t))/_*(e^{c^2} \cdot_* T(t))) = (X^{**}(x))/_*(X(x)).$$

Therefore, there exists a multiplicative constant λ such that

$$(T^{**}(t))/_*(e^{c^2} \cdot_* T(t)) = (X^{**}(x))/_* X(x) = -_*\lambda,$$

whereupon

$$X^{**} +_* \lambda \cdot_* X = 0_*, \quad 1 < x < e^L,$$

$$T^{**} = -_*\lambda \cdot_* e^{c^2} \cdot_* T, \quad t > 1. \tag{4.21}$$

By the boundary condition (4.18), we get

$$u_x^*(1, t) = X^*(1) \cdot_* T(t)$$

$$= 1,$$

$$u_x^*(e^L, t) = X^*(e^L) \cdot_* T(t)$$

$$= 1.$$

Since u is a nontrivial solution, it follows that

$$X^*(1) = X^*(e^L) = 1.$$

Therefore, $X(x)$ should be a solution of the eigenvalue problem:

$$X^{**} +_* \lambda \cdot_* X = 1, \quad 1 < x < e^L, \tag{4.22}$$

$$X^*(1) = X^*(e^L) = 1. \tag{4.23}$$

For the general solution of the first equation of (4.21) we have the following:

1.

$$X(x) = c_1 \cdot_* \cosh_*((-_*\lambda)^{\frac{1}{2}} \cdot_* \cdot_* x) +_* c_2 \cdot_* \sinh_*((-_*\lambda \cdot_* x))^{\frac{1}{2}} \cdot_*$$

for $\lambda < 1$.

2.

$$X(x) = c_1 +_* c_2 \cdot_* x$$

for $\lambda = 1$.

3.

$$X(x) = c_1 \cdot_* \cos_*((\lambda)^{\frac{1}{2}} \cdot_* \cdot_* x) +_* c_2 \cdot_* \sin_*((\lambda)^{\frac{1}{2}} \cdot_* x)$$

for $\lambda > 1$,

where c_1 and c_2 are arbitrary real constants.

1. When $\lambda < 1$, using (4.23), we find

$$c_1 = c_2 = 1.$$

Therefore,

$$X(x) \equiv 1$$

and the multiplicative eigenvalue problem (4.22), (4.23) does not admit multiplicative negative eigenvalues.

2. When $\lambda = 1$, using (4.23), we find $X(x) = c_1$.

3. When $\lambda > 1$, then $c_2 = 1$ and

$$c_1 \cdot_* (\lambda)^{\frac{1}{2}} \cdot_* \cdot_* \sin_*(L(\lambda)^{\frac{1}{2}} \cdot_*) = 1.$$

If $c_1 = 1$, then

$$X(x) \equiv 1$$

and the multiplicative eigenvalue problem (4.22), (4.23) does not admit multiplicative positive eigenvalues. Thus

$$(\lambda)^{\frac{1}{2}} \cdot_* \cdot_* e^L = e^{n\pi}, \quad n \in \mathbb{N},$$

or

$$\lambda = e^{n^2\pi^2}/_*(e^{L^2}), \quad n \in \mathbb{N}.$$

The associated multiplicative eigenfunction is

$$X(x) = \cos_*((e^{n\pi} \cdot_* x)/_* e^L),$$

which is uniquely determined up to a multiplicative factor.

Hence, the solution of the eigenvalue problem (4.22), (4.23) is an infinite sequence of multiplicative nonnegative simple multiplicative eigenvalues and their associated multiplicative eigenfunctions will be denoted by

$$X_n(x) \quad = \quad \cos_*((e^{n\pi} \cdot_* x)/_* e^L),$$

$$\lambda_n \quad = \quad \left(e^{n\pi^2 L}\right)^{2_*}, \quad n \in \mathbb{N}_0.$$

Consider the second equation of (4.21) for $\lambda = \lambda_n$. The solutions are

$$T_0(t) = \alpha_0 +_* \beta_0 \cdot_* t,$$

$$T_n(t) = \alpha_n \cdot_* \cos_*(c \cdot_* e^{\frac{\pi n}{L}} \cdot_* t) +_* \beta_n \cdot_* \sin_*(c \cdot_* e^{\frac{\pi n}{L}} \cdot_* t), \quad n \in \mathbb{N},$$

where $\alpha_n, \beta_n, n \in \mathbb{N}_0$ are multiplicative real constants. Then the multiplicative product solutions of the initial boundary value problem (4.17)–(4.19) are given by

$$u_0(x,t) = X_0(x) \cdot_* T_0(t)$$

$$= A_0 +_* B_0 \cdot_* t,$$

$$u_n(x,t) = \cos_*(e^{\frac{\pi n}{L}} \cdot_* x) \cdot_* \left(A_n \cdot_* \cos_*(c \cdot_* e^{\frac{\pi n}{L}} \cdot_* t)\right.$$

$$\left. +_* B_n \cdot_* \sin_*(c \cdot_* e^{\frac{\pi n}{L}} \cdot_* t)\right), \quad n \in \mathbb{N}.$$

Applying the multiplicative superposition principle, the function

$$u(x,t) \quad = \quad A_0 +_* B_0 \cdot_* t +_* \sum_{*n=1}^{\infty} \left(A_n \cdot_* \cos_*(c \cdot_* e^{\frac{\pi n}{L}} \cdot_* t)\right.$$

$$\left. +_* B_n \cdot_* \sin_*(c \cdot_* e^{\frac{\pi n}{L}} \cdot_* t)\right) \cdot_* \cos_*(c \cdot_* e^{\frac{\pi n}{L}} \cdot_* x)$$

is a generalized (formal) solution of the problem (4.17)–(4.19). To find the constants A_n and B_n, $n \in \mathbb{N}_0$, we will use the initial condition (4.19). Assume that the initial data ϕ and ψ can be expanded into multiplicative generalized Fourier series with respect to the sequence of the multiplicative eigenfunctions of the problem (4.22), (4.23) and these series are multiplicative uniformly convergent,

$$\phi(x) = a_0 +_* \sum_{*n=1}^{\infty} a_n \cdot_* \cos_*(c \cdot_* e^{\frac{\pi n}{L}} \cdot_* x),$$

$$\psi(x) = b_0 +_* \sum_{*n=1}^{\infty} b_n \cdot_* \cos_*(c \cdot_* e^{\frac{\pi n}{L}} \cdot_* x).$$

We have that

$$a_0 = e^{\frac{1}{L}} \cdot_* \int_{*1}^{e^L} \phi(x) \cdot_* d_* x,$$

$$b_0 = e^{\frac{1}{L}} \cdot_* \int_{*1}^{e^L} \psi(x) \cdot_* d_* x,$$

$$a_m = e^{\frac{2}{L}} \cdot_* \int_{*1}^{e^L} \cos_*(c \cdot_* e^{\frac{\pi m}{L}} \cdot_* x) \cdot_* \phi(x) \cdot_* d_* x,$$

$$b_m = e^{\frac{2}{L}} \cdot_* \int_{*1}^{e^L} \cos_*(c \cdot_* e^{\frac{\pi m}{L}} \cdot_* x) \cdot_* \psi(x) \cdot_* d_* x, \quad m \in \mathbb{N}.$$

Hence,

$$\begin{aligned} u(x,1) &= A_0 +_* \sum_{*n=1}^{\infty} A_n \cdot_* \cos_*(c \cdot_* e^{\frac{\pi n}{L}} \cdot_* x) \\ &= a_0 +_* \sum_{*n=1}^{\infty} a_n \cdot_* \cos_*(c \cdot_* e^{\frac{\pi n}{L}} \cdot_* x), \end{aligned}$$

whereupon

$$a_n = A_n, \quad n \in \mathbb{N}_0.$$

Also,

$$\begin{aligned} u_t^*(x,1) &= B_0 +_* \sum_{*n=1}^{\infty} B_n \cdot_* c \cdot_* e^{\frac{\pi n}{L}} \cdot_* \cos_*(c \cdot_* e^{\frac{\pi n}{L}} \cdot_* x) \\ &= b_0 +_* \sum_{*n=1}^{\infty} b_n \cdot_* \cos_*(c \cdot_* e^{\frac{\pi n}{L}} \cdot_* x). \end{aligned}$$

Therefore,

$$B_0 = b_0,$$

$$B_n = (e^L \cdot_* b_n)/_* c \cdot_* e^{\frac{\pi n}{L}}, \quad n \in \mathbb{N}.$$

Thus the problem (4.17)–(4.19) is formally solved.

Example 4.5 We will find a formal solution to the following problem:

$$u_{tt}^{**} -_* u_{xx}^* = 1, \quad 1 < x < e^\pi, \quad t > 1,$$

$$u_x^*(1, t) = u_x^*(e^\pi, t) = 1, \quad t \geq 1,$$

$$u(x, 1) = (\cos_* x)^{2*},$$

$$u_t^*(x, 1) = (\sin_* x)^{2*}, \quad 1 \leq x \leq e^\pi.$$

Here,

$$L = \pi,$$

$$\phi(x) = (\cos_* x)^{2*},$$

$$\psi(x) = (\sin_* x)^{2*}.$$

Then,

$$X_n(x) = \cos_*(e^n \cdot_* x),$$

$$\lambda_n = e^{n^2},$$

$$a_0 = (e/_* e^\pi) \cdot_* \int_{*1}^{e^\pi} (\cos_* x)^{2*} \cdot_* d_* x$$

$$= e^{\frac{1}{2}},$$

$$b_0 = e^{\frac{1}{\pi}} \cdot_* \int_{*1}^{e^\pi} (\sin_* x)^{2*} \cdot_* d_* x$$

$$= e^{\frac{1}{2}},$$

$$a_m = e^{\frac{2}{\pi}} \cdot_* \int_{*1}^{e^\pi} \cos_*(e^m \cdot_* x)(\cos_* x)^{2*} \cdot_* d_* x$$

$$= \begin{cases} 1 & m \neq 2 \\ e^{\frac{1}{2}} & m = 2, \end{cases}$$

$$b_m = e^{\frac{2}{\pi}} \cdot_* \int_{*1}^{e^{\pi}} \cos_*(e^m \cdot_* x)(\sin_* x)^{2*} \cdot_* d_* x$$

$$= \begin{cases} 1 & m \neq 2 \\ -_* e^{\frac{1}{2}} & m = 2. \end{cases}$$

Therefore,

$$A_0 = e,$$

$$B_0 = e,$$

$$A_2 = e^{\frac{1}{2}},$$

$$B_2 = -_* e^{\frac{1}{4}},$$

$$A_m = B_m$$

$$= 1, \quad m \neq 2, \quad m \in \mathbb{N},$$

and

$$u(x,t) = e^{\frac{1}{2}} +_* e^{\frac{t}{2}} +_* \left(e^{\frac{1}{2}} \cdot_* \cos_*(e^2 \cdot_* t) -_* e^{\frac{1}{4}} \cdot_* \sin_*(e^2 \cdot_* t) \right) \cdot_* \cos_*(e^2 \cdot_* x).$$

Exercise 4.5 Find a formal solution to the following problem:

$$u_{tt}^{**} -_* u_{xx}^{**} = 1, \quad 1 < x < e^{\pi}, \quad t > 1,$$

$$u_x^*(1,t) = u_x^*(e^{\pi},t) = 1, \quad t \geq 1,$$

$$u(x,1) = 1,$$

$$u_t^*(x,1) = (\sin_* x)^{3*}, \quad 1 \leq x \leq e^{\pi}.$$

Answer

$$u(x,t) = e^{\frac{4}{3\pi}} \cdot_* t +_* \sum_{*n=1}^{\infty} e^{\frac{12}{\pi n(4n^2-1)(4n^2-9)}} \cdot_* \sin_*(e^{2n} \cdot_* t) \cdot_* \cos_*(e^{2n} \cdot_* x).$$

Now we consider the nonhomogeneous initial boundary value problem:

$$u_{tt}^{**} -_* e^{c^2} u_{xx}^{**} = f(x,t), \quad 1 < x < e^{L}, \quad t > 1, \tag{4.24}$$

$$u_x^*(1,t) = u_x^*(e^{L},t) = 1, \quad t \geq 1, \tag{4.25}$$

$$u(x,1) = \phi(x),$$

$$u_t^*(x,1) = \psi(x), \quad 1 \leq x \leq e^{L}, \tag{4.26}$$

where

$$f(x,t) = e^{C_0} +_* e^{D_0} \cdot_* t +_* \sum_{*n=1}^{\infty} \left(e^{C_n} \cdot_* \cos_* (e^{\frac{c\pi n}{L}} \cdot_* t) \right.$$

$$\left. +_* e^{D_n} \cdot_* \sin_* (e^{\frac{c\pi n}{L}} \cdot_* t) \right) \cdot_* \cos_* (e^{\frac{\pi n}{L}} \cdot_* x), \quad 1 \le x \le e^L, \quad t \ge 1,$$

$$\phi(x) = e^{a_0} +_* \sum_{*n=1}^{\infty} e^{a_n} \cdot_* \cos_* (e^{\frac{\pi n}{L}} \cdot_* x),$$

$$\psi(x) = e^{b_0} +_* \sum_{*n=1}^{\infty} e^{b_n} \cdot_* \cos_* (e^{\frac{\pi n}{L}} \cdot_* x), \quad 1 \le x \le e^L, \quad t \ge 1,$$

$$\phi^*(1) = \phi^*(e^L)$$

$$= \psi^*(1)$$

$$= \psi^*(e^L)$$

$$= 1,$$

C_n, D_n, a_n and b_n, $n \in \mathbb{N}_0$ are given constants. Let $u(x,t)$ be a solution to the problem (4.24)–(4.26) in the following form:

$$u(x,t) = T_0(t) +_* \sum_{*n=1}^{\infty} T_n(t) \cdot_* \cos_* (e^{\frac{\pi n}{L}} \cdot_* x). \tag{4.27}$$

Substituting (4.27) in (4.24), we get

$$T_0^{**} +_* \sum_{*n=1}^{\infty} \left(T_n^{**} +_* e^{\frac{c^2\pi^2 n^2}{L^2}} \cdot_* T_n \right) \cdot_* \cos_* (e^{\frac{\pi n}{L}} \cdot_* x)$$

$$= e^{C_0} +_* e^{D_0} \cdot_* t +_* \sum_{*n=1}^{\infty} \left(e^{C_n} \cdot_* \cos_* (e^{\frac{c\pi n}{L}} \cdot_* t) \right.$$

$$\left. +_* e^{D_n} \cdot_* \sin_* (e^{\frac{c\pi n}{L}} \cdot_* t) \right) \cdot_* \cos_* (e^{\frac{\pi n}{L}} \cdot_* x).$$

Hence,

$$T_0^{**} = e^{C_0} +_* e^{D_0} \cdot_* t,$$

$$T_n^{**} +_* e^{\frac{c^2\pi^2 n^2}{L^2}} \cdot_* T_n = e^{C_n} \cdot_* \cos_* (e^{\frac{c\pi n}{L}} \cdot_* t) +_* e^{D_n} \cdot_* \sin_* (e^{\frac{c\pi n}{L}} \cdot_* t).$$

Therefore,

$$T_0(t) = e^{\frac{C_0}{2}} \cdot_* t^{2*} +_* e^{\frac{D_0}{6}} \cdot_* t^{3*} +_* e^{\beta_0} \cdot_* t +_* e^{\alpha_0},$$

$$T_n(t) = e^{\alpha_n} \cdot_* \cos_*(e^{\frac{c\pi n}{L}} \cdot_* t) +_* e^{\beta_n} \cdot_* \sin_*(e^{\frac{c\pi n}{L}} \cdot_* t)$$

$$-_* e^{\frac{D_n L}{2c\pi n}} \cdot_* t \cos_*(e^{\frac{c\pi n}{L}} \cdot_* t) +_* e^{\frac{C_n L}{2c\pi n}} \cdot_* t \cdot_* \sin_*(e^{\frac{c\pi n}{L}} \cdot_* t), \quad n \in \mathbb{N}.$$

We will find the constants e^{α_0}, e^{α_n} and e^{β_n}, $n \in \mathbb{N}$, using the initial condition (4.26). We have

$$u(x, 1) = e^{\alpha_0} +_* \sum_{*n=1}^{\infty} e^{\alpha_n} \cdot_* \cos_*(e^{\frac{\pi n}{L}} \cdot_* x)$$

$$= e^{a_0} +_* \sum_{*n=1}^{\infty} e^{a_n} \cdot_* \cos_*(e^{\frac{\pi n}{L}} \cdot_* x),$$

whereupon $\alpha_n = a_n$, $n \in \mathbb{N}_0$. Also,

$$u_t^*(x, 1) = e^{\beta_0} +_* \sum_{*n=1}^{\infty} \left(e^{\beta_n} \cdot_* e^{\frac{c\pi n}{L}} -_* e^{\frac{D_n L}{2c\pi n}} \right) \cdot_* \cos_*^(e^{\frac{\pi n}{L}} \cdot_* x)$$

$$= e^{b_0} +_* \sum_{*n=1}^{\infty} e^{b_n} \cdot_* \cos_*(e^{\frac{\pi n}{L}} \cdot_* x),$$

whereupon

$$\beta_0 = b_0,$$

$$\beta_n = \frac{L}{c\pi n} \left(b_n + \frac{D_n L}{2c\pi n} \right).$$

Therefore,

$$u(x, t) = e^{\frac{C_0}{2}} \cdot_* t^{2*} +_* e^{\frac{D_0}{6}} \cdot_* t^{3*} +_* e^{b_0} \cdot_* t +_* e^{a_0} \cdot_*$$

$$+_* \sum_{*n=1}^{\infty} \left(e^{a_n} \cdot_* \cos_*(e^{\frac{c\pi n}{L}} \cdot_* t) \right.$$

$$+_* e^{\frac{L}{c\pi n}} \cdot_* \left(e^{b_n} +_* e^{\frac{D_n L}{2c\pi n}} \right) \cdot_* \sin_*(e^{\frac{c\pi n}{L}} \cdot_* t)$$

$$-_* e^{\frac{D_n L}{2c\pi n}} \cdot_* t_* \cos_*(e^{\frac{c\pi n}{L}} \cdot_* t) +_* e^{\frac{C_n L}{2c\pi n}} \cdot_* t \cdot_* \sin_*(e^{\frac{c\pi n}{L}} \cdot_* t) \right)$$

$$\cdot_* \cos_*(e^{\frac{\pi n}{L}} \cdot_* x).$$

Example 4.6 We will find a formal solution to the following problem:

$$u_{tt}^{**} -_* u_{xx}^{**} = \cos_*(e^{2\pi} \cdot_* x) \cdot_* \cos_*(e^{2\pi} \cdot_* t),$$

$$1 < x < c, \quad t > 1,$$

$$u_x^*(1,t) = u_x^*(e,t) = 1, \quad t \geq 1,$$

$$u(x,1) = (\cos_*(\pi x))^{2_*}, \quad u_t^*(x,1) = e^2 \cdot_* \cos_*(e^{2\pi} \cdot_* x), \quad 1 \leq x \leq e.$$

Here,

$$c = e,$$

$$f(x,t) = \cos_*(e^{2\pi} \cdot_* x) \cdot_* \cos_*(e^{2\pi} \cdot_* t), \quad 1 \leq x \leq e, \quad t \geq 1,$$

$$\phi(x) = e^{\frac{1}{2}} +_* e^{\frac{1}{2}} \cdot_* \cos_*(e^{2\pi} \cdot_* x),$$

$$\psi(x) = e^2 \cdot_* \cos_*(e^{2\pi} \cdot_* x), \quad 1 \leq x \leq e.$$

Then,

$$C_0 = D_0$$

$$= 0,$$

$$C_1 = 0,$$

$$C_2 = 1,$$

$$C_n = 0, \quad n \in \mathbb{N}, \quad n \geq 3,$$

$$D_n = 0, \quad n \in \mathbb{N},$$

$$a_0 = \frac{1}{2},$$

$$a_1 = 0,$$

$$a_2 = \frac{1}{2},$$

$$a_n = 0, \quad n \in \mathbb{N}, \quad n \geq 3,$$

$$b_0 = 0,$$

$$b_1 = 0,$$

$$b_2 = 2,$$

$$b_n = 0, \quad n \in \mathbb{N}, \quad n \geq 3.$$

Therefore,

$$u(x,t) = e^{\frac{1}{2}} +_* \left(e^{\frac{1}{2}} \cdot_* \cos_*(e^{2\pi} \cdot_* t) +_* ((t +_* e^4)/_* e^{4\pi}) \cdot_* \sin_*(e^{2\pi} \cdot_* t) \right)$$

$$\cdot_* \cos_*(e^{2\pi} \cdot_* x).$$

Exercise 4.6 Find a formal solution to the following problem:

$$u_{tt}^{**} -_* e^{c^2} \cdot_* u_{xx}^{**} = 1, \quad 1 < x < e^L, \quad t > 1,$$

$$u(1,t) = u(e^L, t) = 1, \quad t \geq 1,$$

$$u(x,1) = \phi(x),$$

$$u_t^*(x,1) = \psi(x), \quad 1 \leq x \leq e^L,$$

where

$$\phi(1) = \phi^{**}(1) = \psi(1) = 1,$$

$$\phi(e^L) = \phi^{**}(e^L) = \psi(e^L) = 1.$$

Answer

$$u(x,t) = \sum_{*n=1}^{\infty} \left(e^{A_n} \cdot_* \cos_*(e^{\frac{c\pi n}{L}} \cdot_* t) +_* e^{B_n} \cdot_* \sin_*(e^{\frac{c\pi n}{L}} \cdot_* t) \right) \cdot_* \sin_*(e^{\frac{n\pi}{L}} \cdot_* x),$$

$$e^{A_n} = e^{\frac{2}{L}} \cdot_* \int_{*1}^{e^L} \cdot_* \phi(x) \cdot_* \sin_*(e^{\frac{n\pi}{L}} \cdot_* x) \cdot_* d_* x,$$

$$B_n = e^{\frac{2}{c n \pi}} \cdot_* \int_{*1}^{e^L} \psi(x) \cdot_* \sin_*(e^{\frac{n\pi}{L}} \cdot_* x) \cdot_* d_* x, \quad n \in \mathbb{N}.$$

4.1.4 The energy method: uniqueness

The energy method is a fundamental tool in the theory of MPDEs for proving the uniqueness of the solutions of initial boundary value problems. Consider the problem

$$u_{tt}^{**} -_* e^{c^2} \cdot_* u_{xx}^{**} = f(x,t), \quad 1 < x < e^L, \quad t > 1, \tag{4.28}$$

$$u(1,t) = g(t), \quad u(e^L, t) = h(t), \quad t \geq 1, \tag{4.29}$$

$$u(x,1) = \phi(x), \quad u_t^*(x,1) = \psi(x), \quad 1 \leq x \leq e^L, \tag{4.30}$$

where $f \in \mathcal{C}_*([1, e^L] \times [1, \infty))$, $g, h \in \mathcal{C}_*^2([1, \infty))$, $\phi \in \mathcal{C}_*^2(\mathbb{R}_*)$, $\psi \in \mathcal{C}_*^1(\mathbb{R}_*)$ and

$$g(1) = \phi(1), \quad g^*(1) = \psi(1), \quad h^*(1) = \psi(e^L), \quad h(1) = \phi(e^L).$$

Definition 4.5 The condition (4.29) is called Dirichlet boundary condition. Let u_1 and u_2 be two solutions of the problem (4.28)–(4.30). We set

$$v(x,t) = u_1(x,t) -_* u_2(x,t).$$

Then,

$$v_{tt}^{**} -_* e^{c^2} \cdot_* v_{xx}^{**} = 1, \quad 1 < x < e^L, \quad t > 1,$$

$$v(1,t) = v(e^L, t) = 1, \quad t \geq 1,$$

$$v(x,1) = v_t(x,1) = 1, \quad 1 \leq x \leq e^L.$$

Define the total energy

$$E(t) = e^{\frac{1}{2}} \cdot_* \int_{*1}^{e^L} \left((v_t^*(x,t))^{2*} +_* e^{c^2} (v_x^*(x,t))^{2*} \right) \cdot_* d_* x.$$

Here,

$$e^{\frac{1}{2}} \cdot_* \int_{*1}^{e^L} (v_t^*(x,t))^{2*} d_* x$$

is the multiplicative total kinetic energy, while

$$e^{\frac{c^2}{2}} \cdot_* \int_{*1}^{e^L} (v_x^*(x,t))^{2*} d_* x$$

is the multiplicative total potential energy. We have

$$E^*(t) = \int_1^{e^L} \left(v_t^*(x,t) \cdot_* v_{tt}^{**}(x,t) +_* e^{c^2} \cdot_* v_x^*(x,t) \cdot_* v_{xt}^{**}(x,t) \right) \cdot_* d_* x. \tag{4.31}$$

Note that

$$e^{c^2} \cdot_* v_x^*(x,t) \cdot_* v_{xt}^{**}(x,t)$$

$$= \quad e^{c^2} \cdot_* ((v_x^*(x,t) \cdot_* v_t^*(x,t))_x -_* v_{xx}^{**}(x,t) \cdot_* v_t^*(x,t))$$

$$= \quad e^{c^2} \cdot_* (v_x^*(x,t) \cdot_* v_t^*(x,t))_x -_* e^{c^2} \cdot_* v_{xx}^{**}(x,t) \cdot_* v_t^*(x,t)$$

$$= \quad e^{c^2} \cdot_* (v_x^*(x,t) \cdot_* v_t^*(x,t))_x -_* v_{tt}^{**}(x,t) \cdot_* v_t^*(x,t).$$

Hence and by (4.31), we get

$$E^*(t) \quad = \quad e^{c^2} \cdot_* \int_{*1}^{e^L} (v_x^*(x,t) \cdot_* v_t(x,t))_x \cdot_* d_* x$$

$$= \quad e^{c^2} \cdot_* v_x^*(x,t) \cdot_* v_t^*(x,t) \Big|_{x=1}^{x=e^L}$$

$$= \quad 1,$$

where we have used that

$$v_t^*(1,t) = v_t^*(e^L,t) = 1.$$

Therefore,
$$E(t) = \text{const}.$$

By the initial conditions, we have

$$v_t^*(x,1) = v_x^*(x,1) = 1,$$

$1 \le x \le e^L$. Therefore, $E(1) = 1$ and $E(t) \equiv 1$ on $[1,\infty)$. Consequently,

$$(v_t^*(x,t))^{2*} +_* e^{c^2} \cdot_* (v_x^*(x,t))^{2*} = 1$$

on $[1,e^L] \times [1,\infty)$. Hence,

$$v_t^*(x,t) = v_x^*(x,t) = 1$$

on $[1,e^L] \times [1,\infty)$. From here,

$$v(x,t) = \text{const}$$

on $[1,e^L] \times [1,\infty)$. Using
$$v(x,1) = 1$$
on $[1,e^L]$, we conclude that $v(x,t) = 1$ on $[1,L] \times [1,\infty)$, from where

$$u_1(x,t) = u_2(x,t) \quad \text{on} \quad [1,e^L] \times [1,\infty).$$

Exercise 4.7 Prove that the problem

$$u_{tt}^{**} -_* e^{c^2} \cdot_* u_{xx}^{**} = f(x,t), \quad 1 < x < e^L, \quad t > 1,$$

$$u_x^{**}(1,t) = g(t), \quad u_x^*(e^l, t) = h(l), \quad l \geq 1,$$

$$u(x,1) = \phi(x), \quad u_t^*(x,1) = \psi(x), \quad 1 \leq x \leq e^L,$$

where $f \in \mathcal{C}_*([1, e^L] \times [1, \infty))$, $h, g \in \mathcal{C}_*^1([1, \infty))$, $\phi \in \mathcal{C}_*^2(\mathbb{R}_*)$, $\psi \in \mathcal{C}_*^1(\mathbb{R}_*)$ and

$$h(1) = \phi^*(L),$$

$$g(1) = \phi^*(1),$$

$$h^*(1) = \psi^*(e^L),$$

$$g^*(1) = \psi^*(1),$$

has unique solution.

4.2   The Multiplicative Wave Equation in $\mathbb{R}^3_*$

4.2.1   Radially symmetric solutions

We seek solutions $u(x_1, x_2, x_3, t)$ to the multiplicative wave equation

$$u_{tt}^* -_* c^2 \Delta u = 0_*, \quad (x_1, x_2, x_3) \in \mathbb{R}^3, \quad -_*\infty < t < \infty, \tag{4.32}$$

$\Delta_* u = u_{x_1 x_1}^{**} +_* u_{x_2 x_2}^{**} +_* u_{x_3 x_3}^{**}$, that are of the form $u = u(r,t)$, $r = (x_1^{2*} +_* x_2^{2*} +_* x_3^{2*})^{\frac{1}{2}}_*$. Thus

$$u_{x_i}^* = u_r^* \cdot_* (x_i/_* r),$$

$$u_{x_i x_i}^{**} = u_{rr}^* \cdot_* ((x_i^{2*})/_*(r^{2*})) +_* u_r^* \cdot_* ((r^{2*} -_* x_i^{2*})/_*(r^{3*})), \quad i = 1, 2, 3,$$

and

$$u_{x_1 x_1}^{**} +_* u_{x_2 x_2}^{**} +_* u_{x_3 x_3}^{**} = u_{rr}^{**} +_* (e^2/_* r) \cdot_* u_r^*,$$

i.e., the function $u(r,t)$ satisfies the equation

$$u_{tt}^{**} -_* e^{c^2} \cdot_* (u_{rr}^{**} +_* (e^2/_* r) \cdot_* u_r^*) = 0_*. \tag{4.33}$$

Defining

$$v(r,t) = r \cdot_* u(r,t),$$

we see that v satisfies the equation

$$v_{tt}^{**} -_* e^{c^2} \cdot_* v_{rr}^{**} = 0_*.$$

This is exactly the one-dimensional multiplicative wave equation. Therefore, the general radial solution for equation (4.33) is given by

$$u(r,t) = (e/_*r) \cdot_* \left(f(r +_* e^c \cdot_* t) +_* g(r -_* e^c \cdot_* t) \right),$$

where $f, g \in \mathcal{C}^2_*(\mathbb{R}_*)$. We can solve the Cauchy problem that consists of (4.33) for $t > 1$ and the initial conditions

$$u(r,1) = \phi(r),$$

$$u_t^*(r,1) = \psi(r), \quad 1 \le r \le \infty. \tag{4.34}$$

Note that the initial conditions are only given along the ray $r \ge 1$ and not for all r. If we suppose that $\phi, \psi \in \mathcal{C}^1_*([1,\infty))$ and

$$\phi^*(1) = \psi^*(1) = 1,$$

then we can extend ϕ and ψ to the whole line $0 < r < \infty$ by defining them to be multiplicative even extensions of the given ϕ and ψ. Hence, the initial conditions for v are multiplicative odd functions, and therefore $v(r,t)$ is multiplicative odd, which implies that $u(r,t)$ is a multiplicative even function. If we denote the multiplicative even extensions of ϕ and ψ by $\tilde{\phi}$ and $\tilde{\psi}$, respectively, we thus obtain the following radially symmetric solution for the three-dimensional radial multiplicative wave equation:

$$u(r,t) = (e/_*(e^2 \cdot_* r)) \cdot_* \Big((r +_* e^c \cdot_* t)\tilde{\phi}(r +_* e^c \cdot_* t) +_* (r -_* e^c \cdot_* t)$$

$$\tilde{\phi}(r -_* e^c \cdot_* t)\Big)$$

$$+_* (e/_*(e^{2c} \cdot_*)) \cdot_* r \cdot_* \int\limits_{*r -_* e^c t}^{r +_* e^c t} s \cdot_* \tilde{\psi}(s) \cdot_* d_*s.$$

4.2.2   The Cauchy problem

Let $(x,t) = (x_1, x_2, x_3, t) \in \mathbb{R}^3_* \times [1, \infty)$ be arbitrarily chosen. With S we will denote the sphere

$$|y -_* x|^{2*}_* = t^{2*},$$

where $y = (y_1, y_2, y_3)$, $|x -_* y|_*$ is the distance between the points x and y. Let also μ be an arbitrary real-valued twice continuously multiplicative differentiable function defined on S. We will prove that the function

$$u(x,t) = (e/_*t) \cdot_* \int_{*S} \mu(y_1, y_2, y_3) \cdot_* d_* s_y \qquad (4.35)$$

is a solution to equation (4.32) in the case when $c = 1$. If $c \neq 1$, then we make the change

$$t = (e/_* e^c) \cdot_* \tau.$$

Really, the change of variables

$$y_i -_* x_i = t \cdot_* \xi_i, \quad i = 1, 2, 3,$$

brings the expression (4.35) in the form

$$u(x,t) = t \cdot_* \int_{*\sigma} \mu(x_1 +_* t \cdot_* \xi_1, x_2 +_* t \cdot_* \xi_2, x_3 +_* t \cdot_* \xi_3) \cdot_* d_* \sigma_\xi, \qquad (4.36)$$

where σ is the unit sphere $|\xi|_* = e$ and

$$d_* \sigma_\xi = (d_* s_y)/_* (t^{2*})$$

is an element of the unit sphere. By (4.36), we get

$$\Delta_* u(x,t) = t \cdot_* \int_{*\sigma} \sum_{*i=1}^{3} \mu_{y_i y_i}^{**}(x_1 +_* t \cdot_* \xi_1, x_2 +_* t \cdot_* \xi_2, x_3 +_* t \cdot_* \xi_3) \cdot_* d_* \sigma_\xi$$

and

$$u_t^*(x,t)$$
$$= \int_{*\sigma} \mu(x_1 +_* t \cdot_* \xi_1, x_2 +_* t \cdot_* \xi_2, x_3 +_* t \cdot_* \xi_3) \cdot_* d_* \sigma_\xi$$

$$+_* t \cdot_* \int_{*\sigma} \sum_{*i=1}^{3} \xi_i \cdot_* \mu_{y_i}^*(x_1 +_* t \cdot_* \xi_1, x_2 +_* t \cdot_* \xi_2, x_3 +_* t \cdot_* \xi_3) \cdot_* d_* \sigma_\xi$$

$$= (u(x,t)/_*t) +_* t \cdot_* \int_{*\sigma} \sum_{*i=1}^{3} \xi_i \cdot_* \mu_{y_i}^*(x_1 +_* t \cdot_* \xi_1, x_2 +_* t \cdot_* \xi_2, x_3 +_* t \cdot_* \xi_3)$$

$$\cdot_* d_* \sigma_\xi.$$

We set

$$I = \int_{*S} \sum_{*i=1}^{3} \mu_{y_i}^{*}(y) \cdot_* \nu_i(y) \cdot_* d_* s_y,$$

where

$$\nu(y) = (\nu_1(y), \nu_2(y), \nu_3(y))$$

is the outer multiplicative normal vector to S at the point y. Then,

$$t \cdot_* \int_{*\sigma} \sum_{*i=1}^{3} \xi_i \cdot_* \mu_{y_i}^{*}(x +_* t \cdot_* \xi) \cdot_* d_* \sigma_\xi = (e/_* t^{2*}) \cdot_* I$$

and

$$u_t^{*}(x,t) = u(x,t)/_* t +_* (e/_* t) \cdot_* I.$$

We multiplicative differentiate the last equation with respect to t and we find

$$u_{tt}^{**}(x,t) = u_t^{*}(x,t)/_* t -_* ((u(x,t))/_*(t^{2*})) -_* (e/_*(t^{2*})) \cdot_* I +_* (e/_* t) \cdot_* I_t^{*}$$

$$= (e/_* t) \cdot_* (u(x,t)/_* t +_* (e/_* t) \cdot_* I)$$

$$-_* (e/_*(t^{2*})) \cdot_* u(x,t) -_* (e/_*(t^{2*})) \cdot_* I +_* (e/_* t) \cdot_* I_t^{*}$$

$$= (e/_* t) \cdot_* I_t^{*}.$$

(4.37)

By the Gauss–Ostrogradsky formula, we have that

$$I = \int_{*1}^{t} \int_{*1}^{e^{\pi}} \int_{*1}^{e^{2\pi}} \Delta_* \mu \cdot_* \rho^{2*} \cdot_* \sin_* \theta \cdot_* d_* \phi \cdot_* d_* \theta \cdot_* d_* \rho.$$

Hence,

$$I_t^{*} = t^2 \int_{*1}^{e^{\pi}} \int_{*1}^{e^{2\pi}} \Delta_* \mu \cdot_* \sin_* \theta \cdot_* d_* \phi \cdot_* d_* \theta$$

$$= t^{2*} \cdot_* \int_{*\sigma} \Delta_* \mu \cdot_* d_* \sigma_\xi$$

$$= t \cdot_* \Delta_* u.$$

We substitute the last formula in (4.37) and we obtain that u satisfies equation (4.32).

Example 4.7 Consider the function

$$u(x,t) = (e/_* t) \cdot_* \int_{*S} (y_1 +_* y_2 +_* y_3) \cdot_* d_* s_y.$$

Then,

$$u(x,t) = (e/_*t) \cdot_* \int_{*1}^{e^{2\pi}} \int_{*1}^{e^{\pi}} \Big(x_1 +_* x_2 +_* x_3 +_* t \cdot_* \cos_* \phi \cdot_* \sin_* \theta$$

$$+_* t \cdot_* \sin_* \phi \cdot_* \sin_* \theta +_* t \cdot_* \cos_* \theta \Big) \cdot_* t^{2*} \cdot_* \sin_* \theta \cdot_* d_*\theta \cdot_* d_*\phi$$

$$= e^{4\pi} \cdot_* t \cdot_* (x_1 +_* x_2 +_* x_3)$$

and

$$u_{tt}^{**} -_* u_{x_1 x_1}^{**} -_* u_{x_2 x_2}^{**} -_* u_{x_3 x_3}^{**} = 0_*.$$

Exercise 4.8 Check that the function

$$u(x,t) = (e/_*t) \cdot_* \int_{*S} (y_1 -_* e^2 \cdot_* y_2 +_* e^3 \cdot_* y_3) d_*s_y$$

satisfies equation (4.32).

We set

$$M(\mu) = (e/_*t^{2*}) \cdot_* \int_{*S} \mu(y_1, y_2, y_3) \cdot_* d_*s_y.$$

Then,

$$u(x,t) = t \cdot_* M(\mu)$$

is a solution to equation (4.32) for $c = 1$. Now we consider the following Cauchy problem:

$$u_{tt}^{**} -_* \Delta_* u = 1, \quad (x_1, x_2, x_3) \in \mathbb{R}^3_*, \quad t > 1, \tag{4.38}$$

$$u(x_1, x_2, x_3, 1) = \phi(x_1, x_2, x_3),$$

$$u_t^*(x_1, x_2, x_3, 1) = \psi(x_1, x_2, x_3), \quad (x_1, x_2, x_3) \in \mathbb{R}^3_*, \tag{4.39}$$

where $\phi \in C^3_*(\mathbb{R}^3_*)$, $\psi \in C^2_*(\mathbb{R}^3_*)$. We will prove that

$$u(x,t) = e^{\frac{1}{4\pi}} \cdot_* t \cdot_* M(\psi) +_* e^{\frac{1}{4\pi}} \cdot_* (\partial_* /_* (\partial_* t)) (t \cdot_* M(\phi)) \tag{4.40}$$

is its solution.

Definition 4.6 The equality (4.40) is called the multiplicative Kirchhoff formula, shortly the Kirchoff formula.

We have that the function (4.40) satisfies equation (4.38). We will prove that it satisfies the initial condition (4.39). Indeed,

$$u(x_1, x_2, x_3, 0) = e^{\frac{1}{4\pi}} \cdot_* M(\phi)\Big|_{t=1}$$

$$= e^{\frac{1}{4\pi}} \cdot_* \int_{|\xi|=e} \phi(x_1, x_2, x_3) \cdot_* d_*\sigma_\xi$$

$$= \phi(x_1, x_2, x_3),$$

$$u_t^*(x_1, x_2, x_3, t) \;=\; (\partial_*/_*\partial_* t) \cdot_* \left(e^{\frac{1}{4\pi}} \cdot_* t \cdot_* M(\psi) \right.$$

$$\left. +_* e^{\frac{1}{4\pi}} \cdot_* (\partial_*/_*(\partial_* t)) \cdot_* (t \cdot_* M(\phi)) \right)$$

$$=\; e^{\frac{1}{4\pi}} \cdot_* M(\psi) +_* e^{\frac{1}{4\pi}} \cdot_* t \cdot_* (\partial_*/_*(\partial_* t)) \cdot_* M(\psi)$$

$$+_* e^{\frac{1}{4\pi}} \cdot_* (\partial_*^2/_*(\partial_* t^{2*})) \cdot_* (t \cdot_* M(\phi))$$

$$=\; e^{\frac{1}{4\pi}} \cdot_* M(\psi) +_* e^{\frac{1}{4\pi}} \cdot_* t \cdot_* (\partial_*/_*(\partial_* t)) \cdot_* M(\psi)$$

$$+_* e^{\frac{1}{4\pi}} \cdot_* t \cdot_* \Delta_* M(\phi),$$

$$u_t^*(x_1, x_2, x_3, 1) \;=\; \left. e^{\frac{1}{4\pi}} \cdot_* M(\psi) \right|_{t=1}$$

$$=\; e^{\frac{1}{4\pi}} \cdot_* \int_{*|\xi|=e} \psi(x_1, x_2, x_3) \cdot_* d_* \sigma_\xi$$

$$=\; \psi(x_1, x_2, x_3).$$

Example 4.8 Consider the following Cauchy problem:

$$u_{tt}^{**} -_* u_{x_1 x_1}^{**} -_* u_{x_2 x_2}^{**} -_* u_{x_3 x_3}^{**} = 1, \quad (x_1, x_2, x_3) \in \mathbb{R}_*^3, \quad t > 1,$$

$$u(x_1, x_2, x_3, 1) = x_1,$$

$$u_t^*(x_1, x_2, x_3, 1) = x_3, \quad (x_1, x_2, x_3) \in \mathbb{R}_*^3.$$

Here,

$$\phi(x_1, x_2, x_3) \;=\; x_1,$$

$$\psi(x_1, x_2, x_3) \;=\; x_3.$$

Hence,

$$\int_{*|x-_*y|_*=t} \phi(y_1, y_2, y_3) ds_y$$

$$=\; \int_{*|x-_*y|_*=t} y_1 \cdot_* d_* s_y$$

$$= \int_1^{e^{2\pi}} \cdot_* \int_{*1}^{e^\pi} (x_1 +_* t \cdot_* \cos_* \phi \cdot_* \sin_* \theta) \cdot_* t^{2*} \cdot_* \sin_* \theta \cdot_* d_* \theta \cdot_* d_* \phi$$

$$= e^{4\pi} \cdot_* t^{2*} \cdot_* x_1,$$

$$\int_{*|x-_*y|_*=t} \psi(y_1, y_2, y_3) \cdot_* d_* s_y$$

$$= \int_{*|x-_*y|_*=t} y_3 \cdot_* d_* s_y$$

$$= \int_{*1}^{e^{2\pi}} \int_{*1}^{e^\pi} (x_3 +_* t \cdot_* \cos_* \theta) \cdot_* t^{2*} \cdot_* \sin_* \theta \cdot_* d_* \theta \cdot_* d_* \phi$$

$$= e^{4\pi} \cdot_* t^{2*} \cdot_* x_3.$$

Then, using the Kirchhoff formula, we get

$$u(x_1, x_2, x_3, t) = e^{\frac{1}{4\pi}} \cdot_* (e/_*t) \left(e^{4\pi} \cdot_* t^{2*} \cdot_* x_3 \right) +_* e^{\frac{1}{4\pi}} \cdot_* (\partial_*/_*(\partial_* t)) \cdot_*$$

$$\left((e/_*t) \cdot_* e^{4\pi} \cdot_* t^{2*} \cdot_* x_1 \right)$$

$$= t \cdot_* x_3 +_* x_1.$$

Exercise 4.9 Find a solution to the Cauchy problem:

$$u_{tt}^{**} -_* u_{x_1 x_1}^{**} -_* u_{x_2 x_2}^{**} -_* u_{x_3 x_3}^{**} = 1, \quad (x_1, x_2, x_3) \in \mathbb{R}_*^3, \quad t > 1,$$

$$u(x_1, x_2, x_3, 1) = e^{\frac{1}{2}} \cdot_* x_1^{2*} +_* e^{\frac{1}{2}} \cdot_* x_2^{2*} +_* x_3^{2*},$$

$$u_t^*(x_1, x_2, x_3, 1) = 1, \quad (x_1, x_2, x_3) \in \mathbb{R}_*^3.$$

Answer

$$u(x_1, x_2, x_3, t) = e^{\frac{1}{2}} \cdot_* x_1^{2*} +_* e^{\frac{1}{2}} \cdot_* x_2^{2*} +_* x_3^{2*} +_* e^2 \cdot_* t^{2*}.$$

Next, we consider the following Cauchy problem:

$$u_{tt}^{**} -_* \Delta_* u = f(x_1, x_2, x_3, t), \quad (x_1, x_2, x_3) \in \mathbb{R}_*^3, \quad t > 1,$$

$$u(x_1, x_2, x_3, 1) = \phi(x_1, x_2, x_3), \tag{4.41}$$

$$u_t^*(x_1, x_2, x_3, 1) = \psi(x_1, x_2, x_3), \quad (x_1, x_2, x_3) \in \mathbb{R}_*^3,$$

where $\phi \in \mathcal{C}_*^3(\mathbb{R}_*^3)$, $\psi \in \mathcal{C}_*^2(\mathbb{R}_*^3)$, $f \in \mathcal{C}_*^2(\mathbb{R}_*^3 \times [1, \infty))$.

We set

$$v(x_1, x_2, x_3, t) = u(x_1, x_2, x_3, t) -_* \phi(x_1, x_2, x_3) -_* t \cdot_* \psi(x_1, x_2, x_3).$$

Then,

$$v(x_1, x_2, x_3, 1) = u(x_1, x_2, x_3, 0) -_* \phi(x_1, x_2, x_3)$$

$$= 1,$$

$$v_t^*(x_1, x_2, x_3, t) = u_t^*(x_1, x_2, x_3, t) -_* \psi(x_1, x_2, x_3),$$

$$v_t^*(x_1, x_2, x_3, 1) = u_t^*(x_1, x_2, x_3, 1) -_* \psi(x_1, x_2, x_3)$$

$$= 1,$$

$$v_{tt}^{**}(x_1, x_2, x_3, t) = u_{tt}^{**}(x_1, x_2, x_3, t),$$

$$v_{x_i x_i}^{**}(x_1, x_2, x_3, t) = u_{x_i x_i}^{**}(x_1, x_2, x_3, t) -_* \phi_{x_i x_i}^{**}(x_1, x_2, x_3)$$

$$-_* t \cdot_* \psi_{x_i x_i}^{**}(x_1, x_2, x_3), \quad i = 1, 2, 3,$$

$$v_{tt}^{**} -_* \Delta_* v = u_{tt}^* -_* \Delta_* u +_* \Delta_* \phi +_* t \cdot_* \Delta_* \psi$$

$$= f +_* \Delta_* \phi +_* t \cdot_* \Delta_* \psi.$$

Therefore, $v(x_1, x_2, x_3, t)$ satisfies the Cauchy problem:

$$v_{tt}^{**} -_* \Delta_* v = g(x_1, x_2, x_3, t), \quad (x_1, x_2, x_3) \in \mathbb{R}_*^3, \quad t > 1,$$

$$v(x_1, x_2, x_3, 1) = v_t(x_1, x_2, x_3, 1) \qquad\qquad (4.42)$$

$$= 1, \quad (x_1, x_2, x_3) \in \mathbb{R}_*^3,$$

where

$$g(x_1, x_2, x_3, t) = f(x_1, x_2, x_3, t) +_* \Delta_* \phi(x_1, x_2, x_3) +_* t \cdot_* \Delta_* \psi(x_1, x_2, x_3).$$

Let $\tau > 1$ be arbitrarily chosen. Consider the following Cauchy problem:

$$w_{tt}^{**} -_* \Delta_* w = 1, \quad (x_1, x_2, x_3) \in \mathbb{R}_*^3, \quad t > \tau,$$

$$w(x_1, x_2, x_3, \tau) = 1, \qquad\qquad (4.43)$$

$$w_t^*(x_1, x_2, x_3, \tau) = g(x_1, x_2, x_3, \tau).$$

Let $z(x_1, x_2, x_3, t, \tau)$ be the solution of the problem (4.43) which is supposed to be continued as identically multiplicative zero for $t \leq \tau$.

We set

$$h(x_1, x_2, x_3, t) = \int_{*1}^{t} z(x_1, x_2, x_3, t, \tau) \cdot_* d_* \tau.$$

Then,

$$h(x_1, x_2, x_3, 1) = 1,$$

$$h_t(x_1, x_2, x_3, t) = z(x_1, x_2, x_3, t, t) +_* \int_{*1}^{t} z_t(x_1, x_2, x_3, t, \tau) \cdot_* d_* \tau$$

$$= \int_{*1}^{t} z_t^*(x_1, x_2, x_3, t, \tau) \cdot_* d_* \tau,$$

$$h_t^*(x_1, x_2, x_3, 1) = 1,$$

$$h_{tt}^{**}(x_1, x_2, x_3, t) = z_t^*(x_1, x_2, x_3, t, t) +_* \int_{*1}^{t} z_{tt}^{**}(x_1, x_2, x_3, t, \tau) \cdot_* d_* \tau$$

$$= g(x_1, x_2, x_3, t) +_* \int_{*1}^{t} z_{tt}^{**}(x_1, x_2, x_3, t, \tau) \cdot_* d_* \tau$$

$$= g(x_1, x_2, x_3, t) +_* \int_{0}^{t} \Delta z(x_1, x_2, x_3, t, \tau) \cdot_* d_* \tau$$

$$= g(x_1, x_2, x_3, t) +_* \Delta_* h(x_1, x_2, x_3, t),$$

i.e., $h(x_1, x_2, x_3, t)$ is a solution to the Cauchy problem (4.42). Now we apply the Kirchhoff formula and we obtain

$$z(x_1, x_2, x_3, t, \tau) = e^{\frac{1}{4\pi}} \cdot_* (t -_* \tau) \cdot_* M_{t-_*\tau}^*(g(x_1, x_2, x_3, \tau)),$$

whereupon

$$v(x_1, x_2, x_3, t) = e^{\frac{1}{4\pi}} \cdot_* \int_{*1}^{t} (t -_* \tau) M_{t-_*\tau}^*(g(x_1, x_2, x_3, \tau)) \cdot_* d_* \tau$$

and

$$u(x_1, x_2, x_3, t) = \phi(x_1, x_2, x_3) +_* t \cdot_* \psi(x_1, x_2, x_3)$$

$$+_* e^{\frac{1}{4\pi}} \cdot_* \int_{*1}^{t} (t -_* \tau) \cdot_* M_{t-_*\tau}^* \left(f(x_1, x_2, x_3, \tau) \right.$$

(4.44)

$$\left. +_* \Delta_* \phi(x_1, x_2, x_3) +_* \tau \cdot_* \Delta_* \psi(x_1, x_2, x_3) \right) \cdot_* d_*\tau$$

is a solution to the Cauchy problem (4.41).

Example 4.9 Consider the following Cauchy problem:

$$u_{tt}^{**} -_* \Delta_* u = x_1, \quad (x_1, x_2, x_3) \in \mathbb{R}_*^3, \quad t > 1,$$

$$u(x_1, x_2, x_3, 1) = x_2,$$

$$u_t^*(x_1, x_2, x_3, 1) = x_3, \quad (x_1, x_2, x_3) \in \mathbb{R}_*^3.$$

Here,

$$f(x_1, x_2, x_3, t) = x_1,$$

$$\phi(x_1, x_2, x_3) = x_2,$$

$$\psi(x_1, x_2, x_3) = x_3.$$

Then, using (4.44),

$$u(x_1, x_2, x_3, t) = x_2 +_* t \cdot_* x_3 +_* e^{\frac{1}{4\pi}} \cdot_* \int_{*1}^{t} (e /_* (t -_* \tau)) \cdot_*$$

$$\int_{*|x-_*y|=t-_*\tau} y_1 \cdot_* d_* s_y \cdot_* d_*\tau$$

$$= x_2 +_* t \cdot_* x_3$$

$$+_* e^{\frac{1}{4\pi}} \cdot_* \int_{*1}^{t} (e /_* (t -_* \tau)) \cdot_*$$

$$\int_{*1}^{e^{2\pi}} \int_{*1}^{e^{\pi}} (x_1 +_* (t -_* \tau) \cdot_* \cos_* \phi \cdot_* \sin_* \theta)$$

$$\cdot_* (t -_* \tau)^{2*} \cdot_* \sin_* \theta \cdot_* d_*\theta \cdot_* d_*\phi \cdot_* d_*\tau$$

$$= x_2 +_* t \cdot_* x_3 +_* e^{\frac{1}{2}} \cdot_* x_1 \cdot_* t^{2*}.$$

Exercise 4.10 Solve the Cauchy problem:

$$u_{tt}^{**} -_* \Delta_* u = -_* e^4 \cdot_* x_1 -_* e^6 \cdot_* x_2, \quad (x_1, x_2, x_3) \in \mathbb{R}^3_*, \quad t > 1,$$

$$u(x_1, x_2, x_3, 1) = x_1^{3*} +_* x_2^{3*},$$

$$u_t^*(x_1, x_2, x_3, 1) = 1, \quad (x_1, x_2, x_3) \in \mathbb{R}^3_*.$$

Answer

$$u(x_1, x_2, x_3, t) = x_1^{3*} +_* x_2^{3*} +_* t^{2*} \cdot_* x_1.$$

4.3 The Two-Dimensional Wave Equation

We consider the Cauchy problem for the wave equation with two spatial variables:

$$u_{tt}^* -_* u_{x_1 x_1}^* -_* u_{x_2 x_2}^* = 0_*, \quad (x_1, x_2) \in \mathbb{R}^2, \quad t > 0_*, \tag{4.45}$$

$$u(x_1, x_2, 1) = \phi(x_1, x_2),$$

$$u_t^*(x_1, x_2, 1) = \psi(x_1, x_2), \quad (x_1, x_2) \in \mathbb{R}^2_*, \tag{4.46}$$

where $\phi \in C^3_*(\mathbb{R}^2_*)$, $\psi \in C^2_*(\mathbb{R}^2_*)$. The solution $u(x_1, x_2, t)$ of the problem (4.45), (4.46) can be derived using Kirchhoff's formula. Then,

$$u(x_1, x_2, t) = (e/_*(e^{4\pi} \cdot_* t)) \cdot_* \int_{|y|_* = t} \psi(x_1 +_* y_1, x_2 +_* y_2) \cdot_* d_* s_y$$

$$+_* e^{\frac{1}{4\pi}} \cdot_* (\partial_* /_* \partial_* t) \left((e/_* t) \cdot_* \int_{|y|_* = t} \phi(x_1 +_* y_1, x_2 +_* y_2) \cdot_* d_* s_y \right), \tag{4.47}$$

which is independent of x_3 and satisfies (4.45), (4.46).

Note that the projection $d_* y_1 \cdot_* d_* y_2$ of the element of the arc $d_* s_y$ of the sphere $|y|_* = t$ on the circle $y_1^{2*} +_* y_2^{2*} \le t^{2*}$ is expressed by the following formula

$$d_* y_1 \cdot_* d_* y_2 = (|y_3|_* /_* t) \cdot_* d_* s_y.$$

To compute the integrals on the right-hand side of the formula (4.47), we should project on the multiplicative circle

$$y_1^{2*} +_* y_2^{2*} \le t^{2*}$$

both the upper hemisphere $y_3 > 1$ and the lower hemisphere $y_3 < 1$ of the sphere $|y|_* = t$. Therefore,

$$u(x_1, x_2, t) = e^{\frac{1}{2\pi}} \cdot_* \int_{*B} (\psi(y_1, y_2)/_* (t^{2*} -_* (y_1 -_* x_1)^{2*} -_* (y_2 -_* x_2)^{2*})^{\frac{1}{2}}_*)$$

$$\cdot_* d_* y_1 \cdot_* d_* y_2$$

$$+_* e^{\frac{1}{2\pi}} \cdot_* (\partial_* /_* \partial_* t)$$

$$\cdot_* \int_{*B} ((\phi(y_1, y_2))/_* (t^{2*} -_* (y_1 -_* x_1)^{2*} -_* (y_2 -_* x_2)^{2*})^{\frac{1}{2}}_*)$$

$$\cdot_* d_* y_1 \cdot_* d_* y_2,$$

(4.48)

where B is the multiplicative circle

$$(y_1 -_* x_1)^{2*} +_* (y_2 -_* x_2)^{2*} \le t^{2*}.$$

Definition 4.7 The equality (4.48) is called Poisson's formula.

Example 4.10 Consider the following Cauchy problem:

$$u_{tt}^{**} -_* u_{x_1 x_1}^{**} -_* u_{x_2 x_2}^{**} = 0_*, \quad (x_1, x_2) \in \mathbb{R}_*^2, \quad t > 0_*,$$

$$u(x_1, x_2, 1) = x_1,$$

$$u_t^*(x_1, x_2, 1) = x_2, \quad (x_1, x_2) \in \mathbb{R}_*^2.$$

Then, using Poisson's formula, we have

$$u(x_1, x_2, t) = e^{\frac{1}{2\pi}} \cdot_* \int_{*B} (y_2/_* (t^{2*} -_* (y_1 -_* x_1)^{2*} -_* (y_2 -_* x_2)^{2*}))$$

$$\cdot_* d_* y_1 \cdot_* d_* y_2$$

$$+_* e^{\frac{1}{2\pi}} (\partial_* /_* \partial_* t) \cdot_* \int_{*B} (y_1/_* (t^{2*} -_* (y_1 -_* x_1)^{2*}$$

$$-_* (y_2 -_* x_2)^{2*})^{\frac{1}{2}}_*) \cdot_* d_* y_1 \cdot_* d_* y_2,$$

$$B : (y_1 -_* x_1)^{2*} +_* (y_2 -_* x_2)^{2*} \le t^{2*}.$$

Let

$$y_1 = x_1 +_* r \cdot_* \cos_* \phi,$$

$$y_2 = x_2 +_* r \cdot_* \sin_* \phi, \quad r \in [1, t], \quad \phi \in [1, e^{2\pi}].$$

Hence,

$$u(x_1, x_2, t) = e^{\frac{1}{2\pi}} \cdot_* \int\limits_{*1}^{t} \int\limits_{*1}^{e^{2\pi}} ((x_2 +_* r \cdot_* \sin_* \phi)/_* (t^{2*} -_* r^{2*})^{\frac{1}{2}}_*)$$

$$\cdot_* r \cdot_* d_* \phi \cdot_* d_* r$$

$$+_* e^{\frac{1}{2\pi}} \cdot_* (\partial_* /_* \partial_* t) \cdot_* \int\limits_{*1}^{t} \int\limits_{*1}^{e^{2\pi}} ((x_1 +_* r \cdot_* \cos_* \phi)/_* (t^{2*} -_* r^{2*})^{\frac{1}{2}}_*$$

$$\cdot_* r \cdot_* d_* \phi \cdot_* d_* r$$

$$= x_2 \cdot_* t +_* x_1.$$

Exercise 4.11 Solve the Cauchy problem:

$$u_{tt}^{**} -_* u_{x_1 x_1}^{**} -_* u_{x_2 x_2}^{***} = 1, \quad (x_1, x_2) \in \mathbb{R}_*^2, \quad t > 1,$$

$$u(x_1, x_2, 1) = x_1 -_* x_2,$$

$$u_t^*(x_1, x_2, 1) = x_2, \quad (x_1, x_2) \in \mathbb{R}_*^2.$$

Answer

$$u(x_1, x_2, t) = x_1 +_* x_2 \cdot_* (t -_* e).$$

Next, we consider the following Cauchy problem:

$$u_{tt}^{**} -_* u_{x_1 x_1}^{**} -_* u_{x_2 x_2}^{**} = f(x_1, x_2, t), \quad (x_1, x_2) \in \mathbb{R}_*^2, \quad t > 1,$$

$$u(x_1, x_2, 1) = \phi(x_1, x_2),$$

$$u_t^*(x_1, x_2, 1) = \psi(x_1, x_2), \quad (x_1, x_2) \in \mathbb{R}_*^2,$$

where $\phi \in \mathcal{C}_*^3(\mathbb{R}_*^2)$, $\psi \in \mathcal{C}_*^2(\mathbb{R}_*^2)$ and $f \in \mathcal{C}_*^2(\mathbb{R}_*^2 \times [1, \infty))$. Using the formula (4.44), as in above, for its solution $u(x_1, x_2, t)$, we have the following representation:

$$u(x_1, x_2, t) = \phi(x_1, x_2) +_* t \cdot_* \psi(x_1, x_2)$$

$$+_* e^{\frac{1}{2\pi}} \cdot_* \int\limits_{*1}^{t} \int\limits_{B_{t-*\tau}} (f(y_1, y_2, \tau) +_* \triangle_* \phi(y_1, y_2) +_* \tau \cdot_* \triangle_* (\psi(y_1, y_2)))$$

$$/_* ((t -_* \tau)^{2*} -_* (y_1 -_* x_1)^{2*} -_* (y_2 -_* x_2)^{2*})^{\frac{1}{2}}_* \cdot_* d_* y_1 \cdot_* d_* y_2 \cdot_* d\tau,$$

where $B_{t-*\tau} : (y_1 -_* x_1)^{2*} +_* (y_2 -_* x_2)^{2*} \le (t -_* \tau)^{2*}.$

Exercise 4.12 Solve the Cauchy problem:

$$u_{tt}^{**} -_* u_{x_1 x_1}^{**} -_* u_{x_2 x_2}^{**} = x_2^{2*}, \quad (x_1, x_2) \in \mathbb{R}_*^2, \quad t > 1,$$

$$u(x_1, x_2, 1) = e,$$

$$u_t^*(x_1, x_2, 1) = x_1, \quad (x_1, x_2) \in \mathbb{R}_*^2.$$

Answer

$$u(x_1, x_2, t) = e +_* t \cdot_* x_1 +_* e^{\frac{1}{2}} \cdot_* x_2^{2*} \cdot_* t^{2*} +_* e^{\frac{1}{12}} \cdot_* t^{4*}.$$

4.4   The $(2n+1)$-Dimensional Wave Equation

We will start this section with the following useful lemma:

Lemma 4.1 *Let $\phi : \mathbb{R}_* \to \mathbb{R}$ be C_*^{k+1}-function. Then for $k \in \mathbb{N}$ we have*

1.

$$(d_*^{2*}/_* d_* r^{2*}) \cdot_* ((e/_* r) \cdot_* (d_*/_* d_* r))^{(k-1)*} \cdot_* \left(r^{(2k-1)*} \cdot_* \phi(r) \right)$$
$$(4.49)$$
$$= ((e/_* r) \cdot_* (d_*/_* d_* r))^{k*} \cdot_* \left(r^{(2k)*} \cdot_* (d_* \phi/_* d_* r)(r) \right),$$

2.

$$((e/_* r) \cdot_* (d_*/_* (d_* r)))^{(k-1)*} \left(r^{(2k-1)*} \cdot_* \phi(r) \right)$$
$$= \sum_{*j=0}^{k-1} e^{\beta_j^k} \cdot_* r^{(j+1)*} \cdot_* ((d_*^{j*} \phi)/_* (d_* r^{j*}))(r),$$
$$(4.50)$$

where $\beta_j^k = \begin{pmatrix} k -_* 1 \\ j \end{pmatrix} (2k-1) \ldots (k+1+j), \quad j = 0, 1, \ldots, k-1.$

Proof

1. We will use induction.

   (a) For $k = 1$, we have

   $$((d_*^{2*})/_* (d_* r^{2*})) \cdot_* (r \cdot_* \phi(r))$$

   $$= (d_*/_* d_* r) \cdot_* (\phi(r) +_* r \cdot_* \phi^*(r))$$

   $$= e^2 \cdot_* \phi^*(r) +_* r \cdot_* \phi^{**}(r),$$

$$((e/{}_*r) \cdot{}_* (d_*/{}_*d_*r)) \left(r^{2*} \cdot{}_* \phi^*(r)\right)$$

$$= (e/{}_*r) \cdot{}_* \left(e^2 \cdot{}_* r \cdot{}_* \phi^*(r) +{}_* r^{2*} \cdot{}_* \phi^{**}(r)\right)$$

$$= e^2 \cdot{}_* \phi^*(r) +{}_* r \cdot{}_* \phi^{**}(r).$$

Therefore, the assertion is valid for $k = 1$.

(b) Assume that (4.49) holds for some $k \in \mathbb{N}$.

(c) We will prove that

$$((d_*^{2*})/{}_*(d_*r^{2*})) ((e/{}_*r) \cdot{}_* (d_*/{}_*d_*r))^{k*} \cdot{}_* \left(r^{(2k+1)*} \cdot{}_* \phi(r)\right)$$

$$= ((e/{}_*r) \cdot{}_* (d_*/{}_*d_*r))^{(k+1)*} \cdot{}_* \left(r^{(2k+2)*} \cdot{}_* (d_*\phi/{}_*d_*r)(r)\right). \tag{4.51}$$

Indeed,

$$((d_*^{2*}/{}_*d_*r^{2*})) \cdot{}_* ((e/{}_*r) \cdot{}_* (d_*/{}_*d_*r))^{k*} \cdot{}_* \left(r^{(2k+1)*} \cdot{}_* \phi(r)\right)$$

$$= ((d_*^{2*}/{}_*d_*r^{2*})) \cdot{}_* ((e/{}_*r) \cdot{}_* (d_*/{}_*d_*r))^{(k-1)*}$$

$$\cdot{}_* ((e/{}_*r) \cdot{}_* (d_*/{}_*d_*r)) \cdot{}_* \left(r^{(2k+1)*} \cdot{}_* \phi(r)\right)$$

$$= ((d_*^{2*}/{}_*d_*r^{2*})) \cdot{}_* ((e/{}_*r)(d_*/{}_*d_*r))^{(k-1)*} \cdot{}_* ((e/{}_*r))$$

$$\cdot{}_* \left(e^{(2k+1)} \cdot{}_* r^{(2k)*} \cdot{}_* \phi(r) +{}_* r^{(2k+1)*} \cdot{}_* \phi^*(r)\right)$$

$$= ((d_*^{2*}/{}_*d_*r^{2*})) \cdot{}_* ((e/{}_*r) \cdot{}_* (d_*/{}_*d_*r))^{(k-1)*}$$

$$\cdot{}_* \left(e^{2k+1} \cdot{}_* r^{(2k-1)*} \cdot{}_* \phi(r) +{}_* r^{(2k)*} \cdot{}_* \phi^*(r)\right)$$

$$= (d_*^{2*}/{}_*d_*r^{2*}) \cdot{}_* ((e/{}_*r) \cdot{}_* (d_*/{}_*d_*r))^{(k-1)*}$$

$$\cdot{}_* \left(r^{(2k-1)*} \cdot{}_* \left(e^{2k+1} \cdot{}_* \phi(r) +{}_* r \cdot{}_* \phi^*(r)\right)\right)$$

$$= ((e/{}_*r) \cdot{}_* (d_*/{}_*d_*r))^{k*}$$

$$\cdot{}_* \left(r^{(2k)*} \cdot{}_* \left(e^{2k+2} \cdot{}_* \phi^*(r) +{}_* r \cdot{}_* \phi^{**}(r)\right)\right) \tag{4.52}$$

and

$$((e/{}_*r) \cdot{}_* (d_*/{}_*d_*r))^{(k+1)*} \cdot{}_* \left(r^{(2k+2)*} \cdot{}_* (d_*\phi/{}_*d_*r)(r)\right)$$

$$= ((e/{}_*r) \cdot{}_* (d_*/{}_*d_*r))^{k*} \cdot{}_* ((e/{}_*r) \cdot{}_* (d_*/{}_*d_*r))$$

$$\cdot{}_* \left(r^{(2k+2)*} \cdot{}_* \phi^*(r)\right)$$

$$= ((e/_*r) \cdot_* (d_*/_*d_*r))^{k_*} \cdot_* ((e/_*r))$$

$$\cdot_* \left(e^{2k+2} \cdot_* r^{(2k+1)_*} \cdot_* \phi^*(r) +_* r^{(2k+2)_*} \cdot_* \phi^{**}(r) \right)$$

$$= ((e/_*r) \cdot_* (d_*/_*d_*r))^{k_*}$$

$$\cdot_* \left(r^{(2k)_*} \cdot_* \left(e^{2k+2} \cdot_* \phi^*(r) +_* r \cdot_* \phi^{**}(r) \right) \right).$$

Hence from (4.52), we get (4.51).

2. We have

$$((e/_*r) \cdot_* (d_*/_*d_*r))^{(k-1)_*} \cdot_* \left(r^{(2k-1)_*} \cdot_* \phi(r) \right)$$

$$= (e/_*r^{(k-1)_*}) \cdot_* (d_*/_*d_*r)^{(k-1)_*} \cdot_* \left(r^{(2k-1)_*} \cdot_* \phi(r) \right)$$

$$= (e/_*r^{(k-1)_*}) \cdot_* \sum_{*j=0}^{k-1} \left(\begin{array}{c} k-1 \\ j \end{array} \right)_* \cdot_* (d_*/_*d_*r)^{(k-1-j)_*}$$

$$\cdot_* \left(r^{(2k-1)_*} \right) \cdot_* ((d_*^{j_*} \phi)/_*(d_* r^{j_*}))(r)$$

$$= (e/_*(r^{(k-1)_*})) \cdot_* \sum_{*j=0}^{k-1} \left(\begin{array}{c} k-1 \\ j \end{array} \right)_*$$

$$\cdot_*(2k-1)_* \cdots (2k-1-k+1+j-1)_*$$

$$\cdot_* r^{(2k-1-k+1+j)_*} \cdot_* ((d_*^{j_*} \phi)/_*(d_* r^{j_*}))(r)$$

$$= (e/_*r^{(k-1)_*}) \cdot_* \sum_{*j=0}^{k-1} \left(\begin{array}{c} k-1 \\ j \end{array} \right)_* \cdot_* (2k-1)_* \cdots (k+1+j)_*$$

$$\cdot_* r^{(k+j)_*} \cdot_* ((d_*^{j_*} \phi)/_*(d_* r^{j_*}))(r)$$

$$= \sum_{*j=0}^{k-1} \left(\begin{array}{c} k-1 \\ j \end{array} \right)_* \cdot_* (2k-1)_* \cdots (k+1+j)_* \cdot_* r^{(j+1)_*}$$

$$\cdot_*((d_*^{j_*} \phi)/_*(d_* r^{j_*}))(r),$$

which completes the proof.

Now we consider the multiplicative Cauchy problem:

$$u_{tt}^{**} -_* \Delta_* u = f(x,t), \quad (x,t) \in \mathbb{R}_*^{2n+1} \times (1,\infty),$$

$$u(x,1) = \phi(x), \tag{4.53}$$

$$u_t^*(x,1) = \psi(x), \quad x \in \mathbb{R}_*^{2n+1},$$

where $\Delta_* u = \sum\limits_{*i=1}^{2n+1} u_{x_i x_i}^{**}$, $x = (x_1, x_2, \ldots, x_{2n+1}^*)$, $\phi \in C_*^m(\mathbb{R}^{2n+1})$, $\psi \in C_*^{m-1}$

$(\mathbb{R}^{2n+1})$, $f \in C_*^{m-2}(\mathbb{R}_*^{2n+1} \times [1,\infty))$, $m \geq 2$. We denote

$$\alpha(n) = ((e^{2(2\pi)^n})/_*(e^{1\cdots3\cdots(2n-1)})),$$

$$\beta(n) = e^{\frac{2n!(4\pi)^n}{(2n+1)!}}$$

$$= e^{\frac{1}{2n+1}} \cdot_* \alpha(n).$$

Then, if $B(x,r)$ is a ball with centerpoint x and radius r, we have

$$\mu_*(B(x,r)) = \beta(n) \cdot_* r^{(2n+1)*},$$

$$\mu_*(\partial_* B(x,r)) = \alpha(n) \cdot_* r^{(2n)*},$$

where $\mu_*(B(x,r))$ and $\mu_*(\partial_* B(x,r))$ are the multiplicative volume of the multiplicative ball $B(x,r)$ and the multiplicative area of the multiplicative sphere $\partial_* B(x,r)$, respectively. Let $u \in C_*^m(\mathbb{R}_*^{2n+1} \times [1,\infty))$ be a solution to the problem (4.53).

For $x \in \mathbb{R}^{2n+1}$, $t > 1$, $r > 1$, we define

$$U(x,r,t) = (e/_*(\alpha(n) \cdot_* r^{(2n)*})) \cdot_* \int_{\cdot_* \partial_* B(x,r)} u(y,t) \cdot_* d_* s_y,$$

$$F(x,r,t) = (e/_*(\alpha(n) \cdot_* r^{(2n)*})) \cdot_* \int_{*\partial_* B(x,r)} f(y,t) \cdot_* d_* s_y,$$

$$\Phi(x,r) = (e/_*(\alpha(n) \cdot_* r^{(2n)*})) \cdot_* \int_{*\partial_* B(x,r)} \phi(y) \cdot_* d_* s_y,$$

$$\Psi(x,r) = (e/_*(\alpha(n) \cdot_* r^{(2n)*})) \cdot_* \int_{*\partial B(x,r)} \psi(y) \cdot_* d_* s_y.$$

With $D_* u(x,t)$, we denote the multiplicative vector $(u_{x_1}^*(x,t), \ldots, u_{x_{2n+1}}^*(x,t))$.

Theorem 4.6 *We have* $U \in C_*^m(\mathbb{R}_* \times [1, \infty))$ *and*

$$U_{tt}^{**} -_* U_{rr}^{**} -_* (e^{2n}/_*r) \cdot_* U_r^* = F(x, r, t), (r, t) \in \mathbb{R}_* \times (1, \infty), \qquad (4.54)$$

$$U(r, 1) = \Phi(r),$$

$$U_t^*(r, 1) = \Psi(r), \quad r \in \mathbb{R}_*, \qquad (4.55)$$

where $\mathbb{R}_* = (1, \infty)$.

Proof We have

$$U(x, r, t) = (e/_*\alpha(n)) \cdot_* \int_{*\partial_* B(1, e)} u(x +_* r \cdot_* z, t) \cdot_* d_* s_z.$$

Then,

$$\lim_{r \to 1+_*} U(x, r, t) = u(x, t)$$

and

$$U_r^*(x, r, t) = (e/_*\alpha(n)) \cdot_* \int_{*\partial_* B(1, e)} D_* u(x +_* r \cdot_* z, t) \cdot_* z \cdot_* d_* s_z$$

$$= (e/_*(\alpha(n) \cdot_* r^{(2n)*})) \cdot_* \int_{*\partial_* B(x, r)} D_* u(y, t) \cdot_* ((y -_* x)/_*r) \cdot_* d_* s_y$$

$$= (e/_*(\alpha(n) \cdot_* r^{(2n)*})) \cdot_* \int_{*\partial_* B(x, r)} (\partial_* u/_*\partial_* \nu) \cdot_* d_* s_y$$

$$= (e/_*(\alpha(n) \cdot_* r^{(2n)*})) \cdot_* \int_{*B(x, r)} \Delta_* u(y, t) \cdot_* \Delta_* y$$

$$= (r/_*((2n + 1) \cdot_* \beta(n) \cdot_* r^{(2n+1)*})) \cdot_* \int_{*B(x, r)} \Delta_* u(y, t) \cdot_* d_* y$$

$$= (r/_*((2n + 1) \cdot_* \beta(n))) \cdot_* \int_{*B(0, 1)} \Delta_* u(x +_* r \cdot_* z, t) \cdot_* d_* z.$$

Therefore,

$$\lim_{r \to 1+_*} U_r^*(x, r, t) = 1$$

and

$$U_{rr}^{**}(x, r, t) = (e/_*((2n + 1)_* \cdot_* \beta(n) \cdot_* r^{(2n+1)+*})) \cdot_* \int\limits_{*B(x,r)} \Delta_* u(y, t) \cdot_* d_* y$$

$$-_*(e/_*(\beta(n) \cdot_* r^{(2n+1)_*})) \cdot_* \int\limits_{*B(x,r)} \Delta_* u(y, t) \cdot_* d_* y$$

$$+_*(e/_*((2n + 1)_* \cdot_* \beta(n) \cdot_* r^{(2n)_*})) \cdot_* \int\limits_{*\partial_* B(x,r)} \Delta_* u(y, t) \cdot_* d_* s_y$$

$$= ((e/_*(2n + 1)_*) -_* e) \cdot_* (e/_*(\beta(n) \cdot_* r^{(2n+1)_*}))$$

$$\cdot_* \int\limits_{*B(x,r)} \Delta_* u(y, t) \cdot_* d_* y$$

$$+_*(e/_*(\alpha(n) \cdot_* r^{(2n)_*})) \cdot_* \int\limits_{*\partial_* B(x,r)} \Delta_* u(y, t) \cdot_* d_* s_y$$

$$= ((e/_*(2n + 1)_*) -_* e) \cdot_* (e/_*\beta(n))$$

$$\cdot_* \int\limits_{B(0,1)} \Delta_* u(x +_* r \cdot_* z, t) \cdot_* d_* z$$

$$+_*(e/_*\alpha(n)) \cdot_* \int\limits_{*\partial_* B(0,1)} \Delta_* u(x +_* r \cdot_* z, t) \cdot_* d_* s_z.$$

$$(4.56)$$

Consequently,

$$\lim_{r \to 1+_*} U_{rr}^{**}(x, r, t) = (e/_*(2n + 1)_*) \cdot_* \Delta_* u(x, t).$$

Using (4.56), we can compute $U_{rrr}^{***}(x, r, t)$ etc. So, $U \in \mathcal{C}_*^m(\mathbb{R}_* \times [1, \infty))$. Also, we have

$$r^{(2n)_*} \cdot_* U_r^*(x, r, t) = (e/_*((2n + 1)_* \cdot_* \beta(n))) \cdot_* \int\limits_{*B(x,r)} \Delta_* u(y, t) \cdot_* d_* y$$

$$= (e/_*((2n + 1)_* \cdot_* \beta(n))) \cdot_* \int\limits_{B(x,r)} (u_{tt}^{**}(y, t) -_* f(y, t)) \cdot_* d_* y$$

and so

$$\left(r^{(2n)*} \cdot_* U_r^*(x,r,t)\right)_r = (e/_*((2n+1)_* \cdot_* \beta(n)))$$

$$\cdot_* \int_{\partial_* B(x,r)} (u_{tt}^{**}(y,t) -_* f(y,t)) \cdot_* d_* s_y$$

$$= ((r^{(2n)*})/_*(\alpha(n) \cdot_* r^{(2n)*}))$$

$$\cdot_* \int_{*\partial_* B(x,r)} (u_{tt}^{**}(y,t) -_* f(y,t)) \cdot_* d_* s_y$$

$$= r^{(2n)*} \cdot_* U_{tt}^{**}(x,r,t) -_* r^{(2n)*} \cdot_* F(x,r,t),$$

i.e., $U(x,r,t)$ satisfies equation (4.54). Moreover,

$$U(x,r,1) = (e/_*(\alpha(n) \cdot_* r^{(2n)*})) \cdot_* \int_{*\partial_* B(x,r)} u(y,1) \cdot_* d_* s_y$$

$$= (e/_*(\alpha(n) \cdot_* r^{(2n)*})) \cdot_* \int_{*\partial_* B(x,r)} \phi(y) \cdot_* d_* s_y$$

$$= \Phi(x,r),$$

$$U_t^*(x,r,t) = (e/_*(\alpha(n) \cdot_* r^{(2n)*})) \cdot_* \int_{*\partial B(x,r)} u_t^*(y,t) \cdot_* d_* s_y,$$

$$U_t^*(x,r,1) = (e/_*(\alpha(n) \cdot_* r^{(2n)*})) \cdot_* \int_{*\partial_* B(x,r)} u_t^*(y,1) ds_y$$

$$= (e/_*(\alpha(n) \cdot_* r^{(2n)*})) \cdot_* \int_{*\partial_* B(x,r)} \psi(y) \cdot_* d_* s_y$$

$$= \Psi(x,r),$$

i.e., $U(x,r,t)$ satisfies the initial condition (4.55).

Definition 4.8 Equation (4.54) is called multiplicative Euler[1]–Poisson[2]–Darboux[3] equation.

Let

$$\tilde{U}(r,t) \;\; \text{---} \;\; ((c/_*r) \,_* (\partial_*/_*\partial_* r))^{(n-1)*} \cdot_* \left(r^{(2n-1)*} \cdot_* U(x,r,t) \right),$$

$$\tilde{F}(r,t) \;\; = \;\; ((e/_*r) \cdot_* (\partial_*/_*\partial_* r))^{(n-1)*} \cdot_* \left(r^{(2n-1)*} \cdot_* F(x,r,t) \right),$$

$$\tilde{\Phi}(r) \;\; = \;\; ((e/_*r) \cdot_* (\partial_*/_*\partial_* r))^{(n-1)*} \cdot_* \left(r^{(2n-1)*} \cdot_* \Phi(x,r) \right),$$

$$\tilde{\Psi}(r) \;\; = \;\; ((e/_*r) \cdot_* (\partial_*/_*\partial_* r))^{(n-1)*} \cdot_* \left(r^{(2n-1)*} \cdot_* \Psi(x,r) \right).$$

Then,

$$\tilde{U}(r,1) \;\; = \;\; \tilde{\Phi}(r),$$

$$\tilde{U}_t^*(r,1) \;\; = \;\; \tilde{\Psi}(r).$$

Theorem 4.7 *We have*

$$\tilde{U}_{tt}^{**} -_* \tilde{U}_{rr}^{**} = \tilde{F}(r,t), \quad (r,t) \in \mathbb{R}_* \times (1,\infty),$$

$$\tilde{U}(r,1) = \tilde{\Phi}(r),$$

$$\tilde{U}_t(r,1) = \tilde{\Psi}(r), \quad r \in \mathbb{R}_*,$$

$$\tilde{U}(1,t) = 1, \quad t \in (1,\infty).$$

Proof If $r > 0$, applying Lemma 4.1, we have

$$\tilde{U}_{rr}^{**}(r,t) \;\; = \;\; ((\partial_*^{2*})/_*(\partial_* r^{2*})) \, ((e/_*r) \cdot_* (\partial_*/_*\partial_* r))^{(n-1)*}$$

$$\left(r^{(2n-1)*} \cdot_* U(x,r,t) \right)$$

$$= \;\; ((e/_*r) \cdot_* (\partial_*/_*\partial_* r))^{n*} \cdot_* \left(r^{(2n)*} \cdot_* U_r^*(x,r,t) \right)$$

---

[1] Leonhard Euler (April 15, 1707–September 18, 1783) was a Swiss mathematician, physicist, astronomer, logician and engineer who made important and influential discoveries in many branches of mathematics.

[2] Simeon Denis Poisson (June 21, 1781–April 25, 1840) was a French mathematician, geometer and physicist.

[3] Jean Gaston Darboux (August 14, 1842–February 23, 1917) was a French mathematician, who made several important contributions to geometry and mathematical analysis.

$$= ((e/_*r) \cdot_* (\partial_*/_*(\partial_*r)))^{(n-1)*}$$

$$\cdot_* \left(r^{(2n-1)*} \cdot_* U_{rr}^{**}(x,r,t) +_* e^{2n} \cdot_* r^{(2n-2)*} \cdot_* U_r^*(x,r,t) \right)$$

$$= ((e/_*r) \cdot_* (\partial_*/_*\partial_*r))^{(n-1)*}$$

$$\left(r^{(2n-1)*} \cdot_* \left(U_{rr}^{**}(x,r,t) +_* (e^{2n}/_*r) \cdot_* U_r^*(x,r,t) \right) \right)$$

$$= ((e/_*r) \cdot_* (\partial_*/_*(\partial_*r)))^{(n-1)*}$$

$$\left(r^{(2n-1)*} \cdot_* U_{tt}^{**}(x,r,t) -_* r^{(2n-1)*} \cdot_* F(x,r,t) \right)$$

$$= \tilde{U}_{tt}^{**}(r,t) -_* \tilde{F}(r,t).$$

Again applying Lemma 4.1, we have

$$\tilde{U}(r,t) = \sum_{*j=0}^{n-1} e^{\beta_j^n} \cdot_* r^{(j+1)*} \cdot_* ((\partial_*^{j*})/_*(\partial_*r^{j*})) \cdot_* U(x,r,t).$$

Hence from Theorem 4.6, we find that

$$\lim_{r \to 1+_*} \tilde{U}(r,t) = 1,$$

which completes the proof.

By Theorem 4.7, we have

$$\tilde{U}(r,t) = e^{\frac{1}{2}} \cdot_* \left(\tilde{\Phi}(r +_* t) -_* \tilde{\Phi}(t -_* r) \right) +_* e^{\frac{1}{2}} \cdot_* \int_{*t-_*r}^{t+_*r} \tilde{\Psi}(y) \cdot_* d_*y$$

$$+_* e^{\frac{1}{2}} \cdot_* \int_{*1}^{t} \int_{*(t-_*\tau)-_*r}^{(t-_*\tau)+_*r} \tilde{F}(\xi,\tau) \cdot_* d\xi \cdot_* d\tau$$

for all $r \in \mathbb{R}_*$, $t \geq 1$. Because

$$u(x,t) = \lim_{r \to 1} U(x,r,t)$$

and

$$\tilde{U}(r,t) = ((e/_*r) \cdot_* (\partial_*/_*\partial_*r))^{(n-1)*} \cdot_* \left(r^{(2n-1)*} \cdot_* U(x,r,t) \right)$$

$$= \sum_{*j=0}^{n-1} e^{\beta_j^n} r^{(j+1)*} \cdot_* (\partial_*^j/_*\partial_*r^{j*}) U(x,r,t),$$

we get

$$\lim_{r\to 1+_*}(\tilde{U}(r,t)/_*(e^{\beta_0^n}\cdot_* r)) = \lim_{r\to 1}U(x,r,t)$$

$$= u(x,t).$$

Therefore,

$$u(x,t) = (e/_*e^{\beta_0^n})\cdot_* \lim_{r\to 1}\left((\tilde{\Phi}(t+_*r)-_*\tilde{\Phi}(t-_*r))/_*(e^2\cdot_* r)\right)$$

$$+_*(e/_*(e^2\cdot_* r))\cdot_* \int_{*t-_*r}^{t+_*r}\tilde{\Psi}(y)\cdot_* d_*y$$

$$+_*(e/_*(e^2\cdot_* r))\cdot_* \int_{*1}^{t}\int_{*(t-_*\tau)-_*r}^{(t-_*\tau)+_*r}\tilde{F}(\xi,\tau)\cdot_* d_*\xi\cdot_* d_*\tau\right)$$

$$= (e/_*e^{\beta_0^n})\cdot_*\left(\tilde{\Phi}^*(t)+_*\tilde{\Psi}(t)+_*\int_{*1}^{t}\tilde{F}(t-_*\tau,\tau)\cdot_* d_*\tau\right),$$

whereupon

$$u(x,t) = (e/_*e^{\beta_0^n})\cdot_*\left(((\partial_*/_*\partial_*t))\cdot_*((e/_*t)\cdot_*(\partial_*/_*\partial_*t))^{(n-1)_*}\right.$$

$$\cdot_*\left((e/_*(\alpha(n)\cdot_* t))\cdot_* \int_{\partial_* B(x,t)}\phi(y)\cdot_* d_*s_y\right)$$

$$+_*((e/_*t)\cdot_*(\partial_*/_*(\partial_*t)))^{(n-1)_*}$$

$$\cdot_*\left((e/_*(\alpha(n)\cdot_* t))\cdot_* \int_{\partial_* B(x,t)}\psi(y)\cdot_* d_*s_y\right)$$

$$+_*\int_{*1}^{t}((e/_*(t-_*\tau))\cdot_*(\partial_*/_*(\partial_*(t-_*\tau))))^{(n-1)_*}$$

$$\left.\cdot_*\left((e/_*(\alpha(n)\cdot_*(t-_*\tau)))\cdot_* \int_{*\partial_* B(x,t-_*\tau)}f(y,\tau)\cdot_* d_*s_y\right)\cdot_* d_*\tau\right).$$

$$(4.57)$$

Exercise 4.13 Prove that the function $u(x,t)$ defined by (4.57) satisfies the problem (4.53).

4.5 The $(2n) - 1$-Dimensional Wave Equation

Now we consider the following Cauchy problem:

$$u_{tt}^* -_* \Delta u = f(x,t), \quad (x,t) \in \mathbb{R}_*^{2n} \times (1,\infty),$$

$$u(x,1) = \phi(x), \quad (4.58)$$

$$u_t^*(x,1) = \psi(x), \quad x \in \mathbb{R}_*^{2n},$$

where $\Delta_* u = \sum_{*i=1}^{2n} u_{x_i x_i}^{**}$, $x = (x_1, \ldots, x_{2n})$, $\phi \in C_*^m(\mathbb{R}^{2n})$, $\psi \in C_*^{m-1}(\mathbb{R}_*^{2n})$ and $f \in C_*^{m-2}(\mathbb{R}_*^{2n} \times [1,\infty))$.

As we have deducted the solution for the two-dimensional wave equation using the solution of the three-dimensional wave equation, one can get

$$u(x,t) = (e/_* e^{\beta_0^n}) \cdot_* \left((\partial_*/_* \partial_* t) \left((e/_* t) \cdot_* (\partial_*/_* \partial_* t) \right)^{(n-1)*} \right.$$

$$\left((e^2/_* \alpha(n)) \cdot_* \int_{*B(x,t)} (\phi(y)/_* (t^{2*} -_* |y -_* x|^{2*})^{\frac{1}{2}} \cdot_* dy \right)$$

$$+_* \left((e/_* t) \cdot_* (\partial_*/_* \partial_* t) \right)^{(n-1)*}$$

$$\left((e^2/_* \alpha(n)) \cdot_* \int_{*B(x,t)} (\psi(y)/_* (t^{2*} -_* |y -_* x|^{2*})^{\frac{1}{2}} \cdot_*) \cdot_* d_* y \right)$$

$$+_* \int_{*1}^{t} \left((e/_* (t -_* \tau)) \cdot_* (\partial_*/_* (\partial_* (t -_* \tau))) \right)^{(n-1)*}$$

$$\left((e^2/_* \alpha(n)) \cdot_* \int_{*B(x,t-_*\tau)} (f(y,\tau)/_* ((t -_* \tau)^{2*} -_* |y -_* x|^{2*})^{\frac{1}{2}} \cdot_*) \cdot_* dy \right) \cdot_* d\tau \right).$$

$$(4.59)$$

Exercise 4.14 Prove that the function $u(x,t)$ defined by (4.59) satisfies the problem (4.58).

4.6 The Cauchy Problem for a Nonlinear Hyperbolic Equation

Consider the equation

$$u_{x_1 x_2}^{**} = f(x_1, x_2, u, u_{x_1}^*, u_{x_2}^*), \tag{4.60}$$

where f is a given function.

Suppose that the curve l is expressed in the form

$$x_2 = g(x_1),$$

where g is a $\mathcal{C}_*^1$-function and $g^*(x_1) > 1$. The Cauchy data for equation (4.60) are given by

$$
\begin{aligned}
u(x_1, g(x_1)) &= \phi(x_1), \\
u_{x_1}^*(x_1, g(x_1)) &= \psi(x_1), \\
u_{x_2}^*(x_1, g(x_1)) &= \kappa(x_1),
\end{aligned}
$$

where ϕ, ψ, κ are $\mathcal{C}_*^1$-functions satisfying the consistency condition:

$$\phi^*(x_1) = \psi(x_1) +_* \kappa(x_1) \cdot_* g^*(x_1).$$

Without loss of generality, we assume that the functions ϕ, ψ and κ vanish identically. Otherwise, we introduce a new dependent variable:

$$v = u -_* \phi(x_1) -_* (x_2 -_* g(x_1)) \cdot_* \kappa(x_1).$$

Then,

$$v(x_1, g(x_1)) = u(x_1, g(x_1)) -_* \phi(x_1)$$

$$= 1,$$

$$v_{x_1}(x_1, x_2) = u_{x_1}^* -_* \phi^*(x_1) +_* g^*(x_1) \cdot_* \kappa(x_1) -_* (x_2 -_* g(x_1)) \cdot_* \kappa^*(x_1),$$

$$v_{x_1}^*(x_1, g(x_1)) = u_{x_1}^*(x_1, g(x_1)) -_* \phi^*(x_1) +_* g^*(x_1) \cdot_* \kappa(x_1)$$

$$= \psi(x_1) -_* \phi^*(x_1) +_* g^*(x_1) \cdot_* \kappa(x_1)$$

$$= -_* \kappa(x_1) \cdot_* g^*(x_1) +_* \kappa(x_1) \cdot_* g^*(x_1)$$

$$= 1,$$

$$v_{x_2}^*(x_1, x_2) = u_{x_2}^*(x_1, x_2) -_* \kappa(x_1),$$

$$v_{x_2}^*(x_1, g(x_1)) = u_{x_2}^*(x_1, g(x_1)) -_* \kappa(x_1)$$

$$= \kappa(x_1) -_* \kappa(x_1)$$

$$= 1,$$

$$v_{x_1 x_2}^{**}(x_1, x_2) = u_{x_1 x_2}^*(x_1, x_2) -_* \kappa^*(x_1).$$

Equation (4.60) is converted to a similar equation:

$$v_{x_1 x_2}^{**} = \kappa^*(x_1)$$

$$+_* f(x_1, x_2, v +_* \phi(x_1) +_* (x_2 -_* g)\kappa, v_{x_1} +_* \phi^* -_* g^*$$

$$\cdot_* \kappa +_* (x_2 -_* g) \cdot_* \kappa^*, v_{x_2}^* +_* \kappa).$$

Therefore, we consider the problem

$$u_{x_1 x_2}^{**} = f(x_1, x_2, u, u_{x_1}^*, u_{x_2}^*)$$

$$u(x_1, g(x_1)) = u_{x_1}^*(x_1, g(x_1))$$

$$= u_{x_2}^*(x_1, g(x_1)) \qquad (4.61)$$

$$= 1.$$

Theorem 4.8 *Let* $f(x_1, x_2, u, u_{x_1}^*, u_{x_2}^*)$ *be a continuous function of its variables in a neighborhood* N *of the point* $(x_{10}^*, x_{20}^*, 0_*, 0_*, 0_*)$, *where* (x_{10}^*, x_{20}^*) *lies on a multiplicative smooth curve* l *whose multiplicative tangent is never multiplicative horizontal or multiplicative vertical. Let also* f *satisfy in* N *a multiplicative Lipschitz condition with respect to each of the variables* u, $u_{x_1}^*$ *and* $u_{x_2}^*$. *Then, in a sufficiently small neighborhood of* (x_{10}^*, x_{20}^*), *there exists a unique solution of the problem* (4.61).

Proof Let $\delta > 0_*$ be chosen so small that the region S defined by the inequalities

$$|x_1 -_* x_{10}^*|_* \leq \delta,$$

$$|x_2 -_* x_{20}^*|_* \leq \delta,$$

$$|u|_* \leq \delta,$$

$$|u_{x_1}^*|_* \leq \delta,$$

$$|u_{x_2}^*|_* \leq \delta,$$

lies entirely in N. We denote $m = \max\limits_{S} |f|_*$ and let k_1, k_2 and k_3 denote the multiplicative Lipschitz constants associated with u, $u^*_{x_1}$, and $u^*_{x_2}$, respectively. We set

$$M = \max\{m, k_1, k_2, k_3\}.$$

Now, we choose a multiplicative positive number p so small that

$$M \cdot_* p^{2*} < \delta,$$

$$M \cdot_* p < \delta,$$

$$M \cdot_* p \cdot_* (p +_* e^2) < e,$$

$$p < e^2 \cdot_* \delta.$$

Let T be the multiplicative square formed by the lines

$$x_1 = x^*_{10} \pm_* \frac{p}{2},$$

$$x_2 = x^*_{20} \pm_* \frac{p}{2}$$

and let $l \cap T = \{A, B\}$. With R, we will denote the multiplicative rectangle formed by the multiplicative horizontal and multiplicative vertical lines through A and B. Let D denote the multiplicative region bounded by l and the multiplicative horizontal and multiplicative vertical multiplicative lines through the point (x_1, x_2). We define within R the sequences $\{u^*_n\}_{n \in \mathbb{N}}$, $\{u^*_{x_1 n}\}_{n \in \mathbb{N}}$, $\{u^*_{x_2 n}\}_{n \in \mathbb{N}}$ as follows:

$$u^*_0(x_1, x_2) = 1,$$

$$u^*_{n+1}(x_1, x_2) = \int\int_{*D} f(\xi_1, \xi_2, u^*_n(\xi_1, \xi_2), u^*_{nx_1}(\xi_1, \xi_2), u^*_{nx_2}(\xi_1, \xi_2))$$

$$\cdot_* d_* \xi_1 \cdot_* d_* \xi_2,$$

$$u^*_{x_1 n} = (\partial_* /_* \partial x_1) u^*_n(x_1, x_2),$$

$$u^*_{x_2 n} = (\partial_* /_* \partial x_2) u^*_n(x_1, x_2), \quad n \geq 0.$$

(4.62)

First, we will prove that the sequences $\{u^*_n\}_{n \in \mathbb{N}}$, $\{u^*_{x_1 n}\}_{n \in \mathbb{N}}$, $\{u^*_{x_2 n}\}_{n \in \mathbb{N}}$ are well-defined on R. Suppose that for some $n \in \mathbb{N}$, we have

$$|u^*_n|_* \leq \delta,$$

$$|u^*_{x_1 n}|_* \leq \delta,$$

$$|u^*_{x_2 n}|_* \leq \delta \quad \text{on} \quad R.$$

Then,

$$|u^*_{n+1}(x_1, x_2)|_*$$

$$= \left| \int\!\!\int_{*D} f(\xi_1, \xi_2, u^*_n(\xi_1, \xi_2), u^*_{nx_1}(\xi_1, \xi_2), u^*_{nx_2}(\xi_1, \xi_2)) d_*\xi_1 \cdot_* d_*\xi_2 \right|_*$$

$$\leq \int\!\!\int_{*D} |f(\xi_1, \xi_2, u^*_n(\xi_1, \xi_2), u^*_{nx_1}(\xi_1, \xi_2), u^*_{nx_2}(\xi_1\xi_2))|_* \cdot_* d_*\xi_1 \cdot_* d_*\xi_2$$

$$\leq M \cdot_* \mu(D)$$

$$\leq M \cdot_* p^{2*}$$

$$\leq \delta,$$

$$|u^*_{x_1 n+1}(x_1, x_2)|_*$$

$$= \left| \int_{g(x_1)}^{x_2} f(x_1, \xi_2, u^*_n(x_1, \xi_2), u^*_{nx_1}(x_1, \xi_2), u^*_{nx_2}(x_1, \xi_2)) \cdot_* d_*\xi_2 \right|_*$$

$$\leq M \cdot_* p$$

$$\leq \delta,$$

$$|u^*_{x_2 n+_*1}(x_1, x_2)|_*$$

$$= \left| \int_{*g^{-*1}(x_2)}^{x_1} f(\xi_1, x_2, u^*_n(\xi_1, x_2), u^*_{nx_1}(\xi_1, x_2), u^*_{nx_2}(\xi_1, x_2)) \cdot_* d_*\xi_1 \right|_*$$

$$\leq M \cdot_* p$$

$$\leq \delta.$$

Therefore, the sequences $\{u^*_n\}_{n\in\mathbb{N}}$, $\{u^*_{x_1 n}\}_{n\in\mathbb{N}}$, $\{u^*_{x_2 n}\}_{n\in\mathbb{N}}$ are well-defined on R. Also,

$$|u^*_{n+2}(x_1, x_2) -_* u^*_{n+1}(x_1, x_2)|_*$$

$$= \left| \int\!\!\int_{*D} \Big(f(\xi_1, \xi_2, u^*_{n+1}(\xi_1, \xi_2), u^*_{n+1x_1}(\xi_1, \xi_2), u^*_{n+1x_2}(\xi_1, \xi_2)) \right.$$

$$\left. -_* f(\xi_1, \xi_2, u^*_n(\xi_1, \xi_2), u^*_{nx_1}(\xi_1, \xi_2), u^*_{nx_2}(\xi_1, \xi_2)) \Big) \cdot_* d_*\xi_1 \cdot_* d_*\xi_2 \right|_*$$

$$\leq \int\int_{*D} \Big(k_1 \cdot_* |u^*_{n+1}(\xi_1,\xi_2) -_* u^*_n(\xi_1,\xi_2)|_*$$

$$+_* k_2 \cdot_* |u^*_{n+1x_1}(\xi_1,\xi_2) -_* u^*_{nx_1}(\xi_1,\xi_2)|_*$$

$$+_* k_3 \cdot_* |u^*_{n+1x_2}(\xi_1,\xi_2) -_* u^*_{nx_2}(\xi_1,\xi_2)|_* \Big) \cdot_* d_*\xi_1 \cdot_* d_*\xi_2$$

$$\leq M \cdot_* S_n \cdot_* p^{2*},$$

where

$$S_n = \max_R \Big(|u^*_{n+1}(x_1,x_2) -_* u^*_n(x_1,x_2)|_* +_* |u^*_{n+1x_1}(x_1,x_2)$$

$$-_* u^*_{nx_1}(x_1,x_2)|_* +_* |u^*_{n+1x_2}(x_1,x_2) -_* u^*_{nx_2}(x_1,x_2)|_* \Big),$$

and

$$|u^*_{x_1 n+2}(x_1,x_2) -_* u^*_{x_1 n+1}(x_1,x_2)|_*$$

$$= \Big| \int_{*g(x_1)}^{x_2} \Big(f(x_1,\xi_2, u^*_{n+1}(x_1,\xi_2), u^*_{n+1x_1}(x_1,\xi_2), u^*_{n+1x_2}(x_1,\xi_2))$$

$$-_* f(x_1,\xi_2, u^*_n(x_1,\xi_2), u^*_{nx_1}(x_1,\xi_2), u^*_{nx_2}(x_1,\xi_2)) \Big) \cdot_* d_*\xi_2 \cdot_* \Big|_*$$

$$\leq \int_{*g(x_1)}^{x_2} \Big(k_1 \cdot_* |u^*_{n+1}(x_1,\xi_2) -_* u^*_n(x_1,\xi_2)|_*$$

$$+_* k_2 \cdot_* |u^*_{n+1x_1}(x_1,\xi_2) -_* u^*_{nx_1}(x_1,\xi_2)|_*$$

$$+_* k_3 \cdot_* |u^*_{n+1x_2}(x_1,\xi_2) -_* u^*_{nx_2}(x_1,\xi_2)|_* \Big) \cdot_* d_*\xi_2$$

$$\leq M \cdot_* p \cdot_* S_n,$$

$$|u^*_{x_2 n+2}(x_1,x_2) -_* u^*_{x_2 n+1}(x_1,x_2)|_*$$

$$= \Big| \int_{*g^{-*1}(x_2)}^{x_1} \Big(f(\xi_1,x_2, u^*_{n+1}(\xi_1,x_2), u^*_{n+1x_1}(\xi_1,x_2), u^*_{n+1x_2}(\xi_1,x_2))$$

$$-_* f(\xi_1,x_2, u^*_n(\xi_1,x_2), u^*_{nx_1}(\xi_1,x_2), u^*_{nx_2}(\xi_1,x_2)) \Big) \cdot_* d_*\xi_1 \Big|_*$$

$$\leq \int_{g^{-*1}(x_2)}^{x_1} \left(k_1 \cdot_* |u_{n+1}^*(\xi_1, x_2) -_* u_n^*(\xi_1, x_2)|_* \right.$$

$$+_* k_2 \cdot_* |u_{n+1x_1}^*(\xi_1, x_2) -_* u_{nx_1}^*(\xi_1, x_2)|_*$$

$$\left. +_* k_3 \cdot_* |u_{n+1x_2}^*(\xi_1, x_2) -_* u_{nx_2}^*(\xi_1, x_2)|_* \right) \cdot_* d_* \xi_1$$

$$\leq M \cdot_* p \cdot_* S_n.$$

Hence,

$$S_{n+1} \quad \leq \quad M \cdot_* p^{2*} \cdot_* S_n +_* e^2 \cdot_* M \cdot_* p \cdot_* S_n$$

$$= \quad M \cdot_* p \cdot_* (p +_* e^2) \cdot_* S_n.$$

Let

$$q = M \cdot_* p \cdot_* (p +_* e^2).$$

Then, $q < e$ and

$$S_{n+1} \quad \leq \quad q \cdot_* S_n$$

$$\leq \quad q^{2*} \cdot_* S_{n-1}$$

$$\cdots$$

$$\leq \quad q^{(n+1)*} \cdot_* S_0.$$

From here, each of the three series

$$\sum_{*n=0}^{\infty} (u_{n+1}^* -_* u_n^*), \quad \sum_{*n=0}^{\infty} (u_{x_1n+1}^* -_* u_{x_1n}^*), \quad \sum_{*n=0}^{\infty} (u_{x_2n+1}^* -_* u_{x_2n}^*)$$

is dominated by the convergent series $\sum_{*n=0}^{\infty} S_0 \cdot_* q^{n+1}$ on R. Therefore, the series $\{u_n^*\}_{n \in \mathbb{N}}$, $\{u_{x_1n}^*\}_{n \in \mathbb{N}}$, $\{u_{x_2n}^*\}_{n \in \mathbb{N}}$ are uniformly convergent on S to continuous functions which we denote by u, r and s, respectively. Passing the limit under the integral sign in (4.62), we get

$$u(x_1, x_2) = \int\int_{*D} f(\xi_1, \xi_2, u(\xi_1, \xi_2), r(\xi_1, \xi_2), s(\xi_1, \xi_2)) \cdot_* d_* \xi_1 \cdot_* d_* \xi_2 \quad (4.63)$$

and

$$r(x_1, x_2) = \int_{*g(x_1)}^{x_2} f(x_1, \xi_2, u(x_1, \xi_2), r(x_1, \xi_2), s(x_1, \xi_2)) \cdot_* d_* \xi_2,$$

$$s(x_1, x_2) = \int_{*g^{-*1}(x_2)}^{x_1} f(\xi_1, x_2, u(\xi_1, x_2), r(\xi_1, x_2) s(\xi_1, x_2)) \cdot_* d_* \xi_1.$$

Multiplicative differentiating (4.63) with respect to x_1 and x_2, we obtain

$$u_{x_1}^* = r,$$

$$u_{x_2}^* = s,$$

$$u_{x_1 x_2}^{**} = f(x_1, x_2, u, r, s)$$

$$= f(x_1, x_2, u, u_{x_1}^*, u_{x_2}^*).$$

Also, u, $u_{x_1}^*$ and $u_{x_2}^*$ vanish if (x_1, x_2) is taken on l because D shrinks to the single point (x_1, x_2). Therefore, u is a solution to the problem (4.61).

If we assume that u and v are two solutions of the problem (4.61), as in above,

$$\max_R \left(|u -_* v|_* +_* |u_{x_1}^* -_* v_{x_1}|_* +_* |u_{x_2}^* -_* v_{x_2}|_* \right)$$

$$\leq M \cdot_* p \cdot_* (p +_* e^2)$$

$$\cdot_* \max_R \left(|u -_* v|_* +_* |u_{x_1}^* -_* v_{x_1}|_* +_* |u_{x_1}^* -_* v_{x_1}|_* +_* |u_{x_2}^* -_* v_{x_2}|_* \right).$$

Since

$$M \cdot_* p \cdot_* (p +_* e^2) < e,$$

we conclude that $u \equiv v$.

4.7 The Riemann Function

Consider the multiplicative hyperbolic differential operator:

$$Lu = u_{x_1 x_2}^* +_* a \cdot_* u_{x_1}^* +_* b \cdot_* u_{x_2}^* +_* c \cdot_* u, \tag{4.64}$$

where a, b and c are $\mathcal{C}_*^1$-functions of x_1 and x_2, $u \in \mathcal{C}_*^2(\mathbb{R}_*^2)$. For $v \in \mathcal{C}_*^2(\mathbb{R}_*^2)$, we have

$$v \cdot_* Lu = v \cdot_* u_{x_1 x_2}^{**} +_* a \cdot_* v \cdot_* u_{x_1}^* +_* b \cdot_* v \cdot_* u_{x_2}^* +_* c \cdot_* u \cdot_* v. \tag{4.65}$$

Note that

$$v \cdot_* u^{**}_{x_1 x_2} = v \cdot_* (u^*_{x_1})^*_{x_2}$$

$$= (v \cdot_* u^*_{x_1})^*_{x_2} -_* v^*_{x_2} \cdot_* u^*_{x_1}$$

$$= (v \cdot_* u^*_{x_1})^*_{x_2} -_* \big((v^*_{x_2} \cdot_* u)^*_{x_1} -_* v^{**}_{x_1 x_2} \cdot_* u\big)$$

$$= (v \cdot_* u^*_{x_1})^*_{x_2} -_* (v^*_{x_2} \cdot_* u)^*_{x_1} +_* v^{**}_{x_1 x_2} \cdot_* u,$$

$$a \cdot_* v \cdot_* u^*_{x_1} = (a \cdot_* v \cdot_* u)^*_{x_1} -_* (a \cdot_* v)^*_{x_1} \cdot_* u,$$

$$b \cdot_* v \cdot_* u^*_{x_2} = (b \cdot_* v \cdot_* u)^*_{x_2} -_* (b \cdot_* v)^*_{x_2} \cdot_* u.$$

Hence from (4.65), we get

$$v \cdot_* Lu = (v \cdot_* u^*_{x_1})^*_{x_2} -_* (v^*_{x_2} \cdot_* u)^*_{x_1}$$

$$+_* v^{**}_{x_1 x_2} \cdot_* u$$

$$+_* (a \cdot_* v \cdot_* u)^*_{x_1} -_* (a \cdot_* v)^*_{x_1} \cdot_* u$$

$$+_* (b \cdot_* v \cdot_* u)^*_{x_2} -_* (b \cdot_* v)^*_{x_2} \cdot_* u +_* c \cdot_* u \cdot_* v$$

$$= \big(v^{**}_{x_1 x_2} -_* (a \cdot_* v)^*_{x_1} -_* (b \cdot_* v)^*_{x_2} +_* c \cdot_* v\big) \cdot_* u$$

$$+_* \left(e^{\frac{1}{2}} \cdot_* v \cdot_* u^*_{x_2} -_* e^{\frac{1}{2}} \cdot_* u \cdot_* v^*_{x_2} +_* a \cdot_* u \cdot_* v\right)^*_{x_1}$$

$$+_* \left(e^{\frac{1}{2}} \cdot_* v \cdot_* u^*_{x_1} -_* e^{\frac{1}{2}} \cdot_* u \cdot_* v^*_{x_1} +_* b \cdot_* u \cdot_* v\right)^*_{x_2}.$$

We denote

$$L^* v = v^{**}_{x_1 x_2} -_* (a \cdot_* v)^*_{x_1} -_* (b \cdot_* v)^*_{x_2} +_* c \cdot_* v.$$

Definition 4.9 The operator L^* is called the adjoint operator of the operator L.

We have

$$v \cdot_* Lu -_* u \cdot_* L^* v = \left(e^{\frac{1}{2}} \cdot_* v \cdot_* u^*_{x_2} -_* e^{\frac{1}{2}} \cdot_* u \cdot_* v^*_{x_2} +_* a \cdot_* u \cdot_* v\right)^*_{x_1}$$

$$+_* \left(e^{\frac{1}{2}} \cdot_* v \cdot_* u^*_{x_1} -_* e^{\frac{1}{2}} \cdot_* u \cdot_* v^*_{x_1} +_* b \cdot_* u \cdot_* v\right)^*_{x_2}.$$

We integrate the last equality on D and we get

$$\int\int_{*D} (v \cdot_* Lu -_* u \cdot_* L^*v) \cdot_* d_*x_1 \cdot_* d_*x_2$$

$$= \int\int_{*D} \left(\left(e^{\frac{1}{2}} \cdot_* v \cdot_* u^*_{x_2} -_* e^{\frac{1}{2}} \cdot_* u \cdot_* v^*_{x_2} +_* a \cdot_* u \cdot_* v\right)_{x_1} \right.$$

$$\left. +_* \left(e^{\frac{1}{2}} \cdot_* v \cdot_* u^*_{x_1} -_* e^{\frac{1}{2}} \cdot_* u \cdot_* v^*_{x_1} +_* b \cdot_* u \cdot_* v\right)^*_{x_2}\right) \cdot_* d_*x_1 \cdot_* d_*x_2$$

$$= \int_{*A}^{B} \left(e^{\frac{1}{2}} \cdot_* v \cdot_* u^*_{x_2} -_* e^{\frac{1}{2}} \cdot_* u \cdot_* v^*_{x_2} +_* a \cdot_* u \cdot_* v\right) \cdot_* d_*x_2$$

$$+_* \int_{*C}^{B} \left(e^{\frac{1}{2}} \cdot_* v \cdot_* u^*_{x_1} -_* e^{\frac{1}{2}} \cdot_* u \cdot_* v^*_{x_1} +_* b \cdot_* u \cdot_* v\right) \cdot_* d_*x_1$$

$$+_* \int_{*CA} \left(-_* \left(e^{\frac{1}{2}} \cdot_* v \cdot_* u^*_{x_1} -_* e^{\frac{1}{2}} \cdot_* u \cdot_* v^*_{x_1} +_* b \cdot_* u \cdot_* v\right) \cdot_* d_*x_1 \right.$$

$$\left. +_* \left(e^{\frac{1}{2}} \cdot_* v \cdot_* u^*_{x_2} -_* e^{\frac{1}{2}} \cdot_* u \cdot_* v_{x_2} +_* a \cdot_* u \cdot_* v\right) \cdot_* d_*x_2\right)$$

$$= \int_{*A}^{B} u \cdot_* (a \cdot_* v -_* v^*_{x_2}) \cdot_* d_*x_2 +_* \int_{*B}^{C} u \cdot_* (b \cdot_* v -_* v^*_{x_1}) \cdot_* d_*x_1$$

$$+_* u(B) \cdot_* v(B) -_* e^{\frac{1}{2}} \cdot_* u(A) \cdot_* v(A) -_* e^{\frac{1}{2}} \cdot_* u(C) \cdot_* v(C)$$

$$+_* \int_{*CA} \left(-_* \left(e^{\frac{1}{2}} \cdot_* v \cdot_* u^*_{x_1} -_* e^{\frac{1}{2}} \cdot_* u \cdot_* v^*_{x_1} +_* b \cdot_* u \cdot_* v\right) \cdot_* d_*x_1 \right.$$

$$\left. +_* \left(e^{\frac{1}{2}} \cdot_* v \cdot_* u^*_{x_2} -_* e^{\frac{1}{2}} \cdot_* u \cdot_* v^*_{x_2} +_* a \cdot_* u \cdot_* v\right) \cdot_* d_*x_2\right).$$

Now we suppose that it is possible to be determined a function v depending on $P = (x_1, x_2)$ and $B = (\xi_1, \xi_2)$ such that the following conditions are satisfied:

$$L^*v = 1$$

$$b \cdot_* v -_* v^*_{x_1} = e \quad \text{on} \quad x_2 = \xi_2$$

$$a \cdot_* v -_* v^*_{x_2} = e \quad \text{on} \quad x_1 = \xi_1 \qquad (4.66)$$

$$v(\xi_1, \xi_2) = v(B)$$

$$= e.$$

Assume that the function u satisfies the equation $Lu = f$, where $f \in C_*(\mathbb{R}^2_*)$. Then,

$$u(x_1, x_2) = e^{\frac{1}{2}} \cdot_* u(A) \cdot_* v(A) +_* e^{\frac{1}{2}} \cdot_* u(C) \cdot_* v(C)$$

$$+_* \int\int_{*D} v \cdot_* f \cdot_* d_*x_1 \cdot_* d_*x_2$$

$$+_* \int_{*CA} \left(\left(e^{\frac{1}{2}} \cdot_* v \cdot_* u^*_{x_1} -_* e^{\frac{1}{2}} \cdot_* u \cdot_* v^*_{x_1} +_* b \cdot_* u \cdot_* v \right) \cdot_* d_*x_1 \right.$$

$$\left. -_* \left(e^{\frac{1}{2}} \cdot_* v \cdot_* u^*_{x_2} -_* e^{\frac{1}{2}} \cdot_* u \cdot_* v^*_{x_2} +_* a \cdot_* u \cdot_* v \right) \cdot_* d_*x_2 \right).$$

Definition 4.10 The function v defined by the system (4.66) is called the multiplicative Riemann[4] function associated with the operator L and it is denoted by $R_L(x_1, x_2, \xi_1, \xi_2)$.

By the system (4.66), we have

$$v(x_1, \xi_2) = e^{*\xi_1}^{\int^{x_1} b(t,\xi_2)\cdot_* d_*t}, \quad v(\xi_1, x_2) = e^{*\xi_2}^{\int^{x_2} a(\xi_1,t)\cdot_* d_*t}.$$

If

$$Lu = L^*u$$

$$= u^{**}_{x_1 x_2},$$

then

$$R_L(x_1, x_2, \xi_1, \xi_2) = e.$$

If

$$Lu = L^*u$$

$$= u^{**}_{x_1 x_2} +_* c \cdot_* u,$$

where c is a multiplicative constant, then

$$R_L(x_1, \xi_1, \xi_2) = J_{*0}\left(e^2 \left(c \cdot_* (x_1 -_* \xi_1) \cdot_* (x_2 -_* \xi_2) \right)^{\frac{1}{2}*} \right),$$

where

$$J_{*0}(t) = e -_* (t^{2*}/_* e^{2^{2*}}) +_* (t^{4*}/_* (e^{2^{2*}} \cdot_* e^{4^2})) -_* \cdots.$$

Exercise 4.15 Let L be defined by (4.64). Prove that

$$(L^*)^* = L.$$

---

[4]Georg Friedrich Bernhard Riemann(September 17, 1826–July 20, 1866) was an influential German mathematician who made lasting and revolutionary contributions to analysis, number theory and differential geometry.

4.8 Advanced Practical Exercises

Problem 4.1 Solve the Cauchy problem:

$$u_{tt}^{**} -_* e^4 \cdot_* u_{xx}^{**} = 1, \quad -_*\infty < x < \infty, \quad t > 1,$$

$$u(x,1) = e^x,$$

$$u_t^*(x,1) = x, \quad -_*\infty < x < \infty.$$

Answer

$$u(x,t) = e^x \cdot_* \cos_* h(e^2 \cdot_* t) +_* x \cdot_* t.$$

Problem 4.2 Solve the Cauchy problem:

$$u_{tt}^{**} -_* e^4 \cdot_* u_{xx}^{**} = e^6 \cdot_* t, \quad -_*\infty < x < \infty, \quad t > 1,$$

$$u(x,1) = x,$$

$$u_t^*(x,1) = 1, \quad -_*\infty < x < \infty.$$

Answer

$$u(x,t) = x +_* t^{3*}.$$

Problem 4.3 Find a formal solution to the following problem:

$$u_{tt}^{**} -_* u_{xx}^{**} = 1, \quad 1 < x < e^\pi, \quad t > 1,$$

$$u_x^*(0,t) = u_x^*(\pi,t)$$

$$= 1, \quad t \geq 1,$$

$$u(x,1) = \sin_* x,$$

$$u_t^*(x,1) = 1, \quad 1 \leq x \leq e^\pi.$$

Answer

$$u(x,t) = e^{\frac{2}{\pi}} -_* e^{\frac{4}{\pi}} \cdot_* \sum_{*n=1}^{\infty} e^{\frac{1}{4n^2-1}} \cdot_* \cos_*(e^2 \cdot_* n \cdot_* t) \cdot_* \cos_*(e^2 \cdot_* n \cdot_* x).$$

Problem 4.4 Find a formal solution to the following problem:

$$u_{tt}^{**} -_* e^{c^2} \cdot_* u_{xx}^{**} = 1, \quad 1 < x < e^L, \quad t > 1,$$

$$u(1,t) = u_x^*(e^L, t)$$

$$= 1, \quad t \geq 1,$$

$$u(x,1) = \phi(x),$$

$$u_t^*(x,1) = \psi(x),$$

where $\phi \in C_*^2([1, e^L])$, $\psi \in C_*^1([1, e^L])$ and

$$\phi(1) = \psi(1) = \phi^*(e^L) = \psi^*(e^L) = 1.$$

Answer

$$u(x,t) = \sum_{*n=0}^{\infty} \left(A_n \cdot_* \cos_*((e^{(2n+1)c\pi} \cdot_* t)/_* e^{2L}) +_* B_n \right.$$

$$\left. \cdot_* \sin_*((e^{(2n+1)c\pi} \cdot_* t)/_* e^{2L}) \right) \cdot_* \sin_*((e^{(2n+1)\pi} \cdot_* x)/_* e^{2L}),$$

$$A_n = e^{\frac{2}{L}} \cdot_* \int_{*1}^{e^L} \psi(x) \cdot_* \sin_*((e^{(2n+1)\pi} \cdot_* x)/_* e^{2L}) \cdot_* d_* x,$$

$$B_n = e^{\frac{4}{(2n+1)\pi c}} \cdot_* \int_{*1}^{e^L} \psi(x) \cdot_* \sin_*((e^{(2n+*1)\pi} \cdot_* x)/_* e^{2L}) \cdot_* d_* x, \quad n \in \mathbb{N}_0.$$

Problem 4.5 Check that the function

$$u(x,t) = (e/_* t) \cdot_* \int_{*S} (y_1^{2*} -_* e^2 \cdot_* y_1 \cdot_* y_2 +_* y_3) \cdot_* d_* s_y,$$

where $S : |x -_* y|_*^{2*} = t^{2*}$ satisfies equation (4.32).

Problem 4.6 Find a solution to the Cauchy problem:

$$u_{tt}^{**} -_* u_{x_1 x_1}^{**} -_* u_{x_2 x_2}^{**} -_* u_{x_3 x_3}^{**} = 1, \quad (x_1, x_2, x_3) \in \mathbb{R}_*^3, \quad t > 1,$$

$$u(x_1, x_2, x_3, 1) = e^2 \cdot_* x_1 -_* e^3 \cdot_* x_2 +_* e^4 \cdot_* x_3,$$

$$u_t^*(x_1, x_2, x_3, 1) = e^3, \quad (x_1, x_2, x_3) \in \mathbb{R}_*^3.$$

Answer

$$u(x_1, x_2, x_3, t) = e^2 \cdot_* x_1 -_* e^3 \cdot_* x_2 +_* e^4 \cdot_* x_3 +_* e^3 \cdot_* t, \quad (x_1, x_2, x_3) \in \mathbb{R}^3_*, \quad t > 1.$$

Problem 4.7 Solve the Cauchy problem:

$$u^*_{tt} -_* u^*_{x_1 x_1} -_* u^*_{x_2 x_2} -_* u^*_{x_3 x_3} = -_* e^4, \quad (x_1, x_2, x_3) \in \mathbb{R}^3_*, \quad t > 1,$$

$$u(x_1, x_2, x_3, 1) = x_1^{2*} +_* x_2^{2*} +_* x_3^{2*},$$

$$u^*_t(x_1, x_2, x_3, 1) = 1, \quad (x_1, x_2, x_3) \in \mathbb{R}^3_*.$$

Answer

$$u(x_1, x_2, x_3, t) = x_1^{2*} +_* x_2^{2*} +_* x_3^{2*} +_* t^{2*}, \quad (x_1, x_2, x_3) \in \mathbb{R}^3_*, \quad t > 1.$$

Problem 4.8 Solve the Cauchy problem:

$$u^*_{tt} -_* u^*_{x_1 x_1} -_* u^*_{x_2 x_2} = 1, \quad (x_1, x_2) \in \mathbb{R}^2_*, \quad t > 1,$$

$$u(x_1, x_2, 1) = e^2 \cdot_* x_1 -_* e^3 \cdot_* x_2,$$

$$u^*_t(x_1, x_2, 1) = e^3 \cdot_* x_1 +_* x_2, \quad (x_1, x_2) \in \mathbb{R}^2_*.$$

Answer

$$u(x_1, x_2, t) = (e^2 +_* e^3 \cdot_* t) \cdot_* x_1 +_* (t -_* e^3) \cdot_* x_2, \quad (x_1, x_2) \in \mathbb{R}^2_*, \quad t > 1.$$

Problem 4.9 Solve the Cauchy problem:

$$u^*_{tt} -_* u^*_{x_1 x_1} -_* u^*_{x_2 x_2} = t, \quad (x_1, x_2) \in \mathbb{R}^2_*, \quad t > 1,$$

$$u(x_1, x_2, 1) = x_1^{2*},$$

$$u^*_t(x_1, x_2, 1) = x_2^{2*}, \quad (x_1, x_2) \in \mathbb{R}^2_*.$$

Answer

$$u(x_1, x_2, t) = x_1^{2*} +_* t \cdot_* x_2^{2*} +_* e^{\frac{1}{2}} \cdot_* t^{3*} +_* t^{2*}, \quad (x_1, x_2) \in \mathbb{R}^2_*, \quad t > 1.$$

Problem 4.10 Let $L = L^*$. Prove that

$$R_L(x_1, x_2, \xi_1, \xi_2) = R_{L^*}(\xi_1, \xi_2, x_1, x_2).$$

5

The Heat Equation

5.1 The Weak Maximum Principle

Consider the multiplicative heat equation for a function $u(x_1, \ldots, x_n, t)$ in an n-dimensional bounded domain D,

$$u_t^* = e^k \cdot_* \Delta_* u, \quad (x_1, \ldots, x_n) \in D, \quad t > 1, \tag{5.1}$$

where $\Delta u = \sum_{*i=1}^{n} u_{x_i x_i}^{**}$, k is a positive constant. Define the domain

$$Q_T = \{(x_1, \ldots, x_n, t) : (x_1, \ldots, x_n) \in D, \quad 1 < t \le e^T\},$$

where $T > 1$ is arbitrarily chosen. Define the parabolic boundary of Q_T as follows:

$$\partial_{*p} Q_T = \{D \times \{1\}\} \cup \{\partial_* D \times [1, e^T]\}.$$

Let $\mathcal{C}_{*Q_T}$ be the class of functions that are twice multiplicative continuously differentiable in Q_T with respect to $(x_1, \ldots, x_n)$ and once multiplicative continuously differentiable with respect to t in Q_T, and continuous in $\overline{Q_T}$.

Theorem 5.1 *Let $u \in \mathcal{C}_{*Q_T}$ and*

$$u_t^* -_* e^k \cdot_* \Delta_* u < 0_*$$

in Q_T. Then, u has no local maximum in Q_T and u achieves its maximum in $\partial_{*p} Q_T$.

Proof Assume that u has a local maximum at some point $(x, t) \in Q_T$. Then, $u_t^*(x, t) = 1$ and $\Delta_* u(x, t) \le 1$, which is a contradiction. Since u is continuous in $\overline{Q_T}$, then its maximum is achieved somewhere in ∂Q_T. If the maximum is achieved at a point $(x_0, T) \in D \times \{T\}$, then $u_t^*(x, T) = 1$ and $\Delta_* u(x, T) \le 1$, which is a contradiction.

Exercise 5.1 Let $u \in \mathcal{C}_{*Q_T}$ and

$$u_t^* -_* e^k \cdot_* \Delta_* u > 0_*$$

in Q_T. Prove that u has no local minimum in Q_T and u achieves its minimum in $\partial_p Q_T$.

DOI: 10.1201/9781003440116-5   168

Theorem 5.2 *(Weak Maximum Principle for the Heat Equation)* *Let* $u \in \mathcal{C}_{*Q_T}$ *be a solution to the multiplicative heat equation (5.1). Then* u *achieves its maximum (minimum) on* $\partial_{*p}Q_T$.

   Proof Let $\epsilon > 0$ be arbitrarily chosen and

$$e^M = \max_{\partial_{*p}Q_T} u.$$

Define the function

$$v(x,t) = u(x,t) -_* e^\epsilon \cdot_* t.$$

We have

$$\max_{\partial_{*p}Q_T} v \leq e^M.$$

Hence,

$$v_t^* -_* \Delta_* v = u_t^* -_* e^\epsilon -_* \Delta_* u$$

$$= -_* e^\epsilon$$

$$< 0_* \quad \text{in} \quad Q_T.$$

From here, using Theorem 5.1, we conclude that v achieves its maximum in $\partial_{*p}Q_T$. Consequently,

$$v \leq e^M \quad \text{in} \quad Q_T$$

or

$$u \leq e^M +_* e^\epsilon \cdot_* t$$

$$\leq e^M +_* e^{\epsilon T} \quad \text{in} \quad Q_T.$$

Because $\epsilon > 0$ was arbitrarily chosen, we conclude that

$$u \leq e^M$$

in Q_T, which completes the proof.

Corollary 5.1 *Let* $u \in \mathcal{C}_{*Q_T}$ *be a solution to the heat equation (5.1). If*

$$u(x,t) \geq (\leq)0_*$$

on $\partial_{*p}Q_T$, *then*

$$u(x,t) \geq (\leq)0_*$$

on $\overline{Q_T}$.

Theorem 5.3 *Let* $u \in C_{*Q_T}$ *be a solution to the equation*

$$u_t^* -_* e^k \cdot_* \Delta_* u = f(x,t), \quad (x_1, \ldots, x_n) \in D, \quad 0_* < t \le e^T, \tag{5.2}$$

where $f \in C_*(\overline{Q_T})$ *and*

$$|f|_* \le e^N$$

on $\overline{Q_T}$. *Let also*

$$|u(x,t)|_* \le e^m$$

on $\partial_{*p}\overline{Q_T}$. *Then,*

$$|u(x,t)|_* \le e^N \cdot_* t +_* e^m \quad on \quad \overline{Q_T}. \tag{5.3}$$

Proof Let $\epsilon > 0$ be arbitrarily chosen and

$$w_{\pm_*}(x,t) = (e^N +_* e^\epsilon) \cdot_* t +_* e^m \pm_* u(x,t).$$

Note that

$$w_{\pm_*}(x,t) \ge 1$$

on $\partial_{*p}Q_T$. Also,

$$
\begin{aligned}
(w_{\pm_*})_t^* -_* e^k \cdot_* \Delta_*(w_{\pm_*}) &= e^N +_* e^\epsilon \pm_* u_t^* \mp_* e^k \cdot_* \Delta_* u \\[2mm]
&= e^N +_* e^\epsilon \pm_* \left(u_t^* -_* e^k \cdot_* \Delta_* u \right) \\[2mm]
&= e^N +_* e^\epsilon \pm_* f \\[2mm]
&> 0_* \quad on \quad Q_T.
\end{aligned}
$$

Hence from Exercise 5.1, we conclude that $w_{\pm_*}$ achieves its maximum on $\overline{Q_T}$. Because $w_{\pm_*} \ge 1$ on $\partial_{*p}Q_T$, using Corollary 5.1, we obtain $w_{\pm_*} \ge 1$ on $\overline{Q_T}$, i.e.,

$$\pm_* u(x,t) \le (e^N +_* e^\epsilon) \cdot_* t +_* m \quad on \quad \overline{Q_T}.$$

Because $\epsilon > 0$ was arbitrarily chosen, we get (5.3).

Theorem 5.4 *Let* $u_1, u_2 \in C_{*Q_T}$ *be solutions to equation* (5.2) *with initial condition*

$$u_i(x,1) = \phi_i(x), \quad x \in D, \quad i = 1,2,$$

and boundary condition

$$u_i(x,t) = \psi_i(x,t), \quad x \in \partial D, \quad 0_* \le t \le e^T, \quad i = 1,2,$$

respectively, where $\phi_i \in C_*^2(D)$, $\psi_i \in C_*^2(\overline{Q_T})$, $i = 1,2$. *Set*

$$\delta = \max_D |\phi_1 -_* \phi_2|_* +_* \max_{\partial_* D \times \{1\}} |\psi_1 -_* \psi_2|_*.$$

Then

$$|u_1 -_* u_2|_* \le \delta, \quad on \quad \overline{Q_T}. \tag{5.4}$$

Proof Let

$$w = u_1 -_* u_2.$$

Then w is a solution to the problem

$$w_t^* -_* e^k \cdot_+ *\Delta_* w = 1 \quad \text{on} \quad D \times (1, e^T],$$

$$w(x, 1) = \phi_1(x) -_* \phi_2(x), \quad x \in D,$$

$$w(x, t) = \psi_1(x, t) -_* \psi_2(x, t), \quad x \in \partial_* D, \quad 1 \le x \le e^T.$$

Hence from Theorem 5.2, we obtain

$$\min_{\partial_{*p} Q_T} w(x, t) \le w(x, t) \le \max_{\partial_{*p} Q_T} w(x, t) \quad \text{in} \quad \overline{Q_T},$$

whereupon we get (5.4).

5.2 The Strong Maximum Principle

Theorem 5.5 *(The Stong Maximum Principle for the Multiplicative Heat Equation) Let $u \in \mathcal{C}_{*Q_T}$ be a solution to equation (5.1). Let also*

$$u(x^0, t^0) = \max_{\overline{Q_T}} u(x, t) = m > 0_*$$

in some point $(x^0, t^0) \in \overline{Q_T} \backslash \partial_{*p} Q_T$. Then, $u(x, t) = m$ for all $(x, t) \in \overline{Q_{t^0}}$ for which there exists a continuous curve that connects (x, t) and (x^0, t^0) and lies in $\overline{Q_{t^0}}$.

Proof Suppose the contrary. Then there exists a point (x^1, t^1) so that $x^1 \in \overline{D}$, $0_* \le t^1 < e^{t^0+}$ and $u(x^1, t^1) < e^{m_1} < e^m$.
Let Q^1 be the cylinder:

$$Q^1 = \left\{ (x, t) : \left(\sum_{*i=1}^{n} (x_i -_* x_i^1)^{2*} \right)^{\frac{1}{2}*} \le e^\rho, \quad t^1 \le t \le t^2 \right\},$$

where $1 > \rho > 0$ and $t^1 < t^2 < t^0$ are chosen so that $\overline{Q^1} \subset \overline{Q_{t^0}}$ and $u(x, t^1) < e^{m_1}$ for all x for which $\left(\sum_{*i=1}^{n} (x_i -_* x_i^1)^{2*} \right)^{\frac{1}{2}*} \le e^\rho$.

Let $\alpha > 0$ be chosen so that

$$4k^2(n+2)^2 - 8k\alpha\rho^2 < 0.$$

We consider the function

$$w(x,t) = e^m -_* e^{(m-m_1)} \cdot_* \left(e^{\rho^2} -_* \sum_{*i=1}^{n} (x_i -_* x_i^1)^{2*} \right)^{2*}$$

$$\cdot_* e^{-_*\alpha \cdot_* (t-_* t^1)} -_* u(x,t).$$

Then,

$$\left(\left(e^{\rho^2} -_* \sum_{*i=1}^{n} (x_i -_* x_i^1)^{2*} \right)^{2*} \right)^*_{x_j}$$

$$= e^2 \cdot_* \left(e^{\rho^2} -_* \sum_{*i=1}^{n} (x_i -_* x_i^1)^{2*} \right) \cdot_* (-_* e^2 \cdot_* (x_j -_* x_j^1))$$

$$= -_* e^4 \cdot_* \left(e^{\rho^2} -_* \sum_{*i=1}^{n} (x_i -_* x_i^1)^{2*} \right) \cdot_* (x_j -_* x_j^1),$$

$$\left(\left(e^{\rho^2} -_* \sum_{*i=1}^{n} (x_i -_* x_i^1)^{2*} \right)^{2*} \right)^{**}_{x_j x_j}$$

$$= -_* e^4 \cdot_* (-_* e^2 \cdot_* (x_j -_* x_j^1)) \cdot_* (x_j -_* x_j^1)$$

$$-_* e^4 \cdot_* \left(e^{\rho^2} -_* \sum_{*i=1}^{n} (x_i -_* x_i^1)^{2*} \right)$$

$$= -_* e^4 \cdot_* \left(e^{\rho^2} -_* \sum_{*i=1}^{n} (x_i -_* x_i^1)^{2*} \right) +_* e^8 \cdot_* (x_j -_* x_j^1)^{2*},$$

$$\Delta_* \left(\left(e^{\rho^2} -_* \sum_{*i=1}^{n} (x_i -_* x_i^1)^{2*} \right)^{2*} \right)$$

$$= -_* e^{4n} \cdot_* \left(e^{\rho^2} -_* \sum_{*i=1}^{n} (x_i -_* x_i^1)^{2*} \right)$$

$$+_* e^8 \cdot_* \sum_{*i=1}^{n} (x_i -_* x_i^1)^{2*},$$

from where

$$\Delta_* w(x,t) = -_* e^{(m-m_1)} \cdot_* \Delta_* \left(\left(e^{\rho^2} -_* \sum_{*i=1}^{n} (x_i -_* x_i^1)^{2*} \right)^{2*} \right)$$

$$\cdot_* e^{-_* e^{\alpha} \cdot_* (t-_* t^1)} -_* \Delta_* u(x,t)$$

$$= \left(e^{4n(m-m_1)} \cdot_* \left(e^{\rho^2} -_* \sum_{*i=1}^{n} (x_i -_* x_i^1)^{2*}\right)\right.$$

$$\left. -_* e^{8(m-m_1)} \cdot_* \sum_{*i=1}^{n} (x_i -_* x_i^1)^{2*}\right)$$

$$\cdot_* e^{-_* e^{\alpha} \cdot_* (t-_* t^1)} -_* \Delta_* u(x,t),$$

$$-_* e^{k} \cdot_* \Delta_* w(x,t) = \left(-_* e^{4kn(m-m_1)} \cdot_* \left(e^{\rho^2} -_* \sum_{*i=1}^{n} (x_i -_* x_i^1)^{2*}\right)\right.$$

$$\left. +_* e^{8k(m-m_1)} \cdot_* \sum_{*i=1}^{n} (x_i -_* x_i^1)^{2*}\right)$$

$$\cdot_* e^{-_* e^{\alpha} \cdot_* (t-_* t^1)} +_* e^{k} \cdot_* \Delta_* u(x,t)$$

$$w_t^*(x,t) = e^{\alpha(m-m_1)} \cdot_* \left(e^{\rho^2} -_* \sum_{*i=1}^{n} (x_i -_* x_i^1)^{2*}\right)^{2*}$$

$$e^{-_* e^{\alpha} \cdot_* (t-_* t^1)} -_* u_t^*$$

and

$$w_t(x,t) -_* e^{k} \cdot_* \Delta_* w(x,t) = e^{(m-m_1)} \cdot_* \left(e^{\alpha} \cdot_* \left(e^{\rho^2} -_* \sum_{*i=1}^{n} (x_i -_* x_i^1)^{2*}\right)^{2*}\right.$$

$$-_* e^{4kn} \cdot_* \left(e^{\rho^2} -_* \sum_{*i=1}^{n} (x_i -_* x_i^1)^{2*}\right) -_* e^{8k} \cdot_* \left(e^{\rho^2} -_* \sum_{*i=1}^{n} (x_i -_* x_i^1)^{2*}\right)$$

$$\left. +_* e^{8k\rho^2}\right) \cdot_* e^{-_* e^{\alpha} \cdot_* (t-_* t^1)}$$

$$= e^{(m-m_1)} \cdot_* \left(e^{\alpha} \cdot_* \left(e^{\rho^2} -_* \sum_{*i=1}^{n} (x_i -_* x_i^1)^{2*}\right)^{2*}\right.$$

$$\left. -_* e^{4k(n+2)} \cdot_* \left(e^{\rho^2} -_* \sum_{*i=1}^{n} (x_i -_* x_i^1)^{2*}\right) +_* e^{8k\rho^2}\right) e^{-_* e^{\alpha} \cdot_* (t-_* t^1)}$$

$$\geq 1 \quad \text{in} \quad Q^1.$$

When

$$\left(\sum_{*i=1}^{n} (x_i -_* x_i^1)^{2*} \right)^{\frac{1}{2}*} = e^\rho$$

and $t^1 \le t \le t^2$, we have

$$w(x,t) = e^m -_* u(x,t)$$

$$\ge 0_*.$$

When $t = t^1$ and

$$\left(\sum_{*i=1}^{n} (x_i -_* x_i^1)^{2*} \right)^{\frac{1}{2}*} \le e^\rho,$$

we have

$$w(x,t) = e^m -_* e^{(m-m_1)} \cdot_* \left(e^{\rho^2} -_* \sum_{*i=1}^{n} (x_i -_* x_i^1)^{2*} \right)^{2*} -_* u(x,t)$$

$$\ge e^m -_* e^{(m-m_1)\rho^4} -_* e_1^m$$

$$> e^{m-(m-m_1)-m_1}$$

$$= 1.$$

Hence from Corollary 5.1, we conclude that $w(x,t) \ge 1$ in $\overline{Q^1}$. Let $(x^2, t^2) \in \overline{Q^1}$ and

$$\sum_{*i=1}^{n} (x_i^2 -_* x_i^1)^{2*} < e^{\rho^2}.$$

Then, $w(x^2, t^2) \ge 0_*$ and

$$w(x^2, t^2) = e^m -_* e^{(m-m_1)} \cdot_* \left(e^{\rho^2} -_* \sum_{*i=1}^{n} (x_i^2 -_* x_i^1)^{2*} \right)^{2*}$$
$$\cdot_* e^{-_* e^\alpha \cdot_* (t^2 -_* t^1)} -_* u(x^2, t^2)$$

$$\ge 1,$$

whereupon

$$u(x^2, t^2) \le e^m -_* e^{(m-m_1)} \cdot_* \left(e^{\rho^2} -_* \sum_{*i=1}^{n} (x_i^2 -_* x_i^1)^{2*} \right)^{2*} \cdot_* e^{-_* e^\alpha (t^2 -_* t^1)}$$

$$< e^m.$$

Continuing this process, we obtain the points (x^1, t^1), (x^2, t^2), ..., (x^k, t^k), (x^0, t^0) so that

$$t^1 < t^2 < \cdots < t^0$$

and from

$$u(x^s, t^s) < e^m$$

we get

$$u(x^{s+*e}, t^{s+*e}) < e^m,$$

$s = 1, \ldots, k - 1$. Therefore,

$$u(x^0, t^0) < e^m,$$

which is a contradiction.

5.3 The Cauchy Problem

5.3.1 The fundamental solution

Consider the multiplicative heat equation

$$u_t^* -_* \Delta_* u = 0_*, \qquad x \in \mathbb{R}_*^n, \quad t > 1. \tag{5.5}$$

We will seek a solution $u(x, t)$ of (5.5) that has the special structure:

$$u(x, t) = t^{-_* e^{\alpha}} \cdot_* v\left(t^{-_* e^{\frac{1}{2}}} \cdot_* |x|_* \right), \tag{5.6}$$

where α is a constant and v is a function that must be found,

$$|x|_* = \left(\sum_{*i=1}^n x_i^{2*} \right)^{\frac{1}{2}*}.$$

We set

$$y = t^{-_* e^{\frac{1}{2}}} \cdot_* x$$

and insert (5.6) into (5.5). We get

$$e^{\alpha} \cdot_* v(|y|_*) +_* e^{\frac{1}{2}} \cdot_* |y|_* \cdot_* v^*(|y|_*) +_* (e^{n-1}/_*|y|_*) \cdot_* v^*(|y|_*) +_* v^{**}(|y|_*) = 1.$$

We take

$$\alpha = \frac{n}{2}$$

and set

$$r_1 = |y|_*.$$

Then,

$$e^{\frac{n}{2}} \cdot_* v(r_1) +_* e^{\frac{1}{2}} \cdot_* r_1 \cdot_* v^*(r_1) +_* (e^{n-1}/_*r_1) \cdot_* v^*(r_1) +_* v^{**}(r_1) = 1,$$

which we multiply by $r_1^{(n-1)*}$ and we find

$$e^{\frac{1}{2}n} \cdot_* r_1^{(n-1)*} \cdot_* v(r_1) +_* e^{\frac{1}{2}} \cdot_* r_1^{n*} \cdot_* v^*(r_1) +_* e^{n-1} \cdot_* r_1^{(n-2)*}$$

$$\cdot_* v^*(r_1) +_* r_1^{(n-1)*} \cdot_* v^* * (r_1) = 1,$$

whereupon

$$\left(r_1^{(n-1)*} \cdot_* v^* \right)^* +_* e^{\frac{1}{2}} \cdot_* (r_1^{n*} \cdot_* v)^* = 1.$$

Thus,

$$r_1^{(n-1)*} \cdot_* v^* +_* e^{\frac{1}{2}} \cdot_* r_1^{n*} \cdot_* v = e^{C_1}$$

for some constant C_1. Assuming

$$\lim_{r_1 \to \infty} v(r_1), v^*(r_1) = 1,$$

we get $C_1 = 1$ and

$$r_1^{(n-1)*} \cdot_* v^* +_* e^{\frac{1}{2}} \cdot_* r_1^{n*} \cdot_* v = 1,$$

whence

$$v^* = -_* e^{\frac{1}{2}} \cdot_* r_1 \cdot_* v$$

and

$$v = e^C \cdot_* e^{-*((r_1^{2*})/_*e^4)}, \quad C > 0.$$

From here,

$$u(x,t) = (e/_*t^{\frac{n}{2}*}) \cdot_* v \left((|x|_*/_*t^{\frac{1}{2}*}) \right) = (e^C/_*t^{\frac{n}{2}*}) \cdot_* e^{-*(|x|^{2*}/_*(e^4 \cdot_* t))}$$

solves the multiplicative heat equation (5.5).

Definition 5.1 The function

$$\Phi(x,t) = \begin{cases} \dfrac{1}{(e^{4\pi} \cdot_* t)^{\frac{n}{2}*}} \cdot_* e^{-*(|x|^{2*})/_*(e^4 \cdot_* t)}, & x \in \mathbb{R}^n_*, \quad t > 1, \\[2ex] 1, & x \in \mathbb{R}^n_*, \quad t < 1, \end{cases}$$

is called the fundamental solution of the heat equation.

Theorem 5.6 *For each $t > 1$ we have*

$$\int_{*\mathbb{R}^n_*} \Phi(x,t) \cdot_* d_*x = e.$$

Proof We have

$$
\int_{*\mathbb{R}^n_*} \Phi(x,t) \cdot_* d_*x \;=\; (e/_*(e^{4\pi} \cdot_* t)^{\frac{n}{2}}{}_*) \cdot_* \int_{*\mathbb{R}^n_*} e^{-*(|x|^2_*/_*(e^4 \cdot_* t))} \cdot_* d_*x
$$

$$
=\; (e/_*e^{\pi^{\frac{n}{2}}}) \cdot_* \int_{*\mathbb{R}^n_*} e^{-*|z|^2_*} \cdot_* d_*z
$$

$$
=\; e^{\frac{1}{\pi^{\frac{n}{2}}}} \cdot_* \prod_{i=1}^{n}{}_* \int_{-*\infty}^{\infty} e^{-*z_i^2{}_*} \cdot_* d_*z_i
$$

$$
=\; e.
$$

5.3.2 The Cauchy problem

Consider the following Cauchy problem

$$
u_t^* -_* \Delta_* u = 1 \quad \text{in} \quad \mathbb{R}^n_* \times (1, \infty),
$$

$$
u = \phi \quad \text{on} \quad \mathbb{R}^n_* \times \{t = 1\}.
$$

(5.7)

Theorem 5.7 *Assume* $\phi \in C_*(\mathbb{R}^n_*)$ *and*

$$
|\phi(x)|_* \le e^M
$$

for all $x \in \mathbb{R}^n_*$ *and for some* $M > 1$. *Let* $u(x,t)$ *be defined by*

$$
u(x,t) = (e/_*(e^{4\pi} \cdot_* t)^{\frac{n}{2}}{}_*) \cdot_* \int_{*\mathbb{R}^n_*} e^{-*((|x-_*y|^2_*)/_*(e^4 \cdot_* t))}
$$

$$
\cdot_* \phi(y) \cdot_* d_*y, \quad x \in \mathbb{R}^n_*, \quad t > 1.
$$

Then

1. $u \in C_*^{\infty}(\mathbb{R}^n_* \times (1, \infty))$,
2. $u_t^*(x,t) -_* \Delta_* u(x,t) = 1, \quad x \in \mathbb{R}^n_*, \quad t > 1$,
3. $\displaystyle \lim_{\substack{(x,t) \to (x^0, 1) \\ x \in \mathbb{R}^n_*, t > 1}} u(x,t) = \phi(x^0)$ *for each* $x^0 \in \mathbb{R}^n_*$.

Proof

1. Let $t_1 > 1$ be arbitrarily chosen. Since $(e/_* t^{\frac{n}{2}}{}_*) \cdot_* e^{-*(|x|^2_*/_*(e^4 \cdot_* t))}$ is infinetely many multiplicative differentiable with uniformly bounded multiplicative derivatives of all orders on $\mathbb{R}^n_* \times [t_1, \infty)$ and $t_1 > 1$ was arbitrarily chosen, we conclude that $u \in C_*^{\infty}(\mathbb{R}^n_* \times (1, \infty))$.

2. Note that

$$u_t^*(x,t) -_* \Delta_* u(x,t) \quad = \quad \int\limits_{*\mathbb{R}_*^n} (\Phi_t^* -_* \Delta_{*x}\Phi)\,(x -_* y, t) \cdot_* \phi(y) \cdot_* d_* y$$

$$= \quad 1,$$

because

$$\Phi_t(x -_* y, t) \quad = \quad -_* e^{\frac{n}{2}}(4\pi)^{-*\frac{n}{2}} \cdot_* t^{-*\frac{n}{2}} \cdot_* {}^{-*e} \cdot_* e^{-*(|x-_*y|_*^{2*})/_*(e^4 \cdot_* t)}$$

$$+_* e^{(4\pi)^{\frac{n}{2}}} \cdot_* t^{-*\frac{n}{2}} \cdot_* {}^{-*e^2}$$

$$\cdot_* |x -_* y|_*^{2*} e^{-*(|x-_*y|_*^{2*})/_*(e^4 \cdot_* t)},$$

$$\Phi_{x_j}^*(x -_* y, t) \quad = \quad -_* e^{\frac{1}{2}}(4\pi)^{-*\frac{n}{2}} \cdot_* t^{-*\frac{n}{2}, -*e} \cdot_* (x_j -_* y_j)$$

$$\cdot_* e^{-*(|x-_*y|_*^{2*})/_*(e^4 \cdot_* t)},$$

$$\Phi_{x_j x_j}^{**}(x -_* y, t) \quad = \quad -_* e^{\frac{1}{2}}(4\pi)^{-*\frac{n}{2}} \cdot_* t^{-*\frac{n}{2}, -*e} e^{-*(|x-_*y|_*^{2*})/_*(e^4 \cdot_* t)}$$

$$+_* e^{\frac{1}{4}}(4\pi)^{-*\frac{n}{2}} \cdot_* t^{-*\frac{n}{2}, -*e} \cdot_* (x_j -_* y_j)^{2*}$$

$$\cdot_* e^{-*(|x-_*y|_*^{2*})/_*(e^4 \cdot_* t)},$$

$$\Delta_{*x}\Phi(x -_* y, t) \quad = \quad -_* e^{\frac{n}{2}}(4\pi)^{-*\frac{n}{2}} \cdot_* t^{-*\frac{n}{2}, -*e} e^{-*(|x-_*y|_*^{2*})/_*(e^4 \cdot_* t)}$$

$$+_* e^{\frac{1}{4}}(4\pi)^{-*\frac{n}{2}} \cdot_* t^{-*\frac{n}{2}, -*e} \cdot_* |x -_* y|_*^{2*}$$

$$\cdot_* e^{-*(|x-_*y|_*^{2*})/_*(e^4 \cdot_* t)},$$

i.e., Φ itself solves the heat equation.

3. Fix $x^0 \in \mathbb{R}_*^n$ and $\epsilon > 0$. Then there exists $\delta = \delta(\epsilon) > 0$ such that

$$|\phi(y) -_* \phi(x^0)|_* < e^\epsilon \quad \text{if} \quad |y -_* x^0|_* < e^\delta, \quad y \in \mathbb{R}_*^n.$$

Let

$$|x -_* x^0|_* \leq e^{\frac{\delta}{2}}.$$

Then,

$$|u(x,t) -_* \phi(x^0)|_* = \left| \int_{\mathbb{R}_*^n} \Phi(x -_* y, t) \cdot_* (\phi(y) -_* \phi(x^0)) \cdot_* d_* y \right|_*$$

$$\leq \int\limits_{*\mathbb{R}_*^n} |\Phi(y -_* x, t)|_* \cdot_* |\phi(y) -_* \phi(x^0)|_* d_* y$$

$$= \int_{*B(x^0,\delta)} \Phi(x -_* y, t) \cdot_* |\phi(y) -_* \phi(x^0)|_* \cdot_* d_* y$$

$$+_* \int_{*\mathbb{R}^n_* \backslash B(x^0,\delta)} \Phi(x -_* y, t) \cdot_* |\phi(y)$$

$$-_* \phi(x^0)|_* \cdot_* d_* y$$

$$=: I_1 +_* I_2.$$

Note that

$$I_1 \;\leq\; e^\epsilon \cdot_* \int_{*B(x^0,\delta)} \Phi(x -_* y, t) \cdot_* d_* y$$

$$\leq\; e^\epsilon \cdot_* \int_{*\mathbb{R}^n_*} \Phi(x -_* y, t) \cdot_* d_* y$$

$$=\; e^\epsilon$$

and

$$I_2 \;\leq\; \int_{*\mathbb{R}^n_* \backslash B(x^0,\delta)} \Phi(x -_* y, t) \cdot_* \left(|\phi(y)|_* +_* |\phi(x^0)|_*\right) \cdot_* d_* y$$

$$\leq\; e^{2M} \cdot_* \int_{*\mathbb{R}^n_* \backslash B(x^0,\delta)} \Phi(x -_* y, t) \cdot_* d_* y$$

$$=\; (e^{2M} /_* (e^{4\pi} \cdot_* t)^{\frac{n}{2}}_*) \cdot_* \int_{*\mathbb{R}^n_* \backslash B(x^0,\delta)} e^{-((|x-_*y|^{2*}_*)/_*(e^4 \cdot_* t))} \cdot_* d_* y$$

$$\left(|x -_* y|_* \geq e^{\frac{1}{2}} \cdot_* |y -_* x^0|_*\right)$$

$$\leq\; (e^{2M} /_* (e^{4\pi} \cdot_* t)^{\frac{n}{2}}_*) \cdot_* \int_{*\mathbb{R}^n_* \backslash B(x^0,\delta)} e^{-_*(|x^0 -_* y|^{2*}_* /_* (e^{16} \cdot_* t))} \cdot_* d_* y$$

$$\leq\; (e^{C_1} /_* t^{\frac{n}{2}}_*) \cdot_* \int_{*e^\delta}^{\infty} e^{-_*(r^{2*}/_*(e^{16} \cdot_* t))} \cdot_* r^{(n-1)*} \cdot_* d_* r \to 1$$

$$\text{as} \quad t \to 1+_*,$$

where C_1 is a multiplicative constant independent of t. Therefore, if

$$|x -_* x^0|_* \leq e^{\frac{\delta}{2}}$$

and $t > 1$ is small enough, we have $I_2 \le e^\epsilon$ and

$$|u(x,t) -_* \phi(x^0)|_* \le e^{2\epsilon},$$

which completes the proof.

Example 5.1 Consider the following Cauchy problem:

$$u_t^* -_* u_{xx}^{**} = 0_* \quad \text{in} \quad (-_*\infty, \infty) \times (1, \infty),$$

$$u(x, 1) = x \quad \text{in} \quad (-_*\infty, \infty).$$

Then,

$$u(x,t) \;=\; (e/_*(e^{2\sqrt{\pi}} \cdot_* t^{\frac{1}{2}}_*)) \cdot_* \int_{*-_*\infty}^{\infty} e^{-_*((x-_*y)^{2*}/_*(e^4 \cdot_* t))} \cdot_* y \cdot_* d_*y$$

$$=\; (e/_*(e^{2\sqrt{\pi}} \cdot_* t^{\frac{1}{2}}_*)) \cdot_* \int_{*-_*\infty}^{\infty} e^{-_*(z^{2*}/_*(e^4 \cdot_* t))} \cdot_* (x -_* z) \cdot_* d_*z$$

$$=\; (e/_*(e^{2\sqrt{\pi}} \cdot_* t^{\frac{1}{2}}_*)) \cdot_*$$

$$\left(x \cdot_* \int_{*-_*\infty}^{\infty} e^{-_*(z^{2*}/_*(e^4 \cdot_* t))} \cdot_* d_*z -_* \int_{*-_*\infty}^{\infty} e^{-_*(z^{2*}/_*(e^4 \cdot_* t))} \cdot_* z \cdot_* d_*z \right)$$

$$=\; x.$$

Example 5.2 Consider the following Cauchy problem:

$$u_t^* -_* u_{x_1 x_1}^{**} -_* u_{x_2 x_2}^{**} = 1 \quad \text{in} \quad \mathbb{R}_*^2 \times (1, \infty),$$

$$u(x_1, x_2, 1) = x_1^{2*} +_* x_2^{2*} \quad \text{in} \quad \mathbb{R}_*^2.$$

We have

$$u(x_1, x_2, t)$$

$$=\; (e/_*(e^{4\pi} \cdot_* t)) \cdot_* \int_{*\mathbb{R}_*^2} e^{-_*(|x-_*y|^{2*}/_*(e^4 \cdot_* t))} \cdot_* (y_1^{2*} +_* y_2^{2*}) \cdot_* d_*y_1 \cdot_* d_*y_2$$

$$=\; (e/_*(e^{4\pi} \cdot_* t)) \cdot_*$$

$$\int_{*\mathbb{R}_*^2} e^{-_*((z_1^{2*} +_* z_2^{2*})/_*(e^4 \cdot_* t))} \cdot_* ((x_1 -_* z_1)^{2*} +_* (x_2 -_* z_2)^{2*}) \cdot_* d_*z_1 \cdot_* d_*z_2$$

$$= \ (x_1^{2*} +_* x_2^{2*}) \cdot_* (e/_*(e^{4\pi} \cdot_* t)) \cdot_*$$

$$\int_{*-_*\infty}^{\infty} \int_{*-_*\infty}^{\infty} e^{-_*((z_1^{2*}+_* z_2^{2*})/_*(e^4 \cdot_* t))} \cdot_* d_* z_1 \cdot_* d_* z_2$$

$$-_*(x_1/_*(e^{2\pi} \cdot_* t)) \cdot_* \int_{*-_*\infty}^{\infty} \int_{*-_*\infty}^{\infty} e^{-_*((z_1^{2*}+_* z_2^{2*})/_*(e^4 \cdot_* t))} \cdot_* z_1 \cdot_* d_* z_1 \cdot_* d_* z_2$$

$$-_*(x_2/_*(e^{2\pi} \cdot_* t)) \cdot_* \int_{*-_*\infty}^{\infty} \int_{*-_*\infty}^{\infty} e^{-_*((z_1^{2*}+_* z_2^{2*})/_*(e^4 \cdot_* t))} \cdot_* z_2 \cdot_* d_* z_1 \cdot_* d_* z_2$$

$$+_*(e/_*(e^{4\pi} \cdot_* t)) \cdot_*$$

$$\int_{*-_*\infty}^{\infty} \int_{*-_*\infty}^{\infty} e^{-_*((z_1^{2*}+_* z_2^{2*})/_*(e^4 \cdot_* t))} \cdot_* (z_1^{2*} +_* z_2^{2*}) \cdot_* d_* z_1 \cdot_* d_* z_2$$

$$= \ x_1^{2*} +_* x_2^{2*} +_* e^4 \cdot_* t.$$

Exercise 5.2 Find a solution to the Cauchy problem:

$$u_t^* -_* u_{x_1 x_1}^{**} -_* u_{x_2 x_2}^{**} -_* u_{x_3 x_3}^{**} = 1 \quad \text{in} \quad \mathbb{R}_*^3 \times (1, \infty)$$

$$u(x_1, x_2, x_3, 1) = e^2 \cdot_* x_1 -_* x_2 +_* e^3 \cdot_* x_3 \quad \text{in} \quad \mathbb{R}_*^3.$$

Answer

$$u(x_1, x_2, x_3, t) = e^2 \cdot_* x_1 -_* x_2 +_* e^3 \cdot_* x_3.$$

Next, we consider the following Cauchy problem:

$$u_t^* -_* \Delta_* u = f(x, t) \quad \text{in} \quad \mathbb{R}_*^n \times (1, \infty)$$

$$(5.8)$$

$$u = 0 \quad \text{on} \quad \mathbb{R}_*^n \times \{t = 1\}.$$

Theorem 5.8 Let $f \in C_*^2(\mathbb{R}_*^n, C_*^1([1, \infty)))$, $|f_t^*|_*$, $|f_{x_i}^*|_*$, $|f_{x_i x_i}^{**}|_* \leq e^M$, $i = 1, \ldots, n$, in $\mathbb{R}_*^n \times [1, \infty)$ for some positive constant e^M. Let also

$$u(x, t) = \int_{*1}^{t} \int_{\mathbb{R}_*^n} \Phi(x -_* y, t -_* s) \cdot_* f(y, s) \cdot_* d_* y \cdot_* d_* s.$$

Then

1. $u \in \mathcal{C}^2_*(\mathbb{R}^n_*, \mathcal{C}^1_*((1, \infty)))$,
2. $u^*_t -_* \Delta_* u = f(x, t)$ *in* $\mathbb{R}^n_* \times (1, \infty)$,
3. $\displaystyle\lim_{\substack{(x,t) \to (x^0, 1) \\ x \in \mathbb{R}^n_*, t > 1}} u(x, t) = 1$ *for each* $x^0 \in \mathbb{R}^n_*$.

Proof

1. We change the variables and we get

$$u(x, t) = \int_{*1}^{t} \int_{\mathbb{R}^n_*} \Phi(y, s) \cdot_* f(x -_* y, t -_* s) \cdot_* d_*y \cdot_* d_*s.$$

Since $f \in \mathcal{C}^2_*(\mathbb{R}^n_*, \mathcal{C}^1_*((1, \infty)))$ and $|f|_*, |f^*_t|_*, |f^*_{x_i}|_*, |f^{**}_{x_i x_i}|_* \leq e^M$, $i = 1, \ldots, n$, we conclude that $u \in \mathcal{C}^2_*(\mathbb{R}^n_*, \mathcal{C}^1_*((1, \infty)))$.

2. We have

$$u^*_t(x, t) = \int_{*1}^{t} \int_{*\mathbb{R}^n_*} \Phi(y, s) \cdot_* f^*_t(x -_* y, t -_* s) \cdot_* d_*y \cdot_* d_*s$$

$$+_* \int_{*\mathbb{R}^n_*} \Phi(y, t) \cdot_* f(x -_* y, 1) \cdot_* d_*y$$

$$= -_* \int_{*1}^{t} \int_{*\mathbb{R}^n_*} \Phi(y, s) \cdot_* f^*_s(x -_* y, t -_* s) \cdot_* d_*y \cdot_* d_*s$$

$$+_* \int_{*\mathbb{R}^n_*} \Phi(y, t) \cdot_* f(x -_* y, 1) \cdot_* d_*y,$$

$$\Delta_* u(x, t) = \int_{*1}^{t} \int_{*\mathbb{R}^n_*} \Phi(y, s) \cdot_* \Delta_{*x} f(x -_* y, t -_* s) \cdot_* d_*y \cdot_* d_*s$$

$$= \int_{*1}^{t} \int_{*\mathbb{R}^n_*} \Phi(y, s) \cdot_* \Delta_{*y} f(x -_* y, t -_* s) \cdot_* d_*y \cdot_* d_*s.$$

Therefore,

$$u_t^*(x,t) -_* \Delta_* u(x,t)$$

$$= \int\limits_{*1}^{t} \int\limits_{*\mathbb{R}_*^n} \Phi(y,s) \left(-_* f_s^*(x -_* y, t -_* s) -_* \Delta_{*y} f(x -_* y, t -_* s) \right)$$

$$\cdot_* d_* y \cdot_* d_* s$$

$$+_* \int\limits_{*\mathbb{R}^n} \Phi(y,t) \cdot_* f(x -_* y, 1) \cdot_* d_* y$$

$$= \int\limits_{*e^\epsilon}^{t} \int\limits_{*\mathbb{R}_*^n} \Phi(y,s) \cdot_* \left(-_* f_s^*(x -_* y, t -_* s) -_* \Delta_{*y} f(x -_* y, t -_* s) \right)$$

$$\cdot_* d_* y \cdot_* d_* s$$

$$+_* \int\limits_{*1}^{e^\epsilon} \int\limits_{*\mathbb{R}_*^n} \Phi(y,s) \cdot_* \left(-_* f_s(x -_* y, t -_* s) -_* \Delta_{*y} f(x -_* y, t -_* s) \right)$$

$$\cdot_* d_* y \cdot_* d_* s$$

$$+_* \int\limits_{*\mathbb{R}_*^n} \Phi(y,t) \cdot_* f(x -_* y, 1) \cdot_* d_* y$$

$$= J_1 +_* J_2 +_* J_3.$$

Note that

$$J_1 = \int\limits_{e^\epsilon}^{t} \int\limits_{*\mathbb{R}_*^n} \Phi(y,s) \cdot_* \left(-_* f_s(x -_* y, t -_* s) -_* \Delta_{*y} f(x -_* y, t -_* s) \right)$$

$$\cdot_* d_* y \cdot_* d_* s$$

$$= \int\limits_{e^\epsilon}^{t} \int\limits_{*_-*\mathbb{R}^n} \left(\Phi_s(y,s) -_* \Delta_{*y} \Phi(y,s) \right) \cdot_* f(x -_* y, t -_* s) \cdot_* d_* y \cdot_* d_* s$$

$$+_* \int_{*\mathbb{R}^n_*} \Phi(y,e^\epsilon) \cdot_* f(x -_* y, t -_* e^\epsilon) \cdot_* d_*y$$

$$-_* \int_{*\mathbb{R}^n_*} \Phi(y,t) \cdot_* f(x -_* y, 1) \cdot_* d_*y$$

$$= \int_{*\mathbb{R}^n} \Phi(y,e^\epsilon) \cdot_* f(x -_* y, t -_* e^\epsilon) \cdot_* d_*y -_* J_3.$$

Hence,

$$u_t^*(x,t) -_* \Delta_* u(x,t) = \int_{*\mathbb{R}^n_*} \Phi(y,e^\epsilon) \cdot_* f(x -_* y, t -_* e^\epsilon) \cdot_* d_*y -_* J_3$$

$$+_* \int_{*1}^{e^\epsilon} \int_{*\mathbb{R}^n_*} \Phi(y,s) \cdot_* (-_* f_s(x -_* y, t -_* s) -_* \Delta_{*y} f(x -_* y, t -_* s))$$

$$\cdot_* d_*y \cdot_* d_*s +_* J_3$$

$$= \int_{*\mathbb{R}^n} \Phi(y,e^\epsilon) \cdot_* f(x -_* y, t -_* e^\epsilon) \cdot_* d_*y$$

$$+_* \int_{*1}^{e^\epsilon} \int_{*\mathbb{R}^n_*} \Phi(y,s) \cdot_* (-_* f_s(x -_* y, t -_* s) -_* \Delta_{*y} f(x -_* y, t -_* s))$$

$$\cdot_* d_*y \cdot_* d_*s$$

and

$$u_t^*(x,t) -_* \Delta_* u(x,t) = \lim_{e^\epsilon \to 0_*} \int_{*\mathbb{R}^n_*} \Phi(y,e^\epsilon) \cdot_* f(x -_* y, t -_* e^\epsilon) \cdot_* d_*y.$$

We observe that

$$\left| \int_{*\mathbb{R}^n_*} \Phi(y,e^\epsilon) \cdot_* f(x -_* y, t -_* e^\epsilon) \cdot_* d_*y -_* f(x,t) \right|_*$$

$$\le \int_{*\mathbb{R}^n_*} \Phi(y,e^\epsilon) \cdot_* |f(x -_* y, t -_* e^\epsilon) -_* f(x,t)|_* \cdot_* d_*y$$

$$= \int_{*B(x,\delta)} \Phi(y,e^\epsilon) \cdot_* |f(x -_* y, t -_* e^\epsilon) -_* f(x,t)|_* \cdot_* d_*y$$

$$+_* \int_{*\mathbb{R}^n_* \setminus B(x,\delta)} \Phi(y,e^\epsilon) \cdot_* |f(x -_* y, t -_* \epsilon) -_* f(x,t)|_* \cdot_* d_*y$$

$$= J_4 +_* J_5,$$

$$J_4 \leq e^\epsilon \cdot_* \int_{*\mathbb{R}^n_*} \Phi(y,e^\epsilon) \cdot_* d_*y$$

$$= e^\epsilon,$$

$$J_5 \leq (e^{2M}/_*(e^{4\pi\epsilon})^{\frac{n}{2}}_*) \cdot_* \int_{*\mathbb{R}^n_* \setminus B(x,\delta)} e^{-*(|y|^{2*}_*)/_*(e^{4\epsilon})} \cdot_* d_*y$$

$$\leq e^{\frac{C_2}{\epsilon}}_* \cdot_* \int_{e^\delta}^\infty e^{-*((r^{2*})/_*(e^{4\epsilon}))} \cdot_* r^{(n-1)*} \cdot_* d_*r$$

$$\leq e^{C_3\epsilon},$$

where C_2 and C_3 are positive constants independent of $0 < \epsilon < 1$. Therefore,

$$\left| \int_{*\mathbb{R}^n_*} \Phi(y,e^\epsilon) \cdot_* f(x -_* y, t -_* e^\epsilon) \cdot_* d_*y -_* f(x,t) \right|_* \leq e^{(1+C_3)\epsilon}.$$

From here,

$$\lim_{e^\epsilon \to 0_*} \int_{*\mathbb{R}^n_*} \Phi(y,e^\epsilon) \cdot_* f(x -_* y, t -_* e^\epsilon) \cdot_* d_*y = f(x,t)$$

and

$$u_t^*(x,t) -_* \Delta_* u(x,t) = 1 \quad \text{in} \quad \mathbb{R}^n_* \times (1,\infty).$$

3. From the definition of the function u, we have

$$|u(x,t)|_* = \left| \int_{*1}^t \int_{*\mathbb{R}^n_*} \Phi(y,s) \cdot_* f(x -_* y, t -_* s) \cdot_* d_*y \cdot_* d_*s \right|_*$$

$$\leq \int_{*1}^t \int_{*\mathbb{R}^n_*} \Phi(y,s) \cdot_* |f(x -_* y, t -_* s)|_* \cdot_* d_*y \cdot_* d_*s$$

$$\le \; e^M \cdot_* \int\limits_{*1}^{t} \int\limits_{*\mathbb{R}^n_*} \Phi(y,s) \cdot_* d_*y \cdot_* d_*s$$

$$= \; e^M \cdot_* t.$$

Consequently,

$$\lim_{\substack{(x,t) \to (x^0, 1) \\ x \in \mathbb{R}^n_*, t > 1}} u(x,t) = 1 \quad \text{for} \quad \text{each} \quad x^0 \in \mathbb{R}^n_*.$$

Example 5.3 Consider the following Cauchy problem:

$$u_t^* -_* u_{x_1 x_1}^{**} -_* u_{x_2 x_2}^{**} = x_1 +_* x_2 \quad \text{in} \quad \mathbb{R}^2_* \times (1, \infty)$$

$$u(x_1, x_2, 1) = 1 \quad \text{in} \quad \mathbb{R}^2_*.$$

We have

$$u(x_1, x_2, t) = \int\limits_{*1}^{t} (e/_*(e^{4\pi} \cdot_* (t -_* s))) \cdot_* \int\limits_{*\mathbb{R}^2_*} e^{-_*((|x -_* y|^{2*})/_*(e^4 \cdot_* (t -_* s)))}$$

$$\cdot_* (y_1 +_* y_2) \cdot_* d_*y_1 \cdot_* d_*y_2 \cdot_* d_*s$$

$$= \int\limits_{*1}^{t} (e/_*(e^{4\pi} \cdot_* (t -_* s)))$$

$$\cdot_* \int\limits_{-_*\infty}^{\infty} \int\limits_{-_*\infty}^{\infty} e^{-_*((z_1^{2*} +_* z_2^{2*})/_*(e^4 \cdot_* (t -_* s)))}$$

$$\cdot_* (x_1 +_* x_2 -_* z_1 -_* z_2) \cdot_* d_*z_1 \cdot_* d_*z_2 \cdot_* d_*s$$

$$= \; t \cdot_* (x_1 +_* x_2).$$

Exercise 5.3 Find a solution to the Cauchy problem:

$$u_t^* -_* u_{x_1 x_1}^{**} -_* u_{x_2 x_2}^{**} -_* u_{x_3 x_3}^{**} = e^2 \cdot_* t \cdot_* (x_1 -_* x_2 +_* x_3) \quad \text{in} \quad \mathbb{R}^3_* \times (1, \infty)$$

$$u(x_1, x_2, x_3, 1) = 1 \quad \text{in} \quad \mathbb{R}^3_*.$$

Answer

$$u(x_1, x_2, x_3, t) = t^{2*} \cdot_* (x_1 -_* x_2 +_* x_3).$$

Remark 5.1 If u_1 is a solution to the problem (5.7) and u_2 is a solution to the problem (5.8), then by Theorem 5.7 and Theorem 5.8, it follows that

$$u(x,t) \quad = \quad u_1(x,t) +_* u_2(x,t)$$

$$= \int_{*\mathbb{R}^n_*} \Phi(x -_* y, t) \cdot_* \phi(y) \cdot_* d_* y +_* \int_{*1}^{t} \int_{*\mathbb{R}^n_*} \Phi(x -_* y, t -_* s) \cdot_* f(y,s)$$

$$\cdot_* d_* y \cdot_* d_* s,$$

where f and ϕ satisfy the hypotheses above, is a solution to the problem:

$$u_t^* -_* \Delta_* u = f \quad \text{in} \quad \mathbb{R}^n_* \times (1, \infty)$$

$$u = \phi \quad \text{on} \quad \mathbb{R}^n_* \times \{t = 1\}.$$

Exercise 5.4 Find a solution to the Cauchy problem:

$$u_t^* -_* u_{x_1 x_1}^{**} -_* u_{x_2 x_2}^{**} = e^2 \cdot_* t -_* e^2 \cdot_* t^{3*} +_* x_1 +_* e^3 \cdot_* t^{2*} \cdot_* x_2^{2*}$$

$$\text{in} \quad \mathbb{R}^2_* \times (1, \infty)$$

$$u(x_1, x_2, 1) = x_2 \quad \text{in} \quad \mathbb{R}^2_*.$$

Answer

$$u(x_1, x_2, t) = t^{2*} +_* t \cdot_* x_1 +_* t^{3*} \cdot_* x_2^{2*} +_* x_2.$$

5.4 The Mean Value Formula

For fixed $x \in \mathbb{R}^n_*$, $t \in \mathbb{R}_*$, we define

$$W(x, t, r) = \left\{ (y, s) \in \mathbb{R}^{n+1}_* : s \leq t, \quad \Phi(x -_* y, t -_* s) \geq (e/_* r^{n*}) \right\}.$$

We set

$$W(r) = W(1, 1, r).$$

Example 5.4 For $n = 2$ we will compute

$$\int \int_{*W(1)} (|y|_*^{2*} /_* s^{2*}) \cdot_* d_* y \cdot_* d_* s.$$

We have

$$W(e) = \left\{ (y, s) \in \mathbb{R}^3_* : -_*(e/_*(e^{4\pi} \cdot_* s)) \cdot_* e^{(|y|_*^{2*} /_* (e^4 \cdot_* s))} \geq e, \quad s \leq 1 \right\}.$$

Hence,

$$W(e) = \left\{ (y,s) \in \mathbb{R}^3_* : |y|_* \le e^2 \cdot_* (s \cdot_* \log_*(-_*e^{4\pi} \cdot_* s))^{\frac{1}{2}}_*, \quad -_*e^{\frac{1}{4\pi}} \le s \le 1 \right\}.$$

Therefore,

$$\iint_{*W(e)} (|y|^{2*}/_*s^{2*}) \cdot_* d_*y \cdot_* d_*s$$

$$= \int_{*-_*e^{\frac{1}{4\pi}}}^{1} \int_{*|y|_* \le e^2 \cdot_* (s \cdot_* \log_*(-_*e^{4\pi} \cdot_* s))^{\frac{1}{2}}_*} (|y|^{2*}/_*s^{2*}) \cdot_* d_*y \cdot_* d_*s$$

$$= e^{2\pi} \cdot_* \int_{*-_*e^{\frac{1}{4\pi}}}^{1} (e/_*s^{2*}) \cdot_* \int_{1}^{e^2} (s \cdot_* \log_*(-_*e^{4\pi} \cdot_* s))^{\frac{1}{2}}_* \cdot_* r^{3*} \cdot_* d_*r \cdot_* d_*s$$

$$= e^4.$$

Exercise 5.5 Consider the case $n = 3$ and compute

$$\iint_{*W(e)} (|y|^{2*}/_*s^{2*}) \cdot_* d_*y \cdot_* d_*s.$$

Generalize the result for arbitrary n.

Answer e^4.

Theorem 5.9 *(The Mean Value Formula) Let $u \in C_{*Q_T}$ solve the heat equation:*

$$u_t^* -_* \Delta_* u = 1.$$

Then,

$$u(x,t) = (e/_*(e^4 \cdot_* r^{n*})) \cdot_* \iint_{*W(x,t,r)} u(y,s)(|x -_* y|^{2*}_*/_*(t -_* s)^{2*}) \cdot_* d_*y \cdot_* d_*s.$$

Proof Without loss of generality, we will prove the formula for $x = 1, t = 1$. Otherwise, we translate the space and time coordinates so that $x = 1$, $t = 1$. Let

$$\phi(r) = \iint_{*W(e)} u(r \cdot_* z, r^{2*} \cdot_* \tau) \cdot_* (|z|^{2*}_*/_*\tau^{2*}) \cdot_* d_*z \cdot_* d_*\tau.$$

Then,

$$\phi^*(r) = \iint_{*W(e)} \left(\sum_{*i=1}^{n} u_{y_i}^* \cdot_* z_i(|z|^{2*}/_*\tau^{2*}) +_* e^2 \cdot_* r \cdot_* u_s^* \cdot_* (|z|^{2*}/_*\tau) \right)$$

$$\cdot_* d_*z \cdot_* d_*\tau$$

$$= (e/_*r^{(n+1)*}) \cdot_* \int\!\!\!\int_{*W(r)} \sum_{*i=1}^{n} u^*_{y_i} \cdot_* y_i \cdot_* (|y|^{2*}/_*s^{2*}) \cdot_* d_*y \cdot_* d_*s$$

$$+_*(e^2/_*r^{(n+1)*}) \cdot_* \int\!\!\!\int_{*W(r)} u^*_{s} \cdot_* (|y|^{2*}/_*s) \cdot_* d_*y \cdot_* d_*s$$

$$= A +_* B.$$

We introduce the function

$$\psi(y, r, s) = -_*e^{\frac{n}{2}} \cdot_* \log(-_*e^{4\pi} \cdot_* s) +_* (|y|^{2*}/_*(e^4 \cdot_* s)) +_* e^n \cdot_* \log_* r.$$

We have

$$(e/_*(-_*e^{4\pi} \cdot_* s)^{\frac{n}{2}*}) \cdot_* e^{(|y|^{2*}/_*(e^4 \cdot_* s))} = (e/_*r^{n*}) \quad \text{on} \quad \partial_*W(r).$$

Hence, $\psi = 1$ on $\partial_*W(r)$. Also,

$$\psi^*_{y_i} = (y_i/_*(e^2 \cdot_* s)), \quad \sum_{*i=1}^{n} y_i \cdot_* \psi^*_{y_i}$$

$$= (e/_*(e^2 \cdot_* s)) \cdot_* \sum_{*i=1}^{n} y_i^{2*}$$

$$= (|y|^{2*}/_*(e^2 \cdot_* s)),$$

whereupon

$$(|y|^{2*}/_*s) = e^2 \cdot_* \sum_{*i=1}^{n} y_i \cdot_* \psi^*_{y_i}.$$

Therefore,

$$B = (e^4/_*r^{(n+1)*}) \cdot_* \int\!\!\!\int_{*W(r)} u^*_s \cdot_* \sum_{*i=1}^{n} y_i \cdot_* \psi^*_{y_i} \cdot_* d_*y \cdot_* d_*s$$

$$= (e^4/_*r^{(n+1)*}) \cdot_* \int\!\!\!\int_{*W(r)} u^*_s \cdot_* \sum_{*i=1}^{n} ((y_i \cdot_* \psi)^*_{y_i} -_* \psi) \cdot_* d_*y \cdot_* d_*s$$

$$= (e^4/_*r^{(n+1)*}) \cdot_* \int\!\!\!\int_{*W(r)} \left(-_*nu^*_s \cdot_* \psi +_* u^*_s \cdot_* \sum_{*i=1}^{n} (y_i \cdot_* \psi)^*_{y_i} \right)$$

$$\cdot_* d_*y \cdot_* d_*s \quad (\psi = 1 \quad \text{on} \quad \partial_*W(r))$$

$$= \ (e^4/_* r^{(n+1)*}) \cdot_* \int_{*W(r)} \int \left(-_* n u_s^* \cdot_* \psi -_* \sum_{*i=1}^{n} u_{s y_i}^{**} \cdot_* y_i \cdot_* \psi \right)$$

$$\cdot_* d_* y \cdot_* d_* s.$$

Now, we multiplicative integrate by parts with respect to s and we get

$$B \ = \ (e^4/_* r^{(n+1)*}) \cdot_* \int_{*W(r)} \int \left(-_* n u_s^* \cdot_* \psi +_* \sum_{*i=1}^{n} u_{y_i}^* \cdot_* y_i \cdot_* \psi_s \right)$$

$$\cdot_* d_* y \cdot_* d_* s$$

$$= \ (e^4/_* r^{(n+1)*}) \cdot_* \int_{*W(r)} \int \left(-_* n u_s^* \cdot_* \psi +_* \sum_{*i=1}^{n} u_{y_i}^* \right.$$

$$\left. \cdot_* y_i \left(-_* \left(e^n /_* (e^2 \cdot_* s) \right) -_* \left(|y|^{2*} /_* (e^4 \cdot_* s^{2*}) \right) \right) \right) \cdot_* d_* y \cdot_* d_* s$$

$$= \ -_* (e^4/_* r^{(n+1)*})$$

$$\cdot_* \int_{*W(r)} \int \left(n u_s^* \cdot_* \psi +_* (e^n/_* (e^2 \cdot_* s)) \cdot_* \sum_{*i=1}^{n} u_{y_i}^* \cdot_* y_i \right) \cdot_* d_* y \cdot_* d_* s$$

$$-_* (e/_* r^{(n+1)*}) \cdot_* \int_{*W(r)} \int \sum_{*i=1}^{n} u_{y_i}^* \cdot_* y_i \cdot_* \left(|y|^{2*} /_* s^{2*} \right) \cdot_* d_* y \cdot_* d_* s$$

$$= \ -_* (e^4/_* r^{(n+1)*})$$

$$\cdot_* \int_{*W(r)} \int \left(n u_s^* \cdot_* \psi +_* (e^n/_* (e^2 \cdot_* s)) \cdot_* \sum_{*i=1}^{n} u_{y_i}^* \cdot_* y_i \right)$$

$$\cdot_* d_* y \cdot_* d_* s -_* A.$$

Consequently,

$$\phi^*(r) \ = \ -_* (e^4/_* r^{(n+1)*})$$

$$\cdot_* \int_{*W(r)} \int \left(\nu_s^* \cdot_* \psi +_* (e^n/_* (e^2 \cdot_* s)) \cdot_* \sum_{*i=1}^{n} u_{y_i}^* \cdot_* y_i \right) \cdot_* d_* y \cdot_* d_* s$$

$$= \quad -_*(e^4/_*r^{(n+1)*})$$

$$\cdot_* \int\!\!\int_{*W(r)} \left(e^n \cdot_* \Delta_* u \cdot_* \psi +_* (e^n/_*(e^2 \cdot_* s)) \cdot_* \sum_{*i=1}^{n} u^*_{y_i} \cdot_* y_i \right)$$

$$\cdot_* d_* y \cdot_* d_* s$$

$$(\psi = 1 \quad \text{on} \quad \partial_* W(r))$$

$$= \quad -_*(e^4/_*r^{(n+1)*})$$

$$\cdot_* \int\!\!\int_{*W(r)} \left(-_* e^n \cdot_* \sum_{*i=1}^{n} u^*_{y_i} \cdot_* \psi_{y_i} +_* (e^n/_*(e^2 \cdot_* s)) \cdot_* \sum_{*i=1}^{n} u^*_{y_i} \cdot_* y_i \right)$$

$$\cdot_* d_* y \cdot_* d_* s$$

$$= \quad -_*(e^4/_*r^{(n+1)*}) \cdot_* \int\!\!\int_{*W(r)} \left(-_*(e^n/_*(e^2 \cdot_* s)) \cdot_* \sum_{*i=1}^{n} u^*_{y_i} \cdot_* y_i \right.$$

$$\left. +_*(e^n/_*(e^2 \cdot_* s)) \cdot_* \sum_{*i=1}^{n} u^*_{y_i} \cdot_* y_i \right) \cdot_* d_* y \cdot_* d_* s$$

$$= \quad 1.$$

Thus ϕ is a multiplicative constant. Hence,

$$\phi(r) \quad = \quad \lim_{t \to 1} \phi(t)$$

$$= \quad u(1,1) \cdot_* \int\!\!\int_{*W(1)} (|y|^{2*}/_*s^{2*}) \cdot_* d_* y \cdot_* d_* s$$

$$= \quad e^4 \cdot_* u(1,1),$$

whereupon

$$u(1,1) = (e/_*(e^4 \cdot_* r^{n*})) \cdot_* \int\!\!\int_{*W(r)} u(y,s) \cdot_* (|y|^{2*}/_*s^{2*}) \cdot_* d_* y \cdot_* d_* s.$$

5.5   The Maximum Principle for the Cauchy Problem

Here we consider the following Cauchy problem:

$$u_t^* -_* \Delta_* u = 1 \quad \text{in} \quad \mathbb{R}_*^n \times (1, e^T),$$

$$u = \phi \quad \text{on} \quad \mathbb{R}_*^n \times \{t = 1\}, \tag{5.9}$$

where $T > 0$ is fixed.

Theorem 5.10 *(The Maximum Principle for the Cauchy Problem) Suppose that $\phi \in \mathcal{C}_*(\mathbb{R}_*^n)$ and $u \in \mathcal{C}_*^2(\mathbb{R}_*^n, \mathcal{C}_*^1((1, e^T])) \cap \mathcal{C}_*(\mathbb{R}_*^n \times [1, e^T])$ solves the Cauchy problem* (5.9) *and satisfies the growth estimate*

$$u(x, t) \le e^A \cdot_* e^{e^a \cdot_* |x|_*^{2*}}, \quad x \in \mathbb{R}_*^n, \quad 1 \le t \le e^T,$$

for some constants $A, a > 0$. Then,

$$\sup_{\mathbb{R}_*^n \times [1, e^T]} u = \sup_{\mathbb{R}_*^n} \phi.$$

Proof

1. Let $e^{4aT} < e$. Then there are $\epsilon > 0$ and $\gamma > 0$ such that

$$e^{4a(T+\epsilon)} < e \quad \text{and} \quad e^{\frac{1}{4(T+\epsilon)}} = e^{a+\gamma}.$$

We fix $y \in \mathbb{R}_*^n$ and $\mu > 0$. Define the function

$$v(x, t) = u(x, t) -_* (e^\mu /_* (e^T +_* e^\epsilon -_* t)^{\frac{n}{2}}_*)$$
$$\cdot_* e^{(|x -_* y|_*^{2*}/_*(e^4 \cdot_* (e^T +_* e^\epsilon -_* t)))}, \quad x \in \mathbb{R}_*^n, 1 < t \le e^T.$$

We have

$$v_t^*(x, t) = u_t^*(x, t) -_* e^{\frac{n}{2}\mu} \cdot_* (e^T +_* e^\epsilon -_* t)^{-*\frac{n}{2}} \cdot_{-*\epsilon}$$
$$\cdot_* e^{(|x -_* y|_*^{2*}/_*(e^4 \cdot_*(e^T +_* e^\epsilon -_* t)))}$$

$$-_* e^{\frac{\mu}{4}} \cdot_* (e^T +_* e^\epsilon -_* t)^{-*(\frac{n}{2}-2)*}$$
$$\cdot_* |x -_* y|_*^{2*} e^{(|x -_* y|_*^{2*}/_*(e^4 \cdot_*(e^T +_* e^\epsilon -_* t)))},$$

$$v_{x_i}^*(x, t) = u_{x_i}^*(x, t) -_* e^{\frac{\mu}{2}} \cdot_* (e^T +_* e^\epsilon -_* t)^{-*(\frac{n}{2}-1)*}$$
$$\cdot_* (x_i -_* y_i) \cdot_* e^{(|x -_* y|_*^{2*}/_*(e^4 \cdot_*(e^T +_* e^\epsilon -_* t)))},$$

$$v^{**}_{x_i x_i}(x,t) \;=\; u^{**}_{x_i x_i}(x,t) -_* e^{\frac{\mu}{2}} \cdot_* \left(e^T +_* e^\epsilon -_* t\right)^{-*\left(\frac{n}{2}-1\right)*}$$

$$\cdot_* e^{\left(|x-_*y|^{2*}_*/_*\left(e^4\cdot_*\left(e^T+_*e^\epsilon-_*t\right)\right)\right)}$$

$$-_* e^{\frac{\mu}{4}} \cdot_* \left(e^T +_* e^\epsilon -_* t\right)^{-*\left(\frac{n}{2}+2\right)*} \cdot_* \left(x_i -_* y_i\right)^{2*}$$

$$\cdot_* e^{\left(|x-_*y|^{2*}_*/_*\left(e^4\cdot_*\left(e^T+_*e^\epsilon-_*t\right)\right)\right)}, \qquad i = 1,\ldots,n,$$

$$\Delta_* v(x,t) \;=\; \Delta_* u(x,t) -_* e^{\frac{n}{2}\mu}$$

$$\cdot_* \left(e^T +_* e^\epsilon -_* t\right)^{-*\frac{n}{2}} \cdot_* e^{\left(|x-_*y|^{2*}_*/_*\left(e^4\cdot_*\left(e^T+_*e^\epsilon-_*t\right)\right)\right)}$$

$$-_* e^{\frac{\mu}{4}} \cdot_* \left(e^T +_* e^\epsilon -_* t\right)^{-*\left(\frac{n}{2}+2\right)*}$$

$$\cdot_* |x -_* y|^{2*}_* e^{\left(|x-_*y|^{2*}_*/_*\left(e^4\cdot_*\left(e^T+_*e^\epsilon-_*t\right)\right)\right)},$$

whence

$$v^*_t(x,t) -_* \Delta_* v(x,t) = 0_* \quad \text{in} \quad \mathbb{R}^n_* \times (1, e^T].$$

Let $D = B(y,r)$. Then, by Theorem 5.2, we get

$$\max_{\overline{Q_T}} v = \max_{\partial_{*p} Q_T} v.$$

If $x \in \mathbb{R}^n_*$, then

$$v(x,1) \;=\; u(x,1) -_* e^{\frac{\mu}{(T+\epsilon)\frac{n}{2}}} \cdot_* e^{\left(|x-_*y|^{2*}_*/_*\left(e^4\cdot_*\left(e^T+_*e^\epsilon\right)\right)\right)}$$

$$\leq\; u(x,1)$$

$$=\; \phi(x).$$

If $|x -_* y|_* = r$, $1 \leq t \leq e^T$, then

$$v(x,t) = u(x,t) -_* \left(e^\mu/_*\left(e^T +_* e^\epsilon -_* t\right)\right)^{\frac{n}{2}*} e^{r^{2*}_*/_*\left(e^4\cdot_*\left(e^T+_*e^\epsilon-_*t\right)\right)}$$

$$\leq e^A \cdot_* e^{e^a \cdot_* |x|^{2*}_*} -_* \left(e^\mu/_*\left(e^T +_* e^\epsilon -_* t\right)\right)^{\frac{n}{2}*}$$

$$\cdot_* e^{\left(r^{2*}_*/_*\left(e^4\cdot_*\left(e^T+_*e^\epsilon-_*t\right)\right)\right)}$$

$$\le e^A \cdot_* e^{e^a \cdot_* (|y|_* + _*r)^{2*}} -_* (e^\mu /_* (e^T +_* e^\epsilon)^{\frac{n}{2}*})$$

$$\cdot_* e^{(r^{2*}/_*(e^4 \cdot_*(e^T +_*e^\epsilon)))}$$

$$= e^A \cdot_* e^{e^a \cdot_* (|y|_* + _*r)^{2*}} -_* (e^\mu /_* (e^4 \cdot_* (e^a +_* e^\gamma)))^{\frac{n}{2}*} \cdot_* e^{(e^a +_*e^\gamma)\cdot_* r^{2*}}$$

$$\le \sup_{\mathbb{R}^n_*} \phi$$

for $r > 0_*$ is sufficiently small.

Consequently,

$$v(x,t) \le \sup_{\mathbb{R}^n_*} \phi$$

for all $x \in \mathbb{R}^n_*$, $t \in [1, e^T]$. Let $\mu \to 0$. Then,

$$u(x,t) \le \sup_{\mathbb{R}^n_*} \phi$$

for all $x \in \mathbb{R}^n_*$ and $1 \le t \le e^T$.

2. If $4aT \ge 1$, then we apply the above result on $[1, e^{T_1}]$, $[e^{T_1}, e^{2T_1}]$, ..., for some $T_1 \in \left(0, \dfrac{1}{4a}\right)$.

Exercise 5.6 Suppose that $\phi \in \mathcal{C}_*(\mathbb{R}^n_*)$ and $u \in \mathcal{C}^2_*(\mathbb{R}^n_*, \mathcal{C}^1_*((1, e^T])) \cap \mathcal{C}_*(\mathbb{R}^n_* \times [1, e^T])$ solves the Cauchy problem (5.9) and satisfies the growth estimate

$$u(x,t) \ge -_*e^A \cdot_* e^{e^a \cdot_* |x|^{2*}_*}, \quad x \in \mathbb{R}^n_*, \quad 1 \le t \le e^T,$$

for some constants $A, a > 0$. Prove that

$$\inf_{\mathbb{R}^n_* \times [1, e^T]} u = \inf_{\mathbb{R}^n_*} \phi.$$

Hint 5.1 *Use the function*

$$v(x,t) = u(x,t) +_* (e^\mu /_* (e^T +_* e^\epsilon -_* t)^{\frac{n}{2}*} \cdot_* e^{(|x -_* y|^{2*}_*)/_*(e^4 \cdot_*(e^T +_*e^\epsilon -_*t))},$$

$$x \in \mathbb{R}^n_*, 1 < t \le e^T,$$

for some $\mu > 0$.

Theorem 5.11 *Let $\phi \in \mathcal{C}_*(\mathbb{R}^n_*)$, $f \in \mathcal{C}_*(\mathbb{R}^n_* \times [1, e^T])$. Then there exists at most one solution $u \in \mathcal{C}^2_*(\mathbb{R}^n_*, \mathcal{C}^1_*((1, e^T])) \cap \mathcal{C}_*(\mathbb{R}^n_* \times [1, e^T])$ of the Cauchy problem:*

$$u^*_t -_* \Delta_* u = f(x,t) \quad in \quad \mathbb{R}^n_* \times (1, e^T],$$

$$(5.10)$$

$$u = \phi \quad on \quad \mathbb{R}^n_* \times \{t = 1\},$$

satisfying the growth estimate

$$|u(x,t)|_* \leq e^A \cdot_* e^{e^a \cdot_* |x|^{2*}}, \quad x \in \mathbb{R}^n_*, \quad 1 \leq t \leq e^T, \quad (5.11)$$

for some constants $A, a > 0$.

Proof Assume that $u_1, u_2 \in C^2_*(\mathbb{R}^n_*, C^1_*((1, e^T])) \cap C_*(\mathbb{R}^n_* \times [1, e^T])$ satisfy (5.10), (5.11). Then $u_1 -_* u_2$ satisfies

$$u^*_t -_* \Delta_* u = 0 \quad \text{in} \quad \mathbb{R}^n_* \times (1, e^T],$$

$$u = 1 \quad \text{on} \quad \mathbb{R}^n_* \times \{t = 1\},$$

and (5.11). Hence and Theorem 5.10, applied for $\pm_*(u_1 -_* u_2)$, we get

$$\sup_{\mathbb{R}^n_* \times [1, e^T]} |u_1 -_* u_2|_* = 1,$$

which completes the proof.

5.6 The Method of Separation of Variables

Consider the following initial boundary value problem:

$$u^*_t -_* e^k \cdot_* u^{**}_{xx} = 1, \quad 1 < x < e^L, \quad t > 1, \quad (5.12)$$

$$u(1, t) = 1,$$
$$\qquad\qquad\qquad\qquad\qquad\qquad\qquad\qquad (5.13)$$
$$u(e^L, t) = 1, \quad t \geq 1,$$

$$u(x, 1) = \phi(x), \quad 1 \leq x \leq e^L, \quad (5.14)$$

where ϕ is a given initial condition and $k > 0$ is a given constant. Assume the compatibility condition

$$\phi(1) = \phi(e^L) = 1.$$

We seek solutions of the problem (5.12)–(5.14) that have the following special form:

$$u(x, t) = X(x) \cdot_* T(t). \quad (5.15)$$

We are not interested in the zero solution $u(x, t) = 1$. Therefore, we seek functions X and T that do not vanish identically. We substitute (5.15) into (5.12) and we find

$$(T^*(t)/_*(e^k \cdot_* T(t))) = (X^{**}(x)/_* X(x)).$$

Since x and t are independent variables, from the last equality, it follows that there exists a constant λ, which is called the separation constant, such that

$$(T^*(t)/_*(e^k \cdot_* T(t))) = (X^{**}(x)/_*X(x)) = -_*e^\lambda.$$

Using the boundary condition (5.13), because u is not the multiplicative trivial solution $u = 1$, it follows that

$$X(1) = X(e^L) = 1.$$

Thus, the function X should be a solution to the boundary problem:

$$X^{**} +_* e^\lambda \cdot_* X = 1, \quad 1 < x < e^L,$$
$$X(1) = X(e^L) = 1.$$

As in (4.1.3), we find

$$X(x) = \sin_*((e^{n\pi} \cdot_* x)/_*e^L),$$

$$e^\lambda = \left(e^{\frac{n\pi}{L}}\right)^{2*}, \quad n \in \mathbb{N}.$$

For convenience, we use the notation

$$X_n(x) = \sin_*((e^{n\pi} \cdot_* x)/_*e^L),$$

$$e^{\lambda_n} = \left(e^{\frac{n\pi}{L}}\right)^{2*}, \quad n \in \mathbb{N}.$$

Hence,

$$T_n(t) = e^{B_n} \cdot_* e^{-_*e^k \left(e^{\frac{n\pi}{L}}\right)^{2*} \cdot_* t}, \quad n \in \mathbb{N}.$$

Thus, we obtain the following sequence of separated solutions:

$$u_n(x,t) = X_n(x) \cdot_* T_n(t)$$

$$= e^{B_n} \cdot_* e^{-_*e^k \cdot_* \left(e^{\frac{n\pi}{L}}\right)^{2*} \cdot_* t} \cdot_* \sin_*((e^{n\pi} \cdot_* x)/_*e^L), \quad n \in \mathbb{N},$$

where B_n, $n \in \mathbb{N}$, are constants.

The superposition principle implies that

$$u(x,t) = \sum_{*n=1}^{\infty} e^{B_n} \cdot_* \sin_*((e^{n\pi} \cdot_* x)/_*e^L) \cdot_* e^{-_*e^k \cdot_* \left(e^{\frac{n\pi}{L}}\right)^{2*} \cdot_* t}$$

is a formal solution of the problem (5.12)–(5.13).

We will find the constants B_n using the initial condition (5.14). We have

$$u(x,1) = \phi(x)$$

$$= \sum_{*n=1}^{\infty} e^{B_n} \cdot_* \sin_*((e^{n\pi} \cdot_* x)/_*e^L).$$

Fix $m \in \mathbb{N}$ and multiply by $\sin_*((e^{m\pi} \cdot_* x)/_* e^L)$ the last equality. Then integrating it term-by-term over $[1, e^L]$, we find

$$\sum_{*n=1}^{\infty} e^{B_n} \cdot_* \int_{*1}^{e^L} \sin_*((e^{n\pi} \cdot_* x)/_* e^L) \cdot_* \sin_*((e^{m\pi} \cdot_* x)/_* e^L) \cdot_* d_* x$$

$$= \int_{*1}^{e^L} \sin_*((e^{m\pi} \cdot_* x)/_* e^L) \cdot_* \phi(x) \cdot_* d_* x,$$

whereupon

$$e^{B_m} = e^{\frac{2}{L}} \cdot_* \int_{*1}^{e^L} \sin_*((e^{m\pi} \cdot_* x)/_* e^L) \cdot_* \phi(x) \cdot_* d_* x.$$

Example 5.5 Consider the initial boundary value problem:

$$u_t^* -_* u_{xx}^{**} = 1, \quad 1 < x < e^{\pi}, \quad t > 1,$$

$$u(1, t) = 1,$$

$$u(e^{\pi}, t) = 1, \quad t \geq 1,$$

$$u(x, 1) = x^{2*}, \quad 1 \leq x \leq e^{\pi}.$$

Here $L = \pi$, $\phi(x) = x^{2*}$, $k = e$. Then,

$$e^{B_m} = e^{\frac{2}{\pi}} \cdot_* \int_{*1}^{e^{\pi}} \sin_*(e^{m} \cdot_* x) x^{2*} \cdot_* d_* x$$

$$= e^{\frac{2}{m}\pi}(-_* e)^{(m+1)*} +_* e^{\frac{4}{m^3 \pi}} \cdot_* ((-_* e)^{m*} -_* e).$$

Then,

$$u(x, t) = \sum_{*n=1}^{\infty} \left(e^{\frac{2}{n}\pi}(-_* e)^{(n+1)*} +_* e^{\frac{4}{n^3 \pi}} \cdot_* ((-_* e)^{n*} -_* e) \right)$$

$$\cdot_* \sin_*(e^{n} \cdot_* x) e^{-_* n^{2*} \cdot_* t}$$

is a formal solution of the considered problem.

Exercise 5.7 Find a formal solution to the problem:

$$u_t^* -_* u_{xx}^{**} = 1, \quad 1 < x < e^\pi, \quad t > 1,$$
$$u(1,t) = 1,$$
$$u(\pi,t) = 1, \quad t \geq 1,$$
$$u(x,1) = x +_* \cos_* x, \quad 1 \leq x \leq e^\pi.$$

Answer

$$u(x,t) = e^2 \cdot_* \sin_* x \cdot_* e^{-_* t} +_* \sum_{*n=2}^{\infty} \left(e^{\frac{2}{n}}(-_*e)^{(n+1)_*} +_* e^{\frac{2n}{\pi(n^2-1)}}((-_*e)^n +_* e) \right)$$

$$\cdot_* \sin_* (e^n \cdot_* x) \cdot_* e^{-_* e^{n^2} \cdot_* t}.$$

Next, we consider the following problem:

$$u_t^* -_* e^k \cdot_* u_{xx}^{**} = f(x,t), \quad 1 < x < e^L, \quad t > 1, \tag{5.16}$$

$$u(1,t) = 1,$$
$$u(e^L,t) = 1, \quad t \geq 1, \tag{5.17}$$

$$u(x,1) = \phi(x), \quad 1 \leq x \leq e^L, \tag{5.18}$$

where

$$f(x,t) = \sum_{*n=1}^{\infty} f_n(t) \cdot_* \sin_* ((e^{n\pi} \cdot_* x)/_* e^L),$$
$$\phi(x) = \sum_{*n=1}^{\infty} e^{A_n} \sin_* ((e^{n\pi} \cdot_* x)/_* e^L),$$

f_n are given continuous functions and A_n are given constants, $n \in \mathbb{N}$.

Let $u(x,t)$ be a formal solution to the problem (5.16)–(5.18) that has the form

$$u(x,t) = \sum_{*n=1}^{\infty} T_n(t) \cdot_* \sin_* ((e^{n\pi} \cdot_* x)/_* e^L.) \tag{5.19}$$

Substituting (5.19) into (5.16), we get

$$\sum_{*n=1}^{\infty} \left(T_n^*(t) +_* e^k \cdot_* e^{\frac{n^2\pi^2}{L^2}} \cdot_* T_n(t) \right) \cdot_* \sin_* ((e^{n\pi} \cdot_* x)/_* e^L)$$

$$= \sum_{*n=1}^{\infty} f_n(t) \cdot_* \sin_* ((e^{n\pi} \cdot_* x)/_* e^L),$$

whereupon

$$T_n^* +_* e^k \cdot_* e^{\frac{n^2\pi^2}{L^2}} \cdot_* T_n = f_n$$

and

$$T_n(t) = e^{-_* e^{k\frac{n^2\pi^2}{L^2}} \cdot_* t} \cdot_* \left(e^{B_n} +_* \int_* f_n(t) e^{e^{k\frac{n^2\pi^2}{L^2}} \cdot_* t} dt \right), \quad t > 1,$$

where B_n, $n \in \mathbb{N}$, are constants which will be determined below.
We set

$$g_n(t) = \int_* f_n(t) \cdot_* e^{e^{k\frac{n^2\pi^2}{L^2}} \cdot_* t} \cdot_* d_* t.$$

Therefore,

$$u(x,t) = \sum_{*n=1}^{\infty} e^{-_* e^{k\frac{n^2\pi^2}{L^2}} \cdot_* t} \cdot_* \left(e^{B_n} +_* g_n(t) \right) \cdot_* \sin_*((e^{n\pi} \cdot_* x)/_* e^L).$$

We will find the constants B_n using the initial condition (5.18). We have

$$
\begin{aligned}
u(x,1) &= \sum_{*n=1}^{\infty} \left(e^{B_n} +_* g_n(1) \right) \cdot_* \sin_*((e^{n\pi} \cdot_* x)/_* e^L) \\
&= \sum_{*n=1}^{\infty} e^{A_n} \cdot_+ * \sin_*((e^{n\pi} \cdot_* x)/_* e^L).
\end{aligned}
$$

Hence,

$$e^{B_n} = e^{A_n} -_* g_n(1)$$

and

$$u(x,t) = \sum_{*n=1}^{\infty} e^{-_* e^{k\frac{n^2\pi^2}{L^2}} \cdot_* t} \cdot_* \left(e^{A_n} -_* g_n(0) +_* g_n(t) \right) \cdot_* \sin_*((e^{n\pi} \cdot_* x)/_* e^L).$$

Example 5.6 Consider the following problem:

$$
\begin{aligned}
u_t^* -_* u_{xx}^{**} &= t \cdot_* x, \quad 1 < x < e^\pi, \quad t > 1, \\
u(1,t) &= 1, \\
u(e^\pi, t) &= 1, \quad t \geq 1, \\
u(x,1) &= x^{2*}, \quad 1 \leq x \leq e^\pi.
\end{aligned}
$$

Here,

$$
\begin{aligned}
f(x,t) &= t \cdot_* x, \\
\phi(x) &= x^{2*},
\end{aligned}
$$

$$k = 1,$$

$$L = \pi.$$

Note that

$$\int_{*1}^{e^\pi} x \cdot_* \sin_*(e^n \cdot_* x) \cdot_* d_* x = e^{\frac{\pi}{n}} \cdot_* (-_* e)^{(n+1)_*},$$

$$\int_{*1}^{e^\pi} x^{2_*} \cdot_* \sin_*(e^n \cdot_* x) \cdot_* d_* x = e^{\frac{\pi^2}{n}} \cdot_* (-_* e)^{(n+1)_*} +_* e^{\frac{2}{n^3}} \cdot_* ((-_* e)^{n_*} -_* e).$$

Then,

$$f(x,t) = t \cdot_* \sum_{*n=1}^{\infty} e^{\frac{2}{n}} \cdot_* (-_* e)^{(n+1)_*} \cdot_* \sin_*(e^n \cdot_* x),$$

$$f_n(t) = e^{\frac{2}{n}} \cdot_* (-_* e)^{(n+1)_*} \cdot_* t,$$

$$\phi(x) = \sum_{*n=1}^{\infty} \left(e^{\frac{2\pi}{n}} \cdot_* (-_* e)^{(n+1)_*} +_* e^{\frac{4}{n^3\pi}} \cdot_* ((-_* e)^{n_*} -_* e) \right)$$

$$\cdot_* \sin_*(e^n \cdot_* x),$$

$$e^{A_n} = e^{\frac{2\pi}{n}} \cdot_* (-e)^{(n+1)_*} +_* e^{\frac{4}{n^3\pi}} \cdot_* ((-_* e)^{n_*} -_* e),$$

$$g_n(t) = e^{\frac{2}{n}} \cdot_* (-_* e)^{(n+1)_*} \cdot_* \int_* t \cdot_* e^{e^{n^2} \cdot_* t} \cdot_* d_* t$$

$$= e^{\frac{2}{n^3}} \cdot_* (-_* e)^{(n+1)_*} \cdot_* \left(t -_* e^{\frac{1}{n^2}} \right) \cdot_* e^{e^{n^2} \cdot_* t},$$

$$g_n(1) = e^{\frac{2}{n^5}} \cdot_* (-_* e)^{n_*},$$

$$e^{B_n} = e^{\frac{2\pi}{n}} \cdot_* (-_* e)^{(n+1)_*} +_* e^{\frac{4}{n^3\pi}} \cdot_* ((-_* e)^{n_*} -_* e) -_* e^{\frac{2}{n^5}} \cdot_* (-_* e)^{n_*}.$$

Consequently,

$$u(x,t) = \sum_{*n=1}^{\infty} e^{-_* e^{n^2} \cdot_* t} \cdot_* \left(e^{\frac{2\pi}{n}} \cdot_* (-_* e)^{(n+1)_*} +_* e^{\frac{4}{n^3\pi}} \cdot_* ((-_* e)^{n_*} -_* e) \right.$$

$$-_* e^{\frac{2}{n^5}} \cdot_* (-_* e)^{n_*}$$

$$\left. +_* e^{\frac{2}{n^3}} \cdot_* (-_* e)^{(n+1)_*} \cdot_* \left(t -_* e^{\frac{1}{n^2}} \right) \cdot_* e^{e^{n^2} \cdot_* t} \right) \cdot_* \sin_*(e^n \cdot_* x).$$

Exercise 5.8 Find a formal solution to the problem:

$$u_t^* -_* u_{xx}^{**} = \sin_* x \cdot_* \cos_*(e^3 \cdot_* x), \quad 1 < x < e^\pi, \quad t > 1,$$

$$u(1,t) = 1,$$

$$u(e^\pi, t) = 1, \quad t \geq 1,$$

$$u(x,1) = x, \quad 1 \leq x \leq e^\pi.$$

Answer

$$u(x,t) = \sum_{\substack{n=1 \\ *\ n \neq 2,4}}^{\infty} e^{\frac{2}{n}} \cdot_* (-_*e)^{(n+1)_*} \cdot_* e^{-_*e^{n^2} \cdot_* t} \cdot_* \sin_*(e^n \cdot_* x)$$

$$+_* e^{-_*e^4 \cdot_* t} \left(-_* e^{\frac{7}{8}} -_* e^{\frac{1}{8}} \cdot_* e^{e^4 \cdot_* t} \right) \cdot_* \sin_*(e^2 \cdot_* x)$$

$$+_* e^{-_*e^{16} \cdot_* t} \cdot_* \left(-_* e^{\frac{17}{32}} +_* e^{\frac{1}{32}} \cdot_* e^{e^{16} \cdot_* t} \right) \cdot_* \sin_*(e^4 \cdot_* x).$$

5.7 The Energy Method: Uniqueness

Consider the following Cauchy problem:

$$u_t^* -_* \Delta_* u = f(x,t) \quad \text{in} \quad Q_T,$$

$$(5.20)$$

$$u = \phi \quad \text{on} \quad D \times \{t = 1\},$$

where $f \in C_*(\overline{Q_T})$, $\phi \in C_*(\overline{D})$.

Theorem 5.12 *There exists at most one solution $u \in C_{*Q_T}$ of the problem (5.20).*

Proof Suppose that $u_1, u_2 \in C_{*Q_T}$ are solutions of the problem (5.20). Let

$$v = u_1 -_* u_2.$$

Then v satisfies the problem

$$v_t^* -_* \Delta_* v = 1 \quad \text{in} \quad Q_T$$

$$v = 1 \quad \text{on} \quad D \times \{t = 1\}.$$

Set

$$E(t) = \int_{*D} (v(x,t))^{2_*} \cdot_* d_*x, \quad 1 \leq t \leq e^T.$$

Then,

$$E^*(t) = e^2 \cdot_* \int_{*D} v(x,t) \cdot_* v_t^*(x,t) \cdot_* d_*x$$

$$= e^2 \cdot_* \int_{*D} v(x,t) \cdot_* \Delta_* v(x,t) \cdot_* d_*x$$

$$= -_*e^2 \cdot_* \int_{*D} |\nabla v(x,t)|_*^{2*} \cdot_* d_*x$$

$$\leq 1, \quad 1 \leq t \leq e^T.$$

Hence,

$$E(t) \leq E(1) = 1, \quad 1 \leq t \leq e^T.$$

Consequently, $u_1 = u_2$ in Q_T.

5.8 Advanced Practical Exercises

Problem 5.1 Find a solution to the following Cauchy problem:

$$u_t^* -_* u_{x_1 x_1}^{**} -_* u_{x_2 x_2}^{**} -_* u_{x_3 x_3}^{**} = 1 \quad \text{in} \quad \mathbb{R}_*^3 \times (1, \infty),$$

$$u(x_1, x_2, x_3, 1) = e^2 \cdot_* x_1^{2*} -_* e^2 \cdot_* x_2^{2*} +_* e^3 \cdot_* x_3^{2*} \quad \text{in} \quad \mathbb{R}_*^3.$$

Answer

$$u(x_1, x_2, x_3, t) = e^2 \cdot_* x_1^{2*} -_* e^2 \cdot_* x_2^{2*} +_* e^3 \cdot_* x_3^{2*} +_* e^6 \cdot_* t.$$

Problem 5.2 Find a solution to the following Cauchy problem:

$$u_t^* -_* u_{x_1 x_1}^{**} -_* u_{x_2 x_2}^{**} -_* u_{x_3 x_3}^{**} = e^2 \cdot_* t \cdot_* x_1 +_* x_2 +_* e^6 \cdot_* t^{2*} \cdot_* x_3$$

$$\text{in} \quad \mathbb{R}_*^3 \times (1, \infty),$$

$$u(x_1, x_2, x_3, 1) = 1 \quad \text{in} \quad \mathbb{R}_*^3.$$

Answer

$$u(x_1, x_2, x_3, t) = t^{2*} \cdot_* x_1 +_* t \cdot_* x_2 +_* e^2 \cdot_* t^{3*} \cdot_* x_3.$$

Problem 5.3 Find a solution to the following Cauchy problem:

$$u_t^* -_* u_{x_1 x_1}^{**} -_* u_{x_2 x_2}^{**} -_* u_{x_3 x_3}^{**} = e^2 \cdot_* t \cdot_* (x_1^{2*} +_* x_2^{2*} +_* x_3^{2*}) -_* e^6 \cdot_* t^{2*}$$

$$\text{in} \quad \mathbb{R}_*^3 \times (1, \infty),$$

$$u(x_1, x_2, x_3, 1) = x_1 \quad \text{in} \quad \mathbb{R}_*^3.$$

Answer

$$u(x_1, x_2, x_3, t) = t^{2*} \cdot_* (x_1^{2*} +_* x_2^{2*} +_* x_3^{2*}) +_* x_1.$$

Problem 5.4 Find a formal solution to the following problem:

$$u_t^* -_* u_{xx}^{**} - 1, \quad 1 < x < c^{\pi}, \quad t > 1,$$

$$u(1, t) = 1,$$

$$u(e^{\pi}, t) = 1, \quad t \geq 1,$$

$$u(x, 1) = \begin{cases} x & 1 \leq x \leq e^{\frac{\pi}{2}} \\ e^{\pi} -_* x & e^{\frac{\pi}{2}} \leq x \leq e^{\pi}. \end{cases}$$

Answer

$$u(x, t) = e^{\frac{4}{\pi}} \cdot_* \sum_{*n=1}^{\infty} e^{\frac{(-1)^{n+1}}{(2n-1)^2}} \cdot_* \sin_*(e^{(2n-1)} \cdot_* x) \cdot_* e^{-*e^{(2n-1)^2} \cdot_* t}.$$

Problem 5.5 Find a formal solution to the following problem:

$$u_t^* -_* u_{xx}^{**} = \sin_*(x/_* e^2), \quad 1 < x < e^{\pi}, \quad t > 1,$$

$$u(1, t) = 1,$$

$$u(e^{\pi}, t) = 1, \quad t \geq 1,$$

$$u(x, 1) = x, \quad 1 \leq x \leq e^{\pi}.$$

Answer

$$u(x, t) = \sum_{*n=1}^{\infty} e^{-*e^{n^2} \cdot_* t} \cdot_* \left(e^{\frac{2}{n}(-1)^{n+1} + (-1)^n \frac{8}{n(4n^2-1)\pi}} +_* e^{\frac{8}{n(4n^2-1)\pi}(-1)^{n-1}} \right.$$
$$\left. \cdot_* e^{e^{n^2} \cdot_* t} \right) \cdot_* \sin_*(e^n \cdot_* x).$$

Problem 5.6 Find a formal solution to the following problem:

$$u_t^* -_* e^k \cdot_* u_{xx}^{**} = 1, \quad 1 < x < e^L, \quad t > 1,$$

$$u_x^*(1, t) = 1,$$

$$u_x^*(e^L, t) = 1, \quad t \geq 1,$$

$$u(x, 1) = \phi(x), \quad 1 \leq x \leq e^L,$$

where $\phi \in C_*^2([1, e^L])$, $\phi^*(1) = \phi^*(e^L) = 1$.

Answer

$$u(x, t) = \sum_{*n=0}^{\infty} e^{B_n} \cdot_* e^{-*e^{k \frac{n^2\pi^2}{L^2}} \cdot_* t} \cdot_* \cos_*((e^{n\pi} \cdot_* x)/_* e^L),$$

where

$$B_n = \frac{2}{L} \int_0^L \phi(x) \cos \frac{n\pi x}{L} dx, \quad n \in \mathbb{N},$$

$$B_0 = \frac{1}{L} \int_0^L \phi(x) dx.$$

Problem 5.7 Find a formal solution to the following problem:

$$u_t^* -_* e^k \cdot_* u_{xx}^{**} = 1, \quad 1 < x < e^L, \quad t > 1,$$

$$u(1, t) = 1,$$

$$u_x^*(e^L, t) = 1, \quad t \geq 1,$$

$$u(x, 1) = \phi(x), \quad 1 \leq x \leq e^L,$$

where $\phi \in C_*^2([1, e^L])$, $\phi(1) = \phi^*(e^L) = 1$.

Answer

$$u(x, t) = \sum_{*n=0}^{\infty} e^{B_n} e^{-_* e^{k \frac{(2n+1)^2 \pi^2}{4L^2}} \cdot_* t} \cdot_* \sin_*(e^{\frac{(2n+1)}{2\pi L}}) \cdot_* x),$$

$$B_n = \frac{2}{L} \int_0^L \phi(x) \sin \frac{(2n+1)\pi x}{2L} dx, \quad n \in \mathbb{N}_0.$$

Problem 5.8 Find a formal solution to the following problem:

$$u_t^* -_* e^k \cdot_* u_{xx}^{**} = 1, \quad 1 < x < e^L, \quad t > 1,$$

$$u_x^*(1, t) = 1,$$

$$u(e^L, t) = 1, \quad t \geq 1,$$

$$u(x, 1) = \phi(x), \quad 1 \leq x \leq e^L,$$

where $\phi \in C_*^2([1, e^L])$, $\phi^*(1) = \phi(e^L) = 1$.

Answer

$$u(x, t) = \sum_{*n=0}^{\infty} e^{B_n} \cdot_* e^{-_* e^{k \frac{(2n+1)^2 \pi^2}{4L^2}} \cdot_* t} \cdot_* \cos_*(e^{\frac{(2n+1)\pi}{2l}} \cdot_* x),$$

$$B_n = \frac{2}{L} \int_0^L \phi(x) \cos \frac{(2n+1)\pi x}{2L} dx, \quad n \in \mathbb{N}_0.$$

6

The Laplace Equation

6.1 Basic Properties of Elliptic Problems

Let D be a domain in $\mathbb{R}_*^n$, $n \geq 2$, with sufficiently multiplicative smooth boundary $\partial_* D$. Consider the multiplicative Laplace[1] equation, shortly Laplace equation,

$$\Delta_* u = \sum_{*i=1}^{n} u_{x_i x_i}^{**} = 1, \quad x = (x_1, x_2, \ldots, x_n) \in D. \tag{6.1}$$

Definition 6.1 A $\mathcal{C}_*^2$-function u satisfying (6.1) is called a multiplicative harmonic function, shortly harmonic equation.

The Laplace equation is a special case of a more general equation

$$\Delta_* u = f(x), \quad x \in D. \tag{6.2}$$

where f is a given function.

Definition 6.2 Equation (6.2) is called multiplicative Poisson equation, shortly Poisson equation.

Definition 6.3 The problem defined by Poisson equation and the Dirichlet boundary condition
$$u(x) = \phi(x), \quad x \in \partial D,$$
for a given function ϕ is called the Dirichlet problem.

Definition 6.4 The problem defined by the Poisson equation and the Neumann boundary condition

$$\partial_{*\nu} u(x) = \phi(x), \quad x \in \partial D,$$

where ϕ is a given function, $\nu = (\nu_1, \ldots, \nu_n)$ denotes the unit outward multiplicative normal to $\partial_* D$, and $\partial_{*\nu}$ denotes a multiplicative differentiation in the multiplicative direction of ν, i.e., $\partial_{*\nu} = \nu \cdot_* \nabla_*$, is called the Neumann problem.

---

[1]Pierre-Simon Laplace (23 March 1749- March 1827) was an influental French scholar whose work was important to the development of mathematics, statistcs, physics, and astronomy. He was one of the first scientists to postulate the existence of black holes and the notion of gravitational collapse.

DOI: 10.1201/9781003440116-6

Definition 6.5 The problem defined by the Poisson equation and the Robin[2] boundary condition (the boundary condition of the third kind)

$$u(x) +_* \alpha(x) \cdot_* \partial_{*\nu} u(x) = \phi(x), \quad x \in \partial_* D,$$

where α and ϕ are given functions, is called the Robin problem (problem of the third kind).

Theorem 6.1 *A necessary condition for the existence of a solution to the Neumann problem*

$$\Delta_* u = f(x), \quad x \in D,$$

$$\partial_{*\nu} u(x) = \phi(x), \quad x \in \partial_* D,$$

is

$$\int_{*\partial_* D} \phi(x(s)) \cdot_* d_* s = \int_{*D} f(y) \cdot_* d_* y, \qquad (6.3)$$

where $x(s)$ is a multiplicative parametrization of $\partial_* D$.

Proof Note that

$$\Delta_* u = \nabla_* \cdot_* \nabla_* u.$$

Therefore, we can rewrite the Poisson equation as follows:

$$\nabla_* \cdot_* \nabla_* u = f(x),$$

which we multiplicative integrate over D and using the multiplicative Gauss theorem, we find

$$\int_{*\partial_* D} \nabla_* u \cdot_* \nu \cdot_* d_* s = \int_{*D} f(y) \cdot_* d_* y,$$

i.e., equation (6.3) holds.

Remark 6.1 It is useful to observe that for harmonic functions we have

$$\int_{*\Gamma} \partial_{*\nu} u \cdot_* d_* s = 1 \qquad (6.4)$$

for any closed curve Γ that is fully contained in D.

Theorem 6.2 *Let u and v be harmonic functions defined in D. Then,*

$$\int_{*\partial_* D} \left(v(y) \cdot_* \partial_{*\nu(y)} u(y) -_* u(y) \cdot_* \partial_{*\nu(y)} v(y) \right) \cdot_* d_* s_y = 1. \qquad (6.5)$$

---

[2]Victor Gustave Robin (1855–1897) was a French mathematical analyst and applied mathematician who lectured in mathematical physics at the Sorbonne in Paris and also worked in the area of thermodynamics. He is known especially for the Robin boundary condition.

Proof Note that

$$\sum_{*i=1}^{n} \left(v \cdot_* u_{x_i}^*\right)_{x_i}^* = \sum_{*i=1}^{n} v_{x_i}^* \cdot_* u_{x_i}^* +_* \sum_{*i=1}^{n} v \cdot_* u_{x_i x_i}^{**}$$

$$= \sum_{*i=1}^{n} u_{x_i}^* \cdot_* v_{x_i}^*,$$

$$\sum_{*i=1}^{n} \left(u \cdot_* v_{x_i}^*\right)_{x_i} = \sum_{*i=1}^{n} u_{x_i}^* \cdot_* v_{x_i}.$$

Therefore,

$$\sum_{*i=1}^{n} \left(\left(v \cdot_* u_{x_i}^*\right)_{x_i}^* -_* \left(u \cdot_* v_{x_i}^*\right)_{x_i}^* \right) = 1,$$

which we multiplicative integrate over D and using the Gauss theorem, we obtain the desired result (6.5). This completes the proof.

Example 6.1 (The Hadamard[3] Example) Consider the Laplace equation in the domain $-_*\infty < x_1 < \infty$, $x_2 > 1$, under the Cauchy conditions:

$$u^n(x_1, 1) = 1,$$
$$\quad (6.6)$$
$$u_{x_2}^{n*}(x_1, 1) = \sin_*(e^n \cdot_* x_1)/_* e^n, \quad -_*\infty < x_1 < \infty,$$

where $n \in \mathbb{N}$. Let

$$u^n(x_1, x_2) = e^{\frac{1}{n^2}} \cdot_* \sin_*(e^n \cdot_* x_1) \cdot_* \sinh_*(e^n \cdot_* x_2).$$

Then,

$$u_{x_1 x_1}^{n**} = -_* \sin_*(e^n \cdot_* x_1) \cdot_* \sinh_*(e^n \cdot_* x_2),$$

$$u_{x_2 x_2}^{n**} = \sin_*(e^n \cdot_* x_1) \cdot_* \sinh_*(e^n \cdot_* x_2),$$

whereupon

$$\Delta_* u^n(x_1, x_2) = 1,$$

i.e., $u^n(x_1, x_2)$ is a harmonic function satisfying (6.6).

When n is large enough, the initial condition (6.6) describes an arbitrary small perturbation of the trivial solution $u = 0$. On the other hand, $\sup_{x_1 \in \mathbb{R}_*^n} |u^n(x_1, x_2)|_*$ grows multiplicative exponentially fast as $n \to \infty$ for any $x_2 > 1$. Therefore, the Cauchy problem for the Laplace equation is not stable and hence it is not well-posed with respect to the initial condition (6.6).

---

[3] Jacques Salomon Hadamard (December 8, 1865–October 17, 1963) was a French mathematician who made major contributions in number theory, complex function theory, differential geometry and partial differential equations.

Definition 6.6 We define a harmonic polynomial of degree m to be a harmonic function $P_m(x_1, \ldots, x_n)$ of the form

$$P_m(x_1, \ldots, x_n) = \sum_{*0 \le i_1 + \cdots + i_n \le m} e^{a_{i_1 \ldots i_n}} \cdot_* x_1^{i_1 *} \cdot_* \cdots \cdot_* x_n^{i_n *},$$

where $a_{i_1 \ldots i_n}$ are constants.

Example 6.2 The polynomial

$$P_1(x_1, \ldots, x_n) = x_1 +_* \cdots +_* x_n$$

is a harmonic polynomial of degree 1.

Example 6.3 The polynomial

$$P_3(x_1, x_2) = x_1^{3*} -_* e^3 \cdot_* x_1 \cdot_* x_2^{2*} -_* x_2$$

is a harmonic polynomial of degree 3.

Definition 6.7 The harmonic polynomials

$$P_m^h(x_1, \ldots, x_n) = \sum_{*i_1 + \cdots + i_n = m} e^{a_{i_1 \ldots i_n}} \cdot_* x_1^{i_1 *} \ldots x_n^{i_n *}$$

are called homogeneous harmonic polynomials of order m.

6.2   The Fundamental Solution

We will find a solution u of the Laplace equation (6.1) in $D = \mathbb{R}_*^n$ that has the form

$$u(x) = v(r),$$

where

$$r = |x|_* = (x_1^{2*} +_* \cdots +_* x_n^{2*})^{\frac{1}{2}*}.$$

We have

$$u_{x_i}^* = v^*(r) \cdot_* (x_i /_* r),$$

$$u_{x_i x_i}^{**} = v^{**}(r) \cdot_* (x_i^{2*} /_* r^{2*}) +_* v^*(r) \cdot_* ((r^{2*} -_* x_i^{2*}) /_* r^{3*}), \quad i = 1, \ldots, n,$$

$$\Delta_* u = \sum_{*i=1}^{n} u_{x_i x_i}^{**}$$

$$= \sum_{*i=1}^{n} \left(v^{**}(r) \cdot_* (x_i^{2*} /_* r^{2*}) +_* v^*(r) \cdot_* ((r^{2*} -_* x_i^{2*}) /_* r^{3*}) \right)$$

$$= v^{**}(r) +_* (e^{n-1} /_* r) \cdot_* v^*(r).$$

Then, $\Delta_* u = 1$ if and only if

$$v^{**} +_* (e^{n-1}/_*r) \cdot_* v^* = 1.$$

1. Let $n = 2$. Then,

$$v^{**} +_* (e/_*r) \cdot_* v^* = 1,$$

whence

$$v(r) = e^a \cdot_* \log_* r +_* e^b, \quad a, b = \text{const.}$$

2. Let $n \geq 3$. Then,

$$v(r) = (e^a/_* r^{(n-2)*}) +_* e^b, \quad a, b = \text{const.}$$

Definition 6.8 The function, defined by

$$\Phi(x) = \begin{cases} -_* e^{\frac{1}{2\pi}} \cdot_* \log_* |x|_* & n = 2 \\ e/_*(e^{n(n-2)\kappa(n)} \cdot_* |x|_*^{(n-2)*}) & n \geq 3, \end{cases}$$

defined for $x \in \mathbb{R}^n_*$, $x \neq 1$, is called the fundamental solution of the Laplace equation. Here $\kappa(n)$ denotes the volume of the unit ball in $\mathbb{R}^n$,

$$\kappa(n) = \frac{\pi^{\frac{n}{2}}}{\Gamma\left(\frac{n}{2} +_* 1\right)}.$$

6.3 Integral Representation of Harmonic Functions

Theorem 6.3 *Let u be multiplicative harmonic in D and continuous in $D \cup \partial_* D$ with its multiplicative partial derivatives of first order. Then*

$$u(x) = \int_{*\partial_* D} \Phi(x -_* y) \cdot_* \partial_{*\nu_{*y}} u(y) \cdot_* d_* s_y$$

$$-_* \int_{*\partial_* D} u(y) \cdot_* \partial_{*\nu_{*y}} \Phi(x -_* y) \cdot_* d_* s_y, \quad x \in D.$$

(6.7)

Proof Let $x \in D$ be arbitrarily chosen. We take $\epsilon > 0$ so that $B(x, \epsilon) \subset D$. Denote

$$D_\epsilon = D \backslash B(x, \epsilon).$$

We have

$$\partial_* D_\epsilon = \partial_* D \cup \partial_* B(x, \epsilon).$$

Then, using Theorem 6.2, we have

$$
\begin{aligned}
0 &= \int_{*\partial_* D_\epsilon} \left(\Phi(x -_* y) \cdot_* \partial_{*\nu_{*y}} u(y) -_* u(y) \cdot_* \partial_{*\nu_{*y}} \Phi(x -_* y) \right) \cdot_* d_* s_y \\[2mm]
&= \int_{*\partial_* D} \left(\Phi(x -_* y) \cdot_* \partial_{*\nu_{*y}} u(y) -_* u(y) \cdot_* \partial_{*\nu_{*y}} \Phi(x -_* y) \right) \cdot_* d_* s_y \\[2mm]
&\quad -_* \int_{*\partial_* B(x,\epsilon)} \left(\Phi(x -_* y) \cdot_* \partial_{*\nu_{*y}} u(y) -_* u(y) \cdot_* \partial_{*\nu_{*y}} \Phi(x -_* y) \right) \cdot_* d_* s_y .
\end{aligned}
$$

Hence,

$$
\begin{aligned}
&\int_{*\partial_* D} \left(\Phi(x -_* y) \cdot_* \partial_{*\nu_{*y}} u(y) -_* u(y) \cdot_* \partial_{*\nu_{*y}} \Phi(x -_* y) \right) \cdot_* d_* s_y \\[2mm]
&= \int_{*\partial_* B(x,\epsilon)} \left(\Phi(x -_* y) \cdot_* \partial_{*\nu_{*y}} u(y) -_* u(y) \cdot_* \partial_{*\nu_{*y}} \Phi(x -_* y) \right) \cdot_* d_* s_y \\[2mm]
&= \int_{*\partial_* B(x,\epsilon)} \Phi(x -_* y) \cdot_* \partial_{*\nu_{*y}} u(y) \cdot_* d_* s_y \\[2mm]
&\quad -_* \int_{*\partial_* B(x,\epsilon)} \left(u(y) -_* u(x) \right) \cdot_* \partial_{*\nu_{*y}} \Phi(x -_* y) \cdot_* d_* s_y \\[2mm]
&\quad -_* \int_{*\partial_* B(x,\epsilon)} u(x) \cdot_* \partial_{*\nu_{*y}} \Phi(x -_* y) \cdot_* d_* s_y .
\end{aligned}
$$

$$(6.8)$$

Note that

$$
\Phi(x -_* y) = \begin{cases}
-_* e^{\frac{1}{2\pi}} \cdot_* \log_* |x -_* y|_* & n = 2 \\[3mm]
(e/_*(e^{n(n-2)\kappa(n)} \cdot_* |x -_* y|_*^{(n-2)_*})) & n \geq 3,
\end{cases}
$$

$$
\partial_{*\nu_{*y}} \Phi(x -_* y) \Big|_{\partial_* B(x,\epsilon)} = \begin{cases}
-_* e^{\frac{1}{2\pi\epsilon}} & n = 2 \\[3mm]
-_* e^{\frac{1}{n\kappa(n)\epsilon^{n-1}}} & n \geq 3.
\end{cases}
$$

Therefore,

$$\int_{*\partial_* B(x,\epsilon)} (u(y) -_* u(x)) \cdot_* \partial_{\nu_{*y}} \Phi(x -_* y) \cdot_* d_* s_y$$

$$= \begin{cases} -_* e^{\frac{1}{2\pi\epsilon}} \cdot_* \displaystyle\int_{*\partial_* B(x,\epsilon)} (u(y) -_* u(x)) \cdot_* d_* s_y & n = 2 \\[2em] -_* e^{\frac{1}{n\kappa(n)\epsilon^{n-1}}} \cdot_* \displaystyle\int_{*\partial_* B(x,\epsilon)} (u(y) -_* u(x)) \cdot_* d_* s_y & n \geq 3 \end{cases}$$

(6.9)

$$= \begin{cases} -_* e^{\frac{1}{2\pi}} \cdot_* \displaystyle\int_{*\partial_* B(0,1)} (u(x +_* \epsilon \cdot_* z) -_* u(x)) \cdot_* d_* s_z & n = 2 \\[2em] -_* e^{\frac{1}{n\kappa(n)}} \cdot_* \displaystyle\int_{*\partial_* B(0,1)} (u(x +_* \epsilon \cdot_* z) -_* u(x)) \cdot_* d_* s_z & n \geq 3 \end{cases}$$

$\to 1 \quad$ as $\quad \epsilon \to 1,$

$$\int_{*\partial_* B(x,\epsilon)} u(x) \cdot_* \partial_{*\nu_{*y}} \Phi(x -_* y) \cdot_* d_* s_y$$

$$= \begin{cases} -_* e^{\frac{1}{2\pi\epsilon}} \cdot_* u(x) \cdot_* \displaystyle\int_{*\partial_* B(x,\epsilon)} d_* s_y & n = 2 \\[2em] -_* e^{\frac{1}{n\kappa(n)\epsilon^{n-1}}} \cdot_* u(x) \cdot_* \displaystyle\int_{*\partial_* B(x,\epsilon)} d_* s_y & n \geq 3 \end{cases}$$

(6.10)

$$= \begin{cases} -_* u(x) & n = 2 \\[1em] -_* u(x) & n \geq 3, \end{cases}$$

$$\left| \int_{*\partial_* B(x,\epsilon)} \Phi(x -_* y) \cdot_* \partial_{*\nu_* y} u(y) \cdot_* d_* s_y \right|_*$$

$$\leq \begin{cases} e^C \cdot_* |\log_* e^\epsilon|_* \cdot_* \displaystyle\int_{*\partial_* B(x,\epsilon)} d_* s_y & n = 2 \\[2em] e^{\frac{C}{\epsilon^{n-2}}} \cdot_* \displaystyle\int_{*\partial_* B(x,\epsilon)} d_* s_y & n \geq 3 \end{cases} \tag{6.11}$$

$$\leq \begin{cases} e^{C_1 \epsilon} \cdot_* |\log_* e^\epsilon|_* & n = 2 \\[1em] e^{C_1 \epsilon} & n \geq 3 \end{cases}$$

$$\to 1 \quad \text{as} \quad \epsilon \to 1.$$

Here C_1 and C are constants independent of ϵ. From (6.8)–(6.11), we obtain the desired result (6.7). This completes the proof.

6.4   Mean Value Formulas

Theorem 6.4 *If $u \in \mathcal{C}_*^2(D)$ is multiplicative harmonic, then*

$$u(x) = (e/_*(e^{n\kappa(n)} \cdot_* r^{(n-1)*})) \cdot_* \int_{*\partial_* B(x,r)} u(y) \cdot_* d_* s_y$$

$$= (e/_*(e^{\kappa(n)} \cdot_* r^{n*})) \cdot_* \int_{*B(x,r)} u(y) \cdot_* d_* y \tag{6.12}$$

for each ball $B(x,r) \subset D$.

Proof Note that

$$\Phi(x -_* y) \cdot_* \Big|_{\partial_* B(x,r)} = \begin{cases} -_* e^{\frac{1}{2\pi}} \cdot_* \log_* r & n = 2 \\[1em] (e/_*(e^{n(n-2)\kappa(n)} \cdot_* r^{(n-2)*})) & n \geq 3, \end{cases}$$

$$\partial_{*\nu_* y} \Phi(x -_* y) \Big|_{*\partial_* B(x,r)} = \begin{cases} -_*(e/_*(e^{2\pi} \cdot_* r)) & n = 2 \\[1em] -_*(e/_*(e^{n\kappa(n)} \cdot_* r^{(n-1)*})) & n \geq 3. \end{cases}$$

Hence from (6.7), we get

$$
u(x) =
\begin{cases}
-_* e^{\frac{1}{2\pi}} \cdot_* \log_* r \cdot_* \displaystyle\int_{*\partial_* B(x,r)} \partial_{*\nu_{*y}} u(y) \cdot_* d_* s_y +_* (e/_*(e^{2\pi} \cdot_* r)) \\[2em]
\cdot_* \displaystyle\int_{*\partial_* B(x,r)} u(y) \cdot_* d_* s_y \qquad n = 2 \\[2em]
(e/_*(e^{n(n-2)\kappa(n)} \cdot_* r^{(n-2)_*})) \\[2em]
\cdot_* \displaystyle\int_{*\partial_* B(x,r)} \partial_{*\nu_{*y}} u(y) \cdot_* d_* s_y +_* (e/_*(e^{n\kappa(n)} \cdot_* r^{(n-1)_*})) \\[2em]
\cdot_* \displaystyle\int_{*\partial_* B(x,r)} u(y) \cdot_* d_* s_y \qquad n \geq 3.
\end{cases}
$$

Hence from (6.4), we obtain (6.12). This completes the proof.

Definition 6.9 The formulas in (6.12) are known as multiplicative mean value formulas for multiplicative harmonic functions, for a multiplicative sphere and for a multiplicative ball, respectively.

Theorem 6.5 *(Converse to Multiplicative Mean Value Formula) If $u \in \mathcal{C}_*^2(D)$ satisfies (6.12) for each ball $B(x,r) \subset D$, then u is multiplicative harmonic in D.*

Proof Suppose that

$$
\Delta_* u \not\equiv 1
$$

in D. Then there exists some multiplicative ball $B(x,r) \subset D$ such that, without loss of generality, $\Delta_* u > 1$ within $B(x,r)$. By (6.12), we have

$$
\begin{aligned}
u(x) &= (e/_*(e^{n\kappa(n)} \cdot_* r^{(n-1)_*})) \cdot_* \int_{*\partial_* B(x,r)} u(y) \cdot_* d_* s_y \\[1em]
&= e^{\frac{1}{n\kappa(n)}} \cdot_* \int_{*\partial B(0,1)} u(x +_* r \cdot_* z) \cdot_* d_* s_z,
\end{aligned}
$$

whereupon

$$
\begin{aligned}
1 &= e^{\frac{1}{n\kappa(n)}} \cdot_* \int_{*\partial_* B(0,1)} \nabla_* u(x +_* r \cdot_* z) \cdot_* z \cdot_* d_* s_z \\[1em]
&= (e/_*(e^{n\kappa(n)} \cdot_* r^{(n-1)_*})) \cdot_* \int_{*\partial_* B(x,r)} \nabla_* u(y) \cdot_* ((y -_* x)/_* r) \cdot_* d_* s_y
\end{aligned}
$$

$$= (e/_*(e^{n\kappa(n)} \cdot_* r^{(n-1)*})) \cdot_* \int\limits_{*B(x,r)} \Delta_* u(y) \cdot_* d_* y$$

$$> 1,$$

which is a contradiction. This completes the proof.

6.5   Strong Maximum Principle: Uniqueness

Theorem 6.6 *Suppose* $u \in C_*^2(D) \cap C_*(\overline{D})$ *is multiplicative harmonic in* D.

1. *If there exists a point* $x_0 \in D$ *such that*

$$u(x_0) = \max_{\overline{D}} u,$$

   then

$$u = \text{const} \quad within \quad D.$$

2. *We have*

$$\max_{\overline{D}} u = \max_{*\partial_* D} u. \tag{6.13}$$

Proof

1. Let

$$e^M = \max_{\overline{D}} u.$$

   Suppose $u(x_0) = e^M$ for some $x_0 \in D$. Let $1 < e^\epsilon < e^{\text{dist}(x_0, \partial_* D)}$ be arbitrarily chosen. Assume that there exists $y \in B(x_0, \epsilon)$ so that $u(y) < e^M$. Since $u \in C_*(\overline{D})$, there is a $\delta > 0$ so that $B(y, \delta) \subset B(x_0, \epsilon)$ and $u(\xi) < e^M$ for all $\xi \in B(y, \delta)$. Also, using (6.12), we have

$$e^M = u(x_0) = e^{\frac{1}{\kappa(n)\epsilon^n}} \cdot_* \int\limits_{*B(x_0,\epsilon)} u(y) \cdot_* d_* y < e^M,$$

   which is a contradiction. Therefore, $u(y) = e^M$ for any $y \in B(x_0, \epsilon)$. Let $x \in D$ be arbitrarily chosen and l be a continuous curve lying within D and joining the points x and x_0. We take $1 < e^{\epsilon_1} < e^{\text{dist}(l, \partial_* D)}$. Note that if $y \in l$, we have $u(\eta) = e^M$ for any $\eta \in B(y, \epsilon_1)$. Therefore, $u(x) = e^M$. Because $x \in D$ was arbitrarily chosen, we conclude that $u(x) = e^M$ for all $x \in D$.

2. If there is $x_0 \in D$ so that $u(x_0) = \max_{\overline{D}} u(x)$, then $u(x) = u(x_0)$ for any $x \in D$, whereupon we get (6.13). This completes the proof.

Theorem 6.7 *The Dirichlet problem*

$$\Delta_* u - f \quad in \quad D$$

$$u = \phi \quad on \quad \partial_* D, \tag{6.14}$$

where $f \in \mathcal{C}_*(D)$, $\phi \in \mathcal{C}_*(\partial_* D)$ has at most one solution in $\mathcal{C}^2_*(D) \cap \mathcal{C}_*(\overline{D})$.

Proof Let $u_1, u_2 \in \mathcal{C}^2_*(D) \cap \mathcal{C}_*(\overline{D})$ be two solutions of the Dirichlet problem (6.14). Then,

$$v = u_1 -_* u_2 \in \mathcal{C}^2_*(D) \cap \mathcal{C}_*(\overline{D})$$

satisfies the Dirichlet problem

$$\Delta_* v = 1 \quad in \quad D$$

$$v = 1 \quad on \quad \partial_* D.$$

Hence from Theorem 6.6, we conclude that $v = 1$ in $\overline{D}$, i.e., $u_1 = u_2$ in $\overline{D}$. This completes the proof.

6.6   The Poisson Equation

Theorem 6.8 *Let $f \in \mathcal{C}^2_*(\mathbb{R}^n_*)$ be with compact support. Let also*

$$u(x) = -_* \int_{*\mathbb{R}^n_*} \Phi(x -_* y) \cdot_* f(y) \cdot_* d_* y, \quad x \in \mathbb{R}^n_*.$$

Then,

1. $u \in \mathcal{C}^2_*(\mathbb{R}^n_*),$
2. $\Delta_* u = f \quad in \quad \mathbb{R}^n_*.$

Proof

1. We have

$$u(x) = -_* \int_{*\mathbb{R}^n_*} \Phi(y) \cdot_* f(x -_* y) \cdot_* d_* y.$$

If $e_j = (1, \ldots, 1, e, 1, \ldots, 1)$, $j = 1, \ldots, n$, the e in the jth slot, then

$$((u(x +_* h \cdot_* e_j) -_* u(x))/_* h$$

$$= -_* \int_{*\mathbb{R}^n_*} \Phi(y)((f(x +_* h \cdot_* e_j -_* y) -_* f(x -_* y))/_* h) \cdot_* d_* y.$$

Because

$$\lim_{h \to 1} (f(x +_* h \cdot_* e_j -_* y) -_* f(x-_*y))/_*h) = f_{x_j^*} \cdot_* (x -_* y)$$

uniformly on $\mathbb{R}_*^n$, we get

$$u_{x_j}^*(x) = -_* \int_{*\mathbb{R}_*^n} \Phi(y) \cdot_* f_{x_j}^*(x -_* y) \cdot_* d_*y. \qquad (6.15)$$

As in above,

$$u_{x_i x_j}^{**}(x) = -_* \int_{*\mathbb{R}_*^n} \Phi(y) \cdot_* f_{x_i x_j}^{**}(x -_* y) \cdot_* d_*y. \qquad (6.16)$$

As the expressions on the right-hand side of (6.15) and (6.16) are continuous in the variable x, we conclude that $u \in \mathcal{C}_*^2(\mathbb{R}_*^n)$.

2. Let $x \in \mathbb{R}_*^n$ and $\epsilon > 0$ be arbitrarily chosen and fixed. Then,

$$\Delta_* u(x) = -_* \int_{\mathbb{R}_*^n} \Phi(y) \cdot_* \Delta_{*x} f(x -_* y) \cdot_* d_*y$$

$$= -_* \int_{*B(1,e^\epsilon)} \Phi(y) \cdot_* \Delta_{*x} f(x -_* y) \cdot_* d_*y \qquad (6.17)$$

$$-_* \int_{*\mathbb{R}_*^n \setminus B(1,e^\epsilon)} \Phi(y) \cdot_* \Delta_{*x} f(x -_* y) \cdot_* d_*y.$$

We have

$$\left| \int_{*B(1,e^\epsilon)} \Phi(y) \cdot_* \Delta_{*x} f(x -_* y) \cdot_* d_*y \right|_*$$

$$\leq \int_{*B(1,e^\epsilon)} |\Phi(y)|_* \cdot_* |\Delta_{*x} f(x -_* y)|_* \cdot_* d_*y$$

$$\leq \sup_{\mathbb{R}_*^n} |\Delta_* f|_* \cdot_* \int_{*B(1,e^\epsilon)} |\Phi(y)|_* \cdot_* d_*y \qquad (6.18)$$

$$\leq \begin{cases} e^{C\epsilon^2} \cdot_* |\log_* e^\epsilon|_* & n = 2 \\ e^{C\epsilon^2} & n \geq 3 \end{cases}$$

$$\to 1 \quad \text{as} \quad \epsilon \to 0.$$

Here C is a constant independent of ϵ. Also,

$$-_* \int_{*\mathbb{R}^n_* \setminus B(1,e^\epsilon)} \Phi(y) \cdot_* \Delta_{*x} f(x -_* y) \cdot_* d_* y$$

$$= -_* \int_{*\mathbb{R}^n_* \setminus B(1,e^\epsilon)} \Phi(y) \cdot_* \Delta_{*y} f(x -_* y) \cdot_* d_* y$$

(6.19)

$$= \int_{*\mathbb{R}^n_* \setminus B(1,e^\epsilon)} \nabla_* \Phi(y) \cdot_* \nabla_{*y} f(x -_* y) \cdot_* d_* y$$

$$-_* \int_{*\partial_* B(1,e^\epsilon)} \Phi(y) \cdot_* f_{\nu_{*y}}(x -_* y) \cdot_* d_* s_y,$$

$$\left| \int_{*\partial_* B(1,e^\epsilon)} \Phi(y) \cdot_* f_{\nu_{*y}}(x -_* y) \cdot_* d_* s_y \right|_*$$

$$\leq e^C \cdot_* \int_{*\partial_* B(1,e^\epsilon)} |\Phi(y)|_* \cdot_* d_* s_y$$

(6.20)

$$\leq \begin{cases} e^{C\epsilon} \cdot_* |\log_* e^\epsilon|_* & n = 2 \\ e^{C\epsilon} & n \geq 3 \end{cases}$$

$$\to 1 \quad \text{as} \quad \epsilon \to 0,$$

where C is a constant independent of ϵ. Note that

$$\nabla_* \Phi(y) = -_* e^{\frac{1}{n\kappa(n)}} \cdot_* (y /_* (|y|^n_*)),$$

$$\Phi_{\nu_{*y}}(y) = e^{\frac{1}{n\kappa(n)\epsilon^{n-1}}} \quad \text{on} \quad \partial_* B(1, e^\epsilon).$$

Therefore,

$$\int_{*\mathbb{R}^n_* \setminus B(1,e^\epsilon)} \nabla_* \Phi(y) \cdot_* \nabla_{*y} f(x -_* y) \cdot_* d_* y$$

$$= -_* \int_{*\mathbb{R}^n_* \setminus B(1,e^\epsilon)} \Delta_* \Phi(y) \cdot_* f(x -_* y) \cdot_* d_* y$$

$$+_* \int_{*\partial_* B(1,e^\epsilon)} \Phi_{\nu_{*y}}(y) \cdot_* f(x -_* y) \cdot_* d_* s_y$$

$$= \int_{*\partial_* B(1,e^\epsilon)} \Phi_{\nu_{*y}}(y) \cdot_* f(x -_* y) \cdot_* d_* s_y$$

$$= e^{\frac{1}{n\kappa(n)\epsilon^{n-1}}} \cdot_* \int_{*\partial_* B(1,e^\epsilon)} f(x -_* y) \cdot_* d_* s_y$$

$$= e^{\frac{1}{n\kappa(n)}} \cdot_* \int_{*\partial_* B(1,e)} f(x -_* \epsilon z) \cdot_* d_* s_z$$

$$\to f(x) \quad \text{as} \quad \epsilon \to 0.$$

From here and (6.19) and (6.20), we obtain

$$-_* \int_{*\mathbb{R}^n_* \backslash B(1,e^\epsilon)} \Phi(y) \cdot_* \Delta_{*x} f(x -_* y) \cdot_* d_* y \to f(x) \quad \text{as} \quad \epsilon \to 0.$$

Hence from (6.17) and (6.18), letting $\epsilon \to 0$, we get that

$$\Delta_* u(x) = f(x),$$

which completes the proof.

6.7   The Green Function of the Dirichlet Problem

Definition 6.10 The Green[4] function of the Dirichlet problem for the Laplace equation in a domain D is a function $G(x,y)$ depending on two points $x, y \in \overline{D}$ which possesses the following properties:

1. $G(x,y)$ has the form

$$G(x,y) = \Phi(x -_* y) +_* g(x,y),$$

   where $g(x,y)$ is a multiplicative harmonic function with respect to $x, y \in D$.

2. When $x \in \partial_* D$ or $y \in \partial_* D$, the equality $G(x,y) = 1$ is fulfilled.

Theorem 6.9 *We have*

$$G(x,y) \geq 1 \quad in \quad \overline{D}.$$

---

[4]George Green (July 14, 1793–May 31, 1841) was a British mathematical physicist who introduced several important concepts, among them a theorem similar to the modern Green theorem, the idea of potential functions and the concept of what are now called Green functions.

Proof Let $y \in D$ be arbitrarily chosen. Let also $\delta > 0$ be sufficiently small so that $B(y, \delta) \subset D$. Denote $D_\delta = D \backslash B(y, e^\delta)$. Since $\lim\limits_{x \to y} G(x, y) = \infty$, then we must have, for sufficiently small $\delta > 0$ that $G(x, y) > 1$ when $x \in B(y, e^\delta)$. Therefore, $G(x, y) \geq 1$ on $\partial_* D_\delta$. Hence from the maximum principle, we conclude that $G(x, y) \geq 1$ for all $x \in D_\delta$. Consequently, $G(x, y) \geq 1$ in $\overline{D}$. This completes the proof.

Theorem 6.10 *(Symmetry Property of the Green Function)* *We have*

$$G(x, y) = G(y, x) \quad for \quad any \quad x, y \in D.$$

Proof Let $x, y \in D$, $x \neq y$, be arbitrarily chosen and fixed. Take $\epsilon > 0$ small enough so that $B(x, e^\epsilon) \subset D$, $B(y, e^\epsilon) \subset D$ and $B(x, e^\epsilon) \cap B(y, e^\epsilon) = \varnothing$. Denote

$$D_\epsilon = D \backslash \left(B(x, e^\epsilon) \cup B(y, e^\epsilon) \right).$$

Note that $G(z, y)$ is multiplicative harmonic in $D \backslash B(y, e^\epsilon)$ and $G(z, x)$ is multiplicative harmonic in $D \backslash B(x, e^\epsilon)$. Applying (6.5) to the domain D_ϵ for $G(z, x)$ and $G(z, y)$, we get

$$
1 = \int\limits_{*\partial_* D_\epsilon} \left(G(z, y) \cdot_* G_{\nu_{*z}}(z, x) -_* G(z, x) \cdot_* G_{\nu_{*z}}(z, y) \right) \cdot_* d_* s_z
$$

$$
= \int\limits_{*\partial_* D} \left(G(z, y) \cdot_* G_{\nu_{*z}}(z, x) -_* G(z, x) \cdot_* G_{\nu_{*z}}(z, y) \right) \cdot_* d_* s_z
$$

$$
-_* \int\limits_{*\partial_* B(y, e^\epsilon)} \left(G(z, y) \cdot_* G_{\nu_{*z}}(z, x) -_* G(z, x) \cdot_* G_{\nu_{*z}}(z, y) \right) \cdot_* d_* s_z
$$

$$
-_* \int\limits_{*\partial_* B(x, e^\epsilon)} \left(G(z, y) \cdot_* G_{\nu_{*z}}(z, x) -_* G(z, x) \cdot_* G_{\nu_{*z}}(z, y) \right) \cdot_* d_* s_z,
$$

whereupon

$$
\int\limits_{*\partial_* D} \left(G(z, y) \cdot_* G_{\nu_{*z}}(z, x) -_* G(z, x) \cdot_* G_{\nu_{*z}}(z, y) \right) \cdot_* d_* s_z
$$

$$
= \int\limits_{*\partial_* B(y, e^\epsilon)} \left(G(z, y) \cdot_* G_{\nu_{*z}}(z, x) -_* G(z, x) G_{\nu_{*z}}(z, y) \right) \cdot_* d_* s_z
$$

$$
+_* \int\limits_{*\partial_* B(x, e^\epsilon)} \left(G(z, y) \cdot_* G_{\nu_{*z}}(z, x) -_* G(z, x) \cdot_* G_{\nu_{*z}}(z, y) \right) \cdot_* d_* s_z.
$$

Because

$$G(z, y) = G(z, x) = 1 \quad for \quad z \in \partial D,$$

we get

$$\int\limits_{*\partial_* B(x,e^\epsilon)} (G(z,y) \cdot_* G_{\nu_{*z}}(z,x) -_* G(z,x) \cdot_* G_{\nu_{*z}}(z,y)) \cdot_* d_* s_z$$

$$= \int\limits_{*\partial_* B(y,e^\epsilon)} (-_* G(z,y) \cdot_* G_{\nu_{*z}}(z,x) +_* G(z,x) \cdot_* G_{\nu_{*z}}(z,y)) \cdot_* d_* s_z.$$

$$(6.21)$$

Note that

$$\int\limits_{*\partial_* B(x,e^\epsilon)} (G(z,y) \cdot_* G_{\nu_{*z}}(z,x) -_* G(z,x) \cdot_* G_{\nu_{*z}}(z,y)) \cdot_* d_* s_z$$

$$= \int\limits_{*\partial_* B(x,e^\epsilon)} \left(\left(\Phi(z -_* y) +_* g(z,y) \right) \cdot_* \left(\Phi_{\nu_{*z}}(z -_* x) +_* g_{\nu_{*z}}(z,x) \right) \right.$$

$$\left. -_* \left(\Phi(z -_* x) +_* g(z,x) \right) \cdot_* \left(\Phi_{\nu_{*z}}(z -_* y) +_* g_{\nu_{*z}}(z,y) \right) \right) \cdot_* d_* s_z$$

$$= \int\limits_{*\partial_* B(x,e^\epsilon)} (\Phi(z -_* y) \cdot_* \Phi_{\nu_{*z}}(z -_* x) -_* \Phi(z -_* x)\Phi_{\nu_{*z}}(z -_* y)) \cdot_* d_* s_z$$

$$+_* \int\limits_{*\partial_* B(x,e^\epsilon)} (g(z,y) \cdot_* \Phi_{\nu_{*z}}(z -_* x) -_* g_{\nu_{*z}}(z,y) \cdot_* \Phi(z -_* x)) \cdot_* d_* s_z$$

$$+_* \int\limits_{*\partial_* B(x,e^\epsilon)} (\Phi(z -_* y) \cdot_* g_{\nu_{*z}}(z,x) -_* g(z,x) \cdot_* \Phi_{\nu_{*z}}(z -_* y)) \cdot_* d_* s_z$$

$$+_* \int\limits_{*\partial_* B(x,e^\epsilon)} (g(z,y) \cdot_* g_{\nu_{*z}}(z,x) -_* g(z,x) \cdot_* g_{\nu_{*z}}(z,y)) \cdot_* d_* s_z.$$

$$(6.22)$$

Since $g(z,y)$ and $g(z,x)$ are multiplicative harmonic with respect to z in $B(x,\epsilon)$, using Theorem 6.2, we get

$$\int\limits_{*\partial_* B(x,e^\epsilon)} (g(z,y) \cdot_* g_{\nu_{*z}}(z,x) -_* g(z,x) \cdot_* g_{\nu_{*z}}(z,y)) \cdot_* d_* s_z = 1. \qquad (6.23)$$

Because $\Phi(z -_* y)$ and $g(z,x)$ are multiplicative harmonic with respect to z in $B(x, e^\epsilon)$, using Theorem 6.2, we obtain

$$\int_{*\partial_* B(x,e^\epsilon)} (\Phi(z -_* y) \cdot_* g_{\nu_{*z}}(z,x) -_* g(z,x) \cdot_* \Phi_{\nu_{*z}}(z -_* y)) \cdot_* d_* s_z = .$$

$$(6.24)$$

Since $\Phi(z -_* y)$ is multiplicative harmonic with respect to z in $B(x, \epsilon)$, using Theorem 6.3, we have

$$\int_{*\partial_* B(x,e^\epsilon)} (\Phi(z -_* y) \cdot_* \Phi_{\nu_{*z}}(z -_* x) -_* \Phi(z -_* x) \cdot_* \Phi_{\nu_{*z}}(z -_* y))$$

$$(6.25)$$

$$\cdot_* d_* s_z = \Phi(x -_* y).$$

Because $g(z,y)$ is multiplicative harmonic with respect to z in $B(x, \epsilon)$, using Theorem 6.3, we obtain

$$\int_{*\partial_* B(x,e^\epsilon)} (g(z,y) \cdot_* \Phi_{\nu_{*z}}(z -_* x) -_* g_{\nu_{*z}}(z,y) \cdot_* \Phi(z -_* x)) \cdot_* d_* s_z = g(x,y).$$

$$(6.26)$$

From (6.22)–(6.26), we obtain

$$\int_{*\partial_* B(x,e^\epsilon)} (G(z,y) \cdot_* G_{\nu_{*z}}(z,x) -_* G(z,x) \cdot_* G_{\nu_{*z}}(z,y))$$

$$(6.27)$$

$$\cdot_* d_* s_z \to \Phi(x -_* y) +_* g(x,y) = G(x,y)$$

as $\epsilon \to 0$. Similarly,

$$\int_{*\partial_* B(y,e^\epsilon)} (G(z,x) \cdot_* G_{\nu_{*z}}(z,y) -_* G(z,y) G_{\nu_{*z}}(z,x)) \cdot_* d_* s_z \to G(y,x)$$

$$(6.28)$$

as $\epsilon \to 0$. From (6.21), (6.27) and (6.28), letting $\epsilon \to 0$, we get

$$G(x,y) = G(y,x).$$

This completes the proof.

Now we suppose that u is multiplicative harmonic in D and satisfies the boundary condition

$$u = \phi \quad \text{on} \quad \partial_* D. \qquad (6.29)$$

Then, by Theorem 6.3, we get

$$u(x) = \int_{*\partial_* D} \Phi(x -_* y) \cdot_* \partial_{\nu_{*y}} u(y) \cdot_* d_* s_y$$

$$-_* \int_{*\partial_* D} u(y) \cdot_* \partial_{*\nu_{*y}} \cdot_* \Phi(x -_* y) \cdot_* d_* s_y.$$

Hence, using Theorem 6.2,

$$\Phi(x -_* y) = G(x, y) -_* g(x, y)$$

and the boundary condition (6.29), we obtain

$$u(x) \quad = \quad \int_{*\partial_* D} (G(x, y) -_* g(x, y)) \cdot_* \partial_{*\nu_{*y}} u(y) \cdot_* d_* s_y$$

$$-_* \int_{*\partial_* D} u(y) \cdot_* \left(G_{\nu_{*y}}(x, y) -_* g_{\nu_{*y}}(x, y)\right) \cdot_* d_* s_y$$

$$= \quad \int_{*\partial_* D} G(x, y) \cdot_* \partial_{*\nu_{*y}} u(y) \cdot_* d_* s_y$$

$$-_* \int_{*\partial_* D} u(y) \cdot_* G_{\nu_{*y}}(x, y) \cdot_* d_* s_y$$

$$+_* \int_{*\partial_* D} \left(u(y) \cdot_* g_{\nu_{*y}}(x, y) -_* g(x, y) \cdot_* u^*_{\nu_{*y}}(y)\right) \cdot_* d_* s_y$$

$$= \quad -_* \int_{*\partial_* D} u(y) \cdot_* G_{\nu_{*y}}(x, y) \cdot_* d_* s_y$$

$$= \quad -_* \int_{*\partial_* D} \phi(y) \cdot_* G_{\nu_{*y}}(x, y) \cdot_* d_* s_y,$$

i.e.,

$$u(x) = -_* \int_{*\partial_* D} \phi(y) \cdot_* G_{\nu_{*y}}(x, y) \cdot_* d_* s_y, \quad x \in D. \tag{6.30}$$

The multiplicative harmonicity of the function u expressed by the formula
(6.30) follows from the fact that the Green function $G(x, y)$ is multiplicative
harmonic with respect to x for $x \neq y$. The fact that this function satisfies the
boundary condition (6.29) requires special proof.

6.8   The Poisson Formula for a Multiplicative Ball

Let D be the multiplicative ball $B(0_*, e)$ and let x, y be two interior points of that multiplicative ball.

Theorem 6.11 *The function*

$$G(x, y) = \Phi(x -_* y) -_* \Phi\left(|x|_* \cdot_* y -_* (x/_*|x|_*)\right)$$

is the Green function for the multiplicative ball D.

Proof Here

$$g(x, y) = -_*\Phi\left(|x|_* \cdot_* y -_* (x/_*|x|_*)\right).$$

Note that if

$$y = x/_*(|x|^{2_*}),$$

then

$$|y|_* = (e/_*|x|_*$$
$$> e,$$

i.e.,

$$|x|_* \cdot_* y -_* (x/_*|x|_* \neq 1)$$

for any $x, y \in B(0_*, e)$. Therefore, $g(x, y)$ is multiplicative harmonic with respect to x and y in $B(1, e)$. If $|y|_* = e$, then

$$|y -_* x|_* = \left(|x|^{2_*}_* -_* e^2 \cdot_* x \cdot_* y +_* e\right)^{\frac{1}{2}_*}$$
$$= ||y|_* \cdot_* x -_* (y/_*|y|_*)|_*$$
$$= ||x|_* \cdot_* y -_* (x/_*|x|_*)|_*,$$

i.e.,

$$\Phi(x -_* y) = \Phi\left(|x|_* \cdot_* y -_* (x/_*|x|_*)\right)$$

and $G(x, y) = 1$. Similarly, if $|x|_* = e$, then $G(x, y) = 1$, which completes the proof.

For $|y|_* = e$, we have

$$G_{\nu^*_y}(x, y) = -_* e^{\frac{1}{n\kappa(n)}} \cdot_* \sum_{*i=1}^{n} (((y_i \cdot_* (y_i -_* x_i))/_*|y -_* x|^{n_*}_*)$$

$$-_* |x|_* ((y_i \cdot_* (|x|_* \cdot_* y_i -_* (x_i/_*|x|_*)))/_* ||x|_* \cdot_* y -_* (x/_*|x|_*)|^{n_*}_*))$$

$$= -_* \, e^{\frac{1}{n\kappa(n)}}$$

$$\cdot_* \sum_{*i=1}^{n} (((y_i \cdot_* (y_i -_* x_i))/_* |y -_* x|_*^{n*}) -_* |x|_*$$

$$\cdot_* ((y_i \cdot_* (|x|_* y_i -_* (x_i/_* |x|_*)))/_* |y -_* x|_*^{n*}))$$

$$= -_* \, e^{\frac{1}{n\kappa(n)}} \cdot_* \sum_{*i=1}^{n} ((y_i^{2*} -_* x_i \cdot_* y_i -_* y_i^{2*}$$

$$\cdot_* |x|_*^{2*} +_* x_i \cdot_* y_i)/_* |y -_* x|_*^{n*})$$

$$= -_* \, e^{\frac{1}{n\kappa(n)}} \cdot_* \sum_{*i=1}^{n} ((y_i^{2*} \cdot_* (e -_* |x|_*^{2*}))/_* |y -_* x|_*^{n*})$$

$$= -_* \, e^{\frac{1}{n\kappa(n)}} \cdot_* ((|y|_*^{2*} \cdot_* (e -_* |x|_*^{2*}))/_* |y -_* x|_*^{n*})$$

$$= e^{\frac{1}{n\kappa(n)}} \cdot_* ((|x|_*^{2*} -_* e)/_* |y -_* x|_*^{n*}).$$

Then, applying (6.30), we obtain

$$u(x) = e^{\frac{1}{n\kappa(n)}} \cdot_* \int_{*|y|_*=e} ((e -_* |x|_*^{2*})/_* |y -_* x|_*^{n*}) \cdot_* \phi(y) \cdot_* d_* s_y, \qquad (6.31)$$

where $\phi \in C_*(\partial_* B(1, e))$.

Definition 6.11 Equation (6.31) is known as Poisson's formula.

The Poisson formula for the multiplicative ball $B(x_0, e^r)$ is

$$u(x) = e^{\frac{1}{rn\kappa(n)}} \cdot_* \int_{*|y-_* x_0|_*=e^r} ((e^{r^2} -_* |x -_* x_0|_*^{2*})/_* |y -_* x|_*^{n*}) \cdot_* \phi(y) \cdot_* d_* s_y,$$

where $\phi \in C_*(\partial_* B(x_0, e^r))$.

Exercise 6.1 Prove that

$$e^{\frac{1}{n\kappa(n)}} \cdot_* \int_{*\partial_* B(0,1)} ((e -_* |x|_*^{2*})/_* |y -_* x|_*^{n*}) \cdot_* d_* s_y = e \quad \text{for} \quad x \in B(1, e).$$

Let $x_0 \in \partial B(1, e)$ be arbitrarily chosen and fixed. Now, we will prove that

$$\lim_{\substack{x \to x_0 \\ x \in B(1, e)}} u(x) = \phi(x_0),$$

where $u(x)$ is defined by (6.31). We have

$$u(x) -_* \phi(x_0) = e^{\frac{1}{n\kappa(n)}} \cdot_* \int_{*|y|_*=e} ((e -_* |x|_*^{2*})/_*|y -_* x|_*^{n*}) \cdot_* (\phi(y) -_* \phi(x_0))$$

$$\cdot_* d_* s_y.$$

Let $\epsilon > 0$ be arbitrarily chosen. Then there exists $\delta = \delta(\epsilon) > 0$ so that

$$|\phi(y) -_* \phi(x_0)|_* < e^\epsilon \quad \text{for} \quad y \in B(x_0, e^\delta) \cap \partial_* B(1, e).$$

Set

$$S = \partial_* B(1, e) \backslash \left(B(x_0, e^\delta) \cap \partial_* B(1, e) \right).$$

We choose $\delta > 0$ small enough so that

$$e -_* |x|_*^{2*} \leq e^{\frac{\epsilon n\kappa(n)}{4MQ}} \quad \text{for} \quad x \in B(1, e), \quad |x -_* x_0|_* \leq e^{\frac{\delta}{2}},$$

where

$$Q = \int_{*S} (e/_*|y -_* x_0|_*^{n*}) \cdot_* d_* s_y,$$

$$M = \sup_{\partial_* B(1,e)} |\phi|_*.$$

Then, for

$$|x -_* x_0|_* \leq e^{\frac{\delta}{2}},$$

$x \in B(1, e)$, we have

$|u(x) -_* \phi(x_0)|_*$

$$\leq e^{\frac{1}{n\kappa(n)}} \cdot_* \int_{*\partial_* B(1,e)} ((e -_* |x|_*^{2*})/_*|y -_* x|_*^{n*}) \cdot_* |\phi(y) -_* \phi(x_0)|_* \cdot_* d_* s_y$$

$$= e^{\frac{1}{n\kappa(n)}} \cdot_* \int_{*B(x_0,e^\delta)\cap\partial_* B(1,e)} ((e -_* |x|_*^{2*})/_*|y -_* x|_*^{n*})$$

$$\cdot_* |\phi(y) -_* \phi(x_0)|_* \cdot_* d_* s_y$$

$$+_* e^{\frac{1}{n\kappa(n)}} \cdot_* \int_{*S} ((e -_* |x|_*^{2*})/_*|y -_* x|_*^{n*}) \cdot_* |\phi(y) -_* \phi(x_0)|_* d_* s_y$$

$$= I_1 +_* I_2.$$

$$(6.32)$$

Note that

$$I_1 \leq e^{\epsilon \frac{1}{n\kappa(n)}} \cdot_* \int_{*B(x_0,e^\delta)\cap\partial_*B(1,e)} ((e -_* |x|_*^{2*})/_*|x -_* y|_*^{n*}) \cdot_* d_* s_y$$

$$\leq e^{\epsilon \frac{1}{n\kappa(n)}} \cdot_* \int_{*\partial_*B(1,e)} ((e -_* |x|_*^{2*})/_*|x -_* y|_*^{n*}) \cdot_* d_* s_y \qquad (6.33)$$

$$= e^\epsilon \quad \text{for} \quad |x -_* x_0|_* \leq e^{\frac{\delta}{2}}, \quad x \in B(1,e).$$

For

$$|x -_* x_0|_* \leq e^{\frac{\delta}{2}}, \quad x \in B(1,e),$$

and $|y -_* x_0|_* \geq e^\delta$, $y \in \partial_* B(1,e)$, we have

$$|y -_* x_0|_* \leq |y -_* x|_* +_* |x -_* x_0|_*$$

$$\leq |y -_* x|_* +_* e^{\frac{\delta}{2}}$$

$$\leq |y -_* x|_* +_* e^{\frac{1}{2}} \cdot_* |y -_* x_0|_*,$$

i.e., if

$$|x -_* x_0|_* \leq e^{\frac{\delta}{2}}, \quad x \in B(1,e),$$

and $|y -_* x_0|_* \geq e^\delta$, $y \in \partial B(1,e)$, we have

$$|y -_* x|_* \geq e^{\frac{1}{2}} \cdot_* |y -_* x_0|_*.$$

Hence, for

$$|x -_* x_0|_* \leq e^{\frac{\delta}{2}}, \quad x \in B(1,e),$$

we have

$$I_2 \leq e^{\frac{2M}{n\kappa(n)}} \cdot_* \int_{*S} ((e -_* |x|_*^{2*})/_*|x -_* y|_*^{n*}) \cdot_* d_* s_y$$

$$\leq e^{\frac{4M}{n\kappa(n)}} \cdot_* \int_{*S} ((e -_* |x|_*^{2*})/_*|y -_* x_0|_*^{n*}) \cdot_* d_* s_y$$

$$\leq e^\epsilon.$$

From the last estimate and (6.32) and (6.33), we get

$$|u(x) -_* \phi(x_0)|_* \leq e^{2\epsilon}$$

for $|x -_* x_0|_* \leq e^{\frac{\delta}{2}}$, $x \in B(1,e)$.

Exercise 6.2 Let $D = \{x \in \mathbb{R}^n_* : x_n > 1\}$.

1. Prove that
$$G(x,y) = \Phi(x -_* y) -_* \Phi(x -_* y'),$$

   where $x, y \in \overline{D}$, $y = (y_1, \ldots, y_{n-1}, y_n)$, $y' = (y_1, \ldots, y_{n-1}, -_* y_n)$ is the Green function.

2. Let $\phi \in C_*(\partial_* D) \cap L^\infty_*(\partial_* D)$. Prove that

$$\lim_{\substack{x \to x_0 \\ x \in D}} ((e^2 \cdot_* x_n)/_* (e^{n\kappa(n)}))$$

$$\cdot_* \int_{*\partial_* D} (\phi(y)/_* |x -_* y|^{n*}_*) \cdot_* d_* s_y = \phi(x_0), \quad x_0 \in \partial D.$$

6.9 Theorems of Liouville and Harnack

Theorem 6.12 *Let u be multiplicative harmonic throughout the space $\mathbb{R}^n_*$ and $u(x) \geq (\leq)1$, $x \in \mathbb{R}^n_*$. Then u is identically equal to a constant in $\mathbb{R}^n_*$.*

Proof Suppose that $u(x) \geq 1$, $x \in \mathbb{R}^n_*$. Let $R > 0$ be arbitrarily chosen. Then by (6.31), we have

$$u(x) = e^{\frac{1}{n\kappa(n)R}} \cdot_* \int_{*|y|_* = e^R} ((e^{R^2} -_* |x|^{2*}_*)/_* |y -_* x|^{n*}_*) \cdot_* \phi(y) \cdot_* d_* s_y, \quad |x|_* < e^R,$$

$$(6.34)$$

for $\phi \in C_*(\partial_* B(1, e^R))$, and $u(x) \to \phi(x_0)$ as $x \to x_0$, $x \in B(1, e^R)$, $x_0 \in \partial_* B(1, e^R)$. Hence,

$$
\begin{aligned}
u(1) &= e^{\frac{1}{n\kappa(n)R}} \cdot_* \int_{*|y|_* = e^R} (e^{R^2}/_* |y|^{n*}_*) \cdot_* \phi(y) \cdot_* d_* s_y \\
&= e^{\frac{1}{n\kappa(n)R^{n-1}}} \cdot_* \int_{*|y|_* = e^R} \phi(y) \cdot_* d_* s_y.
\end{aligned}
$$

Since

$$
\begin{aligned}
e^R -_* |x|_* &\leq |y|_* -_* |x|_* \\
&\leq |y -_* x|_* \\
&\leq |y|_* +_* |x|_* \\
&\leq e^R +_* |x|_*
\end{aligned}
$$

for $|y|_* = e^R$, $|x|_* < e^R$, we get

$$u(x) \geq e^{\frac{1}{n\kappa(n)R}} \cdot_* \int_{|y|_*=e^R} ((e^{R^2} -_* |x|_*^{2*})/_*(e^R +_* |x|_*)^{n*}) \cdot_* \phi(y) \cdot_* d_* s_y$$

$$= ((e^{R^{n-2}} \cdot_* (e^{R^2} -_* |x|_*^{2*}))/_*(e^R +_* |x|_*)^{n*}) e^{\frac{1}{n\kappa(n)R^{n-1}}}$$

$$\cdot_* \int_{|y|_*=e^R} \phi(y) \cdot_* d_* s_y$$

$$= ((e^{R^{n-2}} \cdot_* (e^{R^2} -_* |x|_*^{2*}))/_*(e^R +_* |x|_*)^{n*}) \cdot_* u(1),$$

and

$$u(x) \leq e^{\frac{1}{n\kappa(n)R}} \cdot_* \int_{*|y|_*=e^R} ((e^{R^2} -_* |x|_*^{2*})/_*(e^n R -_* |x|_*)^{n*}) \cdot_* \phi(y) \cdot_* d_* s_y$$

$$= ((e^{R^{n-2}} \cdot_* (e^{R^2} -_* |x|_*^{2*}))/_*(e^R -_* |x|_*)^{n*}) \cdot_* e^{\frac{1}{n\kappa(n)R^{n-1}}}$$

$$\cdot_* \int_{|y|_*=e^R} \phi(y) \cdot_* d_* s_y$$

$$= ((e^{R^{n-2}} \cdot_* (e^{R^2} -_* |x|_*^{2*}))/_*(e^R -_* |x|_*)^{n*}) \cdot_* u(1), \quad |x|_* < e^R.$$

Therefore,

$$((e^{R^{n-2}} \cdot_* (e^{R^2} -_* |x|_*^{2*}))/_*(e^R +_* |x|_*)^{n*}) \cdot_* u(1) \leq u(x)$$

$$\leq ((e^{R^{n-2}} \cdot_* (e^{R^2} -_* |x|_*^{2*}))/_*(e^R -_* |x|_*)^{n*}) \cdot_* u(1), \quad |x|_* < e^R.$$

Making R tend to ∞ we get $u(x) = u(1)$ for any $x \in \mathbb{R}_*^n$.

Theorem 6.13 *(The Liouville[5] Theorem) Let u be multiplicative harmonic throughout $\mathbb{R}_*^n$ and it is bounded above (below) in $\mathbb{R}_*^n$. Then u is identically equal to a constant in $\mathbb{R}_*^n$.*

    Proof Let $u(x) \leq e^M$, $x \in \mathbb{R}_*^n$. Set

$$v(x) = e^M -_* u(x), \quad x \in \mathbb{R}_*^n.$$

Then v is multiplicative harmonic throughout $\mathbb{R}_*^n$ and it is multiplicative non-negative in $\mathbb{R}_*^n$. Hence from Theorem 6.12, we conclude that

$$e^M -_* u(x) = e^M -_* u(1), \quad x \in \mathbb{R}_*^n,$$

---

[5] Joseph Liouville (March 24, 1809–September 8, 1882) was a French mathematician who worked in a number of different fields in mathematics, including number theory, complex analysis, differential geometry, topology, mathematical physics and astronomy.

whereupon

$$u(x) = u(1) \quad x \in \mathbb{R}_*^n.$$

Exercise 6.3 Prove that the Dirichlet problem for the half space $x_n > 1$ cannot have more than one solution in the class of bounded functions.

Theorem 6.14 *(The Harnack[6] Theorem) Let $u_m(x)$, $m \in \mathbb{N}$, be multiplicative harmonic functions in a domain D which are continuous in $\overline{D}$ and the series $\sum\limits_{*m=1}^{\infty} u_m(x)$ is uniformly convergent on the boundary $\partial_* D$. Then this series is uniformly convergent in $\overline{D}$ and its sum $u(x) = \sum\limits_{*m=1}^{\infty} u_m(x)$ is a multiplicative harmonic function in D.*

Proof Let $\epsilon > 0$ be arbitrarily chosen. Since $\sum\limits_{*m=1}^{\infty} u_m(x)$ is uniformly convergent on $\partial_* D$, there exists an index $N = N(\epsilon)$ such that

$$\left| \sum\limits_{*i=1}^{p} u_{N+i}(y) \right|_* < e^{\epsilon}$$

holds for all $p \geq 1$ and $y \in \partial_* D$. Note that $\sum\limits_{*i=1}^{p} u_{N+i}(x)$ is multiplicative harmonic in D and continuous in $\overline{D}$. Hence from the maximum principle, we conclude that

$$\left| \sum\limits_{*i=1}^{p} u_{N+i}(x) \right|_* < e^{\epsilon} \quad \text{for} \quad \text{all} \quad x \in \overline{D}.$$

Therefore, $\sum\limits_{*m=1}^{\infty} u_m(x)$ is uniformly convergent in $\overline{D}$. Let now $x_0 \in D$ be an arbitrary point. We take $R > 0$ so that $B(x_0, e^R) \subset D$. Then,

$$u_m(x) = e^{\frac{1}{n\kappa(n)R}} \cdot_* \int\limits_{\partial_* B(x_0, e^R)} ((e^{R^2} -_* |x -_* x_0|_*^{2_*})/_* |y -_* x|_*^{n_*}) \cdot_* u_m(y)$$

$$\cdot_* d_* s_y, \quad m \in \mathbb{N}, \quad |x -_* x_0|_* < e^R.$$

[6] Carl Gustav Axel von Harnack (May 7, 1851–April 3, 1888) was a German mathematician who contributed to potential theory Harnack inequality applied to harmonic functions. He also worked on the real algebraic geometry of plane curves, proving Harnack's curve theorem for real plane algebraic curves.

Hence, using $\sum\limits_{*m=1}^{\infty} u_m(x)$ that is uniformly convergent in $\overline{D}$, we get

$$u(x) = \sum_{*m=1}^{\infty} u_m(x)$$

$$= e^{\frac{1}{n\kappa(n)R}} \cdot_* \int_{*\partial_* B(x_0,e^R)} ((e^{R^2} -_* |x -_* x_0|_*^{2*})/_* |y -_* x|_*^{n*})$$

$$\cdot_* \sum_{*m=1}^{\infty} u_m(y) \cdot_* d_* s_y$$

$$= e^{\frac{1}{n\kappa(n)R}} \cdot_* \int_{*\partial_* B(x_0,e^R)} ((e^{R^2} -_* |x -_* x_0|_*^{2*})/_* |y -_* x|_*^{n*})$$

$$\cdot_* u(y) \cdot_* d_* s_y, \quad |x -_* x_0|_* < e^R.$$

Therefore, u is multiplicative harmonic in $|x -_* x_0|_* < e^R$. Because $x_0 \in D$ was arbitrarily chosen, we conclude that u is multiplicative harmonic everywhere in D.

6.10   Separation of Variables

6.10.1   Multiplicative rectangles

Let u be the solution to the Dirichlet problem in multiplicative rectangular domain

$$\Delta_* u = e, \quad e^a < x_1 < e^b, \quad e^c < x_2 < e^d, \tag{6.35}$$

with the boundary conditions

$$u(e^a, x_2) = \phi_1(x_2),$$

$$u(e^b, x_2) = \phi_2(x_2), \quad e^c \le x_2 \le e^d,$$

$$u(x_1, e^c) = \psi_1(x_1),$$

$$u(x_1, e^d) = \psi_2(x_1), \quad e^a \le x_1 \le e^b, \tag{6.36}$$

where $\phi_1, \phi_2 \in \mathcal{C}_*([e^c, e^d])$, $\psi_1, \psi_2 \in \mathcal{C}_*([e^a, e^b])$, and

$$\phi_1(e^c) = \psi_1(e^a),$$

$$\phi_1(e^d) = \psi_2(e^a),$$

$$\phi_2(e^c) = \psi_1(e^b),$$

$$\phi_2(e^d) = \psi_2(e^b).$$

We split u into the form $u = u_1 +_* u_2$, where u_1 solves

$$\Delta_* u_1 = 1, \quad e^a < x_1 < e^b, \quad e^c < x_2 < e^d,$$

$$u_1(e^a, x_2) = \phi_1(x_2),$$

$$u_1(e^b, x_2) = \phi_2(x_2), \quad e^c \leq x_2 \leq e^d, \tag{6.37}$$

$$u_1(x_1, e^c) = u_1(x_1, e^d)$$

$$= 1, \quad e^a \leq x_1 \leq e^b,$$

and u_2 satisfies

$$\Delta_* u_2 = 1, \quad e^a < x_1 < e^b, \quad e^c < x_2 < e^d,$$

$$u_2(e^a, x_2) = u_2(e^b, x_2)$$

$$= 1, \quad e^c \leq x_2 \leq e^d, \tag{6.38}$$

$$u_2(x_1, e^c) = \psi_1(x_1),$$

$$u_2(x_1, e^d) = \psi_2(x_1), \quad e^a \leq x_1 \leq e^b,$$

under the compatibility condition

$$\phi_1(e^c) = \psi_1(e^a) = \phi_1(e^d) = \psi_2(e^a) = \phi_2(e^c) = \psi_1(e^b) = \phi_2(e^d) = \psi_2(e^b) = 1.$$

We will seek a solution $u_1 = u_1(x_1, x_2)$ of the problem (6.37) in the form

$$u_1(x_1, x_2) = X^1(x_1)Y^1(x_2).$$

Substituting such solution into the multiplicative Laplace equation, we obtain

$$X^{1**}(x_1) -_* e^\lambda \cdot_* X^1(x_1) = 1, \quad e^a < x_1 < e^b, \tag{6.39}$$

$$Y^{1**}(x_2) +_* e^\lambda \cdot_* Y^1(x_2) = 1, \quad e^c < x_2 < e^d. \tag{6.40}$$

Also, we get

$$Y^1(e^c) = Y^1(e^d) = 1.$$

Thus, using (6.40), we obtain the Sturm[7]–Liouville problem for $Y^1(x_2)$:

$$Y^{1**}(x_2) +_* e^{\lambda} \cdot_* Y^1(x_2) = 1 \quad e^c < x_2 < e^d,$$

$$Y^1(e^c) = Y^1(e^d) = 1.$$

(6.41)

Solving (6.41), as in (4.1.3), we derive a sequence of eigenvalues $e^{\lambda_n^1}$ and a sequence of eigenfunctions $Y_n^1(x_2)$. Then we substitute the sequence $e^{\lambda_n^1}$ into (6.39) and we obtain an associated sequence $X_n^1(x_1)$. The formal solution $u_1(x_1, x_2)$ of the problem (6.37) is written as follows:

$$u_1(x_1, x_2) = \sum_{*n=1}^{\infty} X_n^1(x_1) \cdot_* Y_n^1(x_2).$$

The remaining boundary conditions for u_1, $u_1(e^a, x_2) = \phi_1(x_2)$, $u_1(e^b, x_2) = \phi_2(x_2)$ will be used to eliminate the two free parameters associated with $X_n^1(x_1)$ for each $n \in \mathbb{N}$, as is done in (4.1.3). Now, we seek a solution $u_2(x_1, x_2)$ of the problem (6.38) in the form

$$u_2(x_1, x_2) = X^2(x_1) \cdot_* Y^2(x_2).$$

Substituting it in

$$\Delta_* u_2 = 0,$$

we get

$$X^{2**}(x_1) -_* e^{\lambda} \cdot_* X^2(x_1) = 1, \quad e^a < x_1 < e^b, \tag{6.42}$$

$$Y^{2**}(x_2) +_* e^{\lambda} \cdot_* Y^2(x_2) = 1, \quad e^c < x_2 < e^d. \tag{6.43}$$

Using that

$$u_2(e^a, x_2) = u_2(e^b, x_2) = 1,$$

$e^c \leq x_2 \leq e^d$, we get $X^2(e^a) = X^2(e^b) = 1$. Hence from (6.42), we obtain the Sturm–Liouville problem for $X^2(x_1)$.

$$X^{2**}(x_1) -_* e^{\lambda} \cdot_* X^2(x_1) = 0, \quad e^a < x_1 < e^b,$$

$$X^2(e^a) = X^2(e^b) = 1.$$

(6.44)

For the problem (6.44), as in (4.1.3), we get a sequence of eigenvalues $e^{\lambda_n^2}$ and a sequence of eigenfunctions $X_n^2(x_1)$. Substituting $e^{\lambda_n^2}$ into (6.43), we

---

[7] Jacques Charles Francois Sturm (September 29, 1803–December 15, 1855) was a French mathematician who discovered the theorem which bears his name and which concerns the determination of the number and the localization of the real roots of a polynomial equation included between given limits.

obtain an associated sequence $Y_n^2(x_2)$. The formal solution $u_2(x_1, x_2)$ of the problem (6.38) is written as follows:

$$u_2(x_1, x_2) = \sum_{*n-1}^{\infty} X_n^2(x_1) \cdot_* Y_n^2(x_2).$$

We find the two free parameters in $Y_n^2(x_2)$, $n \in \mathbb{N}$, using the boundary conditions

$$u_2(x_1, e^c) = \psi_1(x_1),$$
$$u_2(x_1, e^d) = \psi_2(x_1).$$

The formal solution $u(x_1, x_2)$ of the problem (6.35), (6.36) is written in the following way:

$$u(x_1, x_2) = \sum_{*n=1}^{\infty} \left(X_n^1(x_1) \cdot_* Y_n^1(x_2) +_* X_n^2(x_1) \cdot_* Y_n^2(x_2) \right).$$

Example 6.4 Consider the problem

$$\Delta_* u = 1 \quad \text{in} \quad 1 < x_1, x_2 < e^{\pi},$$

$$u(1, x_2) = e,$$

$$u(e^{\pi}, x_2) = u(x_1, 1)$$

$$= u(x_2, e^{\pi})$$

$$= 1, \quad 1 \le x_1, x_2 \le e^{\pi}.$$

We seek a nontrivial formal solution

$$u(x_1, x_2) = X(x_1) \cdot_* Y(x_2)$$

Then,

$$(X^{**}(x_1)/_* X(x_1)) = -_*(Y^{**}(x_2)/_* Y(x_2)) = e^{\lambda},$$

where λ is a constant. Since u is nontrivial and

$$u(x_1, 1) = u(x_1, e^{\pi})$$

$$= 1,$$

we obtain

$$Y(1) = Y(e^{\pi}) = 0.$$

Thus we get the Sturm–Liouville problem:

$$Y^{**}(x_2) +_* e^{\lambda} \cdot_* Y(x_2) = 1, \quad 1 < x_2 < e^{\pi},$$

$$Y(1) = Y(e^{\pi}) = 1.$$

Hence,

$$Y_n(x_2) = \sin_*(e^n \cdot_* x_2), \quad \lambda_n = n^2,$$

and

$$X_n^{**}(x_1) -_* e^{n^2} \cdot_* X_n(x_1) = 1, \quad e^a < x_1 < e^b.$$

Therefore,

$$X_n(x_1) = e^{A_n} \cdot_* e^{e^n \cdot_* x_1} +_* e^{B_n} \cdot_* e^{-_* e^n \cdot_* x_1}$$

and

$$u(x_1, x_2) = \sum_{*n=1}^{\infty} \left(e^{A_n} \cdot_* e^{e^n \cdot_* x_1} +_* e^{B_n} e^{-_* e^n \cdot_* x_1} \right) \cdot_* \sin_*(e^n \cdot_* x_2),$$

where A_n and B_n are constants which we will determine using the boundary conditions

$$u(1, x_2) = e$$

and

$$u(e^\pi, x_2) = 1,$$

$1 \le x_2 \le e^\pi$. We get

$$u(1, x_2) \quad = \quad \sum_{*n=1}^{\infty} (e^{A_n} +_e^{*B_n}) \cdot_* \sin_*(e^n \cdot_* x_2)$$

$$= \quad e,$$

whereupon multiplying by $\sin_*(e^n x_2)$ the last equality and integrating over $[1, e^\pi]$, we obtain

$$(e^{A_n} +_* e^{B_n}) \cdot_* \int_{*1}^{e^\pi} (\sin_*(e^n \cdot_* x_2))^{2*} \cdot_* d_* x_2 = \int_{*1}^{e^\pi} \sin_*(e^n \cdot_* x_2) \cdot_* d_* x_2$$

or

$$e^{A_n} +_* e^{B_n} = -_* e^{\frac{2}{n\pi}((-1)^n - 1)}. \tag{6.45}$$

Also,

$$u(e^\pi, x_2) = \sum_{*n=1}^{\infty} \left(e^{A_n} \cdot_* e^{n\pi} +_* e^{B_n} \cdot_* e^{-_* n\pi} \right) \cdot_* \sin_*(e^n \cdot_* x_2) = 1,$$

whence

$$e^{A_n} \cdot_* e^{n\pi} +_* e^{B_n} \cdot_* e^{-_* n\pi} = 1.$$

From the last equality and from (6.45), we go to the system

$$e^{A_n} +_* e^{B_n} = -_* e^{\frac{2}{n\pi}((-1)^n - 1)}$$

$$e^{A_n} \cdot_* e^{n\pi} +_* e^{B_n} \cdot_* e^{-_* n\pi} = 1.$$

For its solution, we have

$$A_n = \frac{2}{n\pi}\left((-1)^n - 1\right)\frac{1}{e^{2n\pi} - 1}$$

$$B_n = -\frac{2}{n\pi}\left((-1)^n - 1\right)\frac{e^{2n\pi}}{e^{2n\pi} - 1}.$$

Therefore,

$$e^{A_n} \cdot_* e^{n}{}_{*x_1} +_* e^{B_n} \cdot_* e^{-*n*x_1} = e^{\frac{4}{n\pi}((-1)^n - 1)\frac{1}{e^{2n\pi}-1}e^{n\pi}} \cdot_* \sinh_*(e_*n(x_1 -_* e^\pi))$$

and

$$u(x_1, x_2) = \sum_{*n=1}^{\infty} e^{\frac{4}{n\pi}((-1)^n-1)\frac{1}{e^{2n\pi}-1}e^{n\pi}} \cdot_* \sinh_*(e^n(x_1 -_* e^\pi)) \cdot_* \sin_*(e^n \cdot_* x_2).$$

Exercise 6.4 Find a formal solution to the following problem:

$$\Delta_* u = 1, \quad 1 < x_1, x_2 < e,$$

$$u(1, x_2) = x_2 \cdot_* (e -_* x_2),$$

$$u(e, x_2) = 1, \quad 1 \le x_2 \le e,$$

$$u(x_1, 1) = \sin_*(e^\pi \cdot_* x_1),$$

$$u(x_1, e) = 1, \quad 1 \le x_1 \le e.$$

Answer

$$u(x_1, x_2) = ((\sinh_*(e^\pi \cdot_* (e -_* x_2)))/_*(\sinh_* e^\pi)) \cdot_* \sin_*(e^\pi \cdot_* x_1)$$

$$+_* e^{\frac{8}{\pi^3}} \cdot_* \sum_{*n=0}^{\infty} ((\sinh_*((e^{(2n+1)\pi} \cdot_* (x_1 -_* e))))/_* e^{(2n+1)^3})$$

$$\cdot_* ((\sin_*(e^{(2n+1)\pi} \cdot_* x_2))/_*(\sinh_*(e^{(2n+1)\pi}))).$$

6.10.2 Multiplicative circular domains

Consider the Dirichlet problem:

$$\Delta_* u = 1, \quad (x_1, x_2) \in B(1, e^a),$$

$$u(x_1, x_2) = \phi(x_1, x_2), \quad (x_1, x_2) \in \partial B(1, e^a). \tag{6.46}$$

Introduce multiplicative polar coordinates

$$x_1 = r \cdot_* \cos_* \theta,$$

$$x_2 = r \cdot_* \sin_* \theta, \quad r > 1, \quad \theta \in [1, e^{2\pi}].$$

Then,

$$B(1, e^a) = \{(r, \theta) : 1 \leq r \leq e^a, \quad 1 \leq \theta \leq e^{2\pi}\},$$

$$\phi(x_1, x_2)\Big|_{\partial B(1, e^a)} = \phi(e^a \cdot_* \cos_* \theta, e^a \cdot_* \sin_* \theta)$$

$$= h(\theta).$$

Thus the problem (6.46) takes the following form:

$$u_{rr}^{**} +_* (e/_* r) \cdot_* u_r^* +_* (e/_* r^{2*}) \cdot_* u_{\theta\theta}^{**} = 1 \quad \text{in} \quad B(1, e^a)$$

$$u(e^a, \theta) = h(\theta), \tag{6.47}$$

$$\lim_{r \to 1} u(r, \theta) \qquad \text{exists} \quad \text{and} \quad \text{is} \quad \text{finite.}$$

We will search a formal solution of the problem (6.47) in the form

$$u(r, \theta) = R(r) \cdot_* \Theta(\theta).$$

Substituting this function in (6.47) and using the arguments in (4.1.3), we find

$$r^{2*} \cdot_* R^{**}(r) +_* r \cdot_* R^*(r) -_* e^\lambda \cdot_* R(r) = 1, \quad 1 < r < e^a,$$

$$\Theta^{**}(\theta) +_* e^\lambda \cdot_* \Theta(\theta) = 1, \quad 1 < \theta < e^{2\pi},$$

$$\Theta(1) = \Theta(e^{2\pi}), \quad \Theta^*(1) = \Theta^*(e^{2\pi})$$

$$R(e^a) \cdot_* \Theta(\theta) = h(\theta), \quad 1 \leq \theta \leq e^{2\pi},$$

$$\lim_{r \to 1} u(r, \theta) \qquad \text{exists} \quad \text{and} \quad \text{is} \quad \text{finite.}$$
$$\tag{6.48}$$

Since we search a solution $u(r, \theta)$ of the class C_*^2, we need to impose the periodicity conditions:

$$\Theta(1) = \Theta(e^{2\pi}),$$

$$\Theta^*(1) = \Theta^*(e^{2\pi}).$$

Hence from the second equation of (6.48), we get

$$\Theta_n(\theta) = e^{A_n} \cdot_* \cos_*(e^n \cdot_* \theta) +_* e^{B_n} \cdot_* \sin_*(e^n \cdot_* \theta),$$

$$\lambda_n = n^2, \quad n \in \mathbb{N}_0.$$

Substituting the eigenvalues λ_n into the first equation of (6.48), we find

$$r^{2*} \cdot_* R^{**}(r) +_* r \cdot_* R^*(r) -_* e^{n^2} \cdot_* R(r) = 1,$$

whereupon

$$R_n(r) = e^{\mathcal{U}_n} \cdot_* r^{n*} +_* e^{\mathcal{D}_n} \cdot_* r^{-*n*}, \quad n \in \mathbb{N},$$

$$R_0(r) = C_0 +_* D_0 \cdot_* \log_* r.$$

Since we want $\lim_{r \to 1} u(r, \theta)$ to exist and and to be finite, we get $D_n = 0, n \in \mathbb{N}_0$. Therefore, we obtain a formal solution as follows:

$$u(r, \theta) = \sum_{*n=0}^{\infty} R_n(r) \cdot_* \Theta_n(\theta)$$

$$= e^{\frac{\alpha_0}{2}} +_* \sum_{*n=1}^{\infty} r^{n*} \cdot_* \left(e^{\alpha_n} \cdot_* \cos_*(e^n \cdot_* \theta) +_* e^{\beta_n} \cdot_* \sin_*(e^n \cdot_* \theta) \right).$$

$$(6.49)$$

Formally differentiating this series term-by-term, we verify that (6.49) is indeed multiplicative harmonic. Imposing the boundary condition

$$u(e^a, \theta) = h(\theta), \quad 1 \le \theta \le e^{2\pi},$$

we obtain

$$e^{\alpha_0} = e^{\frac{1}{\pi}} \cdot_* \int_{*1}^{e^{2\pi}} h(\theta) \cdot_* d_*\theta,$$

$$e^{\alpha_n} = e^{\frac{1}{\pi a^n}} \cdot_* \int_{*1}^{e^{2\pi}} \cdot_* h(\theta) \cdot_* \cos_*(e^n \cdot_* \theta) \cdot_* d_*\theta,$$

$$e^{\beta_n} = e^{\frac{1}{\pi a^n}} \cdot_* \int_{*1}^{e^{2\pi}} h(\theta) \cdot_* \sin_*(e^n \theta) \cdot_* d_*\theta, \quad n \in \mathbb{N}.$$

Example 6.5 Consider the Dirichlet problem:

$$\Delta_* u = 1, \quad x_1^{2*} +_* x_2^{2*} < e,$$

$$u(x_1, x_2) = x_2^{2*} \quad \text{on} \quad x_1^{2*} +_* x_2^{2*} = e.$$

Introducing polar coordinates

$$x_1 = r \cdot_* \cos_* \theta,$$

$$x_2 = r \cdot_* \sin_* \theta, \quad r > 1, \quad \theta \in [1, e^{2\pi}],$$

we get the problem

$$u_{rr}^{**} +_* (e/_*r) \cdot_* u_r^* +_* (e/_*r^{2*}) \cdot_* u_{\theta\theta}^{**} = 1 \quad \text{in} \quad B(0,1),$$

$$u(\cos_* \theta, \sin_* \theta) = (\sin_* \theta)^{2*}, \quad \theta \in [1, e^{2\pi}].$$

Here,

$$h(\theta) = (\sin_* \theta)^{2*}.$$

Then,

$$e^{\alpha_0} = e^{\frac{1}{\pi}} \cdot_* \int_{*1}^{e^{2\pi}} (\sin_* \theta)^{2*} \cdot_* d_*\theta$$

$$= e,$$

$$e^{\alpha_n} = e^{\frac{1}{\pi}} \cdot_* \int_{*1}^{e^{2\pi}} (\sin_* \theta)^{2*} \cdot_* \cos_*(e^n \cdot_* \theta) \cdot_* d_*\theta$$

$$= \begin{cases} e^{-_* \frac{1}{2}} & n = 2 \\ \\ 1 & n \neq 2, \quad n \in \mathbb{N}, \end{cases}$$

$$e^{\beta_n} = e^{\frac{1}{\pi}} \cdot_* \int_{*1}^{e^{2\pi}} (\sin_* \theta)^{2*} \cdot_* \sin_*(e^n \cdot_* \theta) \cdot_* d_*\theta$$

$$= 1, \quad n \in \mathbb{N}.$$

Consequently,

$$u(x_1, x_2) = u(r, \theta)$$

$$= e^{\frac{1}{2}} -_* (r^{2*}/_*e^2) \cdot_* \cos_*(e^2 \cdot_* \theta)$$

$$= e^{\frac{1}{2}} -_* ((r^{2*} \cdot_* (\cos_* \theta)^{2*} -_* r^{2*} \cdot_* (\sin_* \theta)^{2*})/_*e^2$$

$$= e^{\frac{1}{2}} \cdot_* (e -_* x_1^{2*} +_* x_2^{2*}).$$

Exercise 6.5 Find a formal solution to the following problem:

$$\Delta_* u = 1 \quad \text{in} \quad x_1^{2*} +_* x_2^{2*} < e,$$

$$u(x_1, x_2) = x_2^{3*} \quad \text{on} \quad x_1^{2*} +_* x_2^{2*} = e.$$

Answer

$$u(r,\theta) = (r/_*e^4) \cdot_* (e^3 \cdot_* \sin_* \theta -_* r^{2*} \cdot_* \sin_*(e^3 \cdot_* \theta))$$

$$+_* (r/_*e^4) \cdot_* (e^3 \cdot_* \cos_* \theta -_* r^{2*} \cdot_* \cos_*(e^3 \cdot_* \theta)).$$

Exercise 6.6 Find a formal solution to the Dirichlet problem:

$$\Delta_* u = 1 \quad \text{in} \quad \mathbb{R}^2_* \backslash \overline{B(1, e^a)},$$

$$u(x_1, x_2) = \phi(x_1, x_2) \quad \text{on} \quad \partial_* B(1, e^a),$$

where $\phi \in \mathcal{C}_*(\partial_* B(1, e^a))$.

Answer

$$u(r,\theta) = e^{\frac{\alpha_0}{2}} +_* \sum_{*n=1}^{\infty} (e^a/_*r)^{n*} \cdot_* \left(e^{\alpha_n} \cdot_* \cos_*(e^n \cdot_* \theta) +_* e^{\beta_n} \cdot_* \sin_*(e^n \cdot_* \theta) \right),$$

where

$$e^{\alpha_0} = e^{\frac{1}{\pi}} \cdot_* \int_{*1}^{e^{2\pi}} h(\theta) \cdot_* d_* \theta,$$

$$e^{\alpha_n} = e^{\frac{1}{\pi}} \cdot_* \int_{*1}^{e^{2\pi}} h(\theta) \cdot_* \cos_*(e^n \cdot_* \theta) \cdot_* d_* \theta,$$

$$e^{\beta_n} = e^{\frac{1}{\pi}} \cdot_* \int_{*1}^{e^{2\pi}} h(\theta) \cdot_* \sin_*(e^n \cdot_* \theta) \cdot_* d_* \theta, \quad n \in \mathbb{N},$$

$$h(\theta) = \phi(e^a \cdot_* \cos_* \theta, e^a \cdot_* \sin_* \theta).$$

6.11   Advanced Practical Exercises

Problem 6.1 Let

$$D = \{x = (x_1, \ldots, x_n) \in \mathbb{R}^n_* : 1 < x_n < e^a\},$$

where a is a positive constant. Prove that

$$G(x, y) = \sum_{*m=-\infty}^{\infty} \left(\Phi(x, y^m) -_* \Phi(x, y'^m) \right),$$

where $y = (y_1, \ldots, y_{n-1}, y_n)$, $y^m = (y_1, \ldots, y_{n-1}, e^{2ma} +_* y_n)$, $y'^m = (y_1, \ldots, y_{n-1}, e^{2ma} -_* y_n)$ is a Green function.

Problem 6.2 Find a formal solution to the following problem:

$$\Delta_* u = 1, \quad 1 < x_1, x_2 < e,$$

$$u(x_1, 1) = e +_* \sin_*(e^\pi \cdot_* x_1),$$

$$u(x_1, e) = e^2, \quad 1 \le x_1 \le e,$$

$$u(1, x_2) = u(e, x_2)$$

$$= e +_* x_2, \quad 1 \le x_2 \le e.$$

Answer

$$u(x_1, x_2) = \sin_*(e^\pi x_1) \cdot_* ((\sinh_*(e^\pi \cdot_* (e -_* x_2)))/_*(\sinh_* e^\pi)) +_* e +_* x_2.$$

Problem 6.3 Find a formal solution to the following problem:

$$\Delta_* u = 1, \quad 1 < x_1 < e, \quad 1 < x_2 < e^2,$$

$$u(x_1, 1) = 1,$$

$$u(x_1, e^2) = e, \quad 1 \le x_1 \le e,$$

$$u^*_{x_1}(1, x_2) = 1,$$

$$u^*_{x_1}(e, x_2) = \sin_*(e^{2\pi} \cdot_* x_2), \quad 1 \le x_2 \le e^2.$$

Answer

$$u(x_1, x_2) = x_2/_* e^2 +_* ((\cosh_*(e^{2\pi} \cdot_* x_1)$$

$$\cdot_* \sin_*(e^{2\pi} \cdot_* x_2))/_*(e^{2\pi} \cdot_* \sinh_*(e^{2\pi}))).$$

Problem 6.4 Find a bounded formal solution to the following problem:

$$\Delta_* u = 1, \quad 1 < x_1 < e, \quad x_2 > 1,$$

$$u(x_1, 1) = e,$$

$$u(1, x_2) = u(e, x_2)$$

$$= 1.$$

Answer

$$u(x_1, x_2) = e^{\frac{4}{\pi}} \cdot_* \sum_{*n=0}^{\infty} e^{\frac{1}{2n+1}} e^{-_* e^{(2n+1)\pi} \cdot_* x_2} \cdot_* \sin_*(e^{(2n+1)\pi} \cdot_* x_1).$$

Problem 6.5 Find a formal solution to the following problem:

$$\Delta_* u = 1, \quad 1 < x_1, x_2 < e^\pi,$$

$$u^*_{x_1}(1, x_2) = u^*_{x_1}(e^\pi, x_2)$$

$$= 1, \quad 1 \le x_2 \le e^\pi,$$

$$u^*_{x_2}(x_1, 1) = 1,$$

$$u^*_{x_2}(x_1, e^\pi) = x_1 -_* e^{\frac{\pi}{2}}, \quad 1 \le x_1 \le e^\pi.$$

Answer

$$u(x_1, x_2) = e^{A_0} -_* e^{\frac{4}{\pi}} \cdot_* \sum_{*n=0}^{\infty} ((\cos_*(e^{(2n-1)}) \cdot_* x_1) \cdot_* \cosh_*(e^{(2n-*1)}) \cdot_* x_2))/_*$$

$$(e^{(2n-1)^2} \cdot_* \sinh_*(e^{(2n-1)\pi}))),$$

where A_0 is a constant.

Problem 6.6 Find a formal solution to the following problem:

$$\Delta_* u = 1 \quad \text{in} \quad x_1^{2*} +_* x_2^{2*} < e,$$

$$u(x_1, x_2) = x_1^{2*} \quad \text{on} \quad x_1^{2*} +_* x_2^{2*} = e.$$

Answer

$$u(x_1, x_2) = e^{\frac{1}{2}} \cdot_* (e +_* x_1^{2*} -_* x_2^{2*}).$$

Problem 6.7 Let $u(x)$ be a multiplicative harmonic function in D. Prove that $u \in C_*^\infty(D)$.

7

The Cauchy–Kovalevskaya Theorem

7.1 Analytic Functions of One Variable

Definition 7.1 A power series is an expression of the form

$$f(z) = \sum_{*n=0}^{\infty} e^{a_n} \cdot_* (z -_* e^c)^{n_*},$$ (7.1)

with coefficients $a_n \in \mathbb{C}$ and centre $c \in \mathbb{C}$.

Assume that the power series (7.1) is convergent. Then,

$$|e^{a_n}|_* \cdot_* |z -_* e^c|_*^{n_*} \to 1 \quad \text{as} \quad n \to \infty.$$

Therefore, there exists a constant $M > 0$ such that

$$|e^{a_n}|_* \cdot_* |z -_* e^c|_*^{n_*} \le e^M \quad \text{for any} \quad n \in \mathbb{N},$$

whereupon

$$|z -_* e^c|_* \le e^R$$

$$= \sup\{r \ge 1 : \quad \sup |e^{a_n}|_* \cdot_* r^{n_*} < \infty\}.$$ (7.2)

Definition 7.2 The above defined $e^R \in [1, \infty]$ is called the convergence radius of the power series (7.1).

We introduce the notation

$$D_R(e^c) = \{z \in \mathbb{C}_* : |z -_* e^c|_* < e^R\}, \quad D_R = D_R(1).$$

Note that if $|z -_* e^c| > e^R$, then the power series (7.1) is divergent.

Theorem 7.1 Let e^R be defined by (7.2). Then the power series (7.1) converges absolutely uniformly on each compact subset of the open disk $D_R(e^c)$ and diverges at every $z \in \mathbb{C}_* \backslash \overline{D_R(e^c)}$. Moreover, e^R can be determined by the Cauchy–Hadamard formula

$$e/_* e^R = \limsup_{n \to \infty} |e^{a_n}|_*^{\frac{1}{n}*}$$

DOI: 10.1201/9781003440116-7

with the conventions $e/_*\infty = 0$ and $e/_*1 = \infty$, and, furthermore, provided that $e^{a_n} = 1$ for only finitely many n, one can estimate e^R by the ratio test

$$\liminf_{n\to\infty} |e^{a_n}|_*/_*|e^{a_{n+1}}|_* \;\le\; e^R$$

$$\le\; \limsup_{n\to\infty} |e^{a_n}|_*/_*|e^{a_{n+1}}|_*.$$

Proof Without loss of generality, we suppose that $c = 0$. Above we have demonstrated the divergence of $\sum_{*n=1}^{\infty} e^{a_n} \cdot_* z^{n*}$ at every $z \in \mathbb{C}_*\backslash\overline{D_R}$. Take $r < R$. Let $z \in D_r$. Then, for any $\rho \in (r, R)$, we have

$$|e^{a_n} \cdot_* z^{n*}|_* < |e^{a_n}|_* \cdot_* r^{n*} = |e^{a_n}|_* \rho^{n*} \cdot_* (r/_*\rho)^{n*} \le M \cdot_* (r/_*\rho)^{n*}$$

for some constant $M > 0$, $M < \infty$. Since $r/_*\rho < 1$, we have that $\sum_{*n=1}^{\infty} e^{a_n} \cdot_* z^{n*}$ converges uniformly in D_r. Because any $z \in D_R$ is in some D_r, $r < R$, the series converges absolutely uniformly on each compact subset of D_R. Let now, ρ be defined with

$$e/_*\rho = \limsup_{n\to\infty} |e^{a_n}|_*^{\frac{1}{n}*}.$$

We will prove that $\rho = R$. For any $\epsilon \in (0, 1)$, we have

$$|e^{a_n}|_* \cdot_* \rho^{n*} \ge (e -_* \epsilon)^{n*}$$

for infinitely many n. Also, there is n_ϵ such that $|e^{a_n}|_* \cdot_* \rho^{n*} \le (e +_* \epsilon)^{n*}$ for any $n > n_\epsilon$. Thus, if $|z|_* > \rho$, then $|e^{a_n} \cdot_* z^{n*}|_* > |e^{a_n}|_* \cdot_* \rho^{n*}$. Hence, $|e^{a_n} \cdot_* z^{n*}|_* > e$ for infinitely many n provided that $|z|_* > \rho$. Therefore, the series $\sum_{*n=1}^{\infty} e^{a_n} \cdot_* z^{n*}$ diverges, which implies that $\rho \ge R$. On the other hand, if $|z|_* < \rho$, then for any $\epsilon > 0$ we have

$$|e^{a_n} \cdot_* z^n|_* = |e^{a_n}|_* \cdot_* \rho^{n*} \cdot_* (|z|_*^{n*}/_*\rho^{n*}) \le (e +_* \epsilon)^{n*}(|z|_*^{n*}/_*\rho^{n*} = k^{n*}$$

for any $n > n_\epsilon$. By choosing $\epsilon > 0$ small enough, one can ensure that $k \in [0, 1)$, and so $\sum_{*n=1}^{\infty} e^{a_n} \cdot_* z^{n*}$ converges. Therefore, $\rho \le R$. Consequently, $\rho = R$. Let

$$\alpha = \liminf_{n\to\infty}(|e^{a_n}|_*/_*|e^{a_{n+1}}|_*).$$

Suppose that $|z|_* < \alpha$. Then for any $\epsilon > 0$, we have $|e^{a_n}|_* \ge (\alpha -_* \epsilon) \cdot_* |e^{a_{n+1}}|_*$ for all sufficiently large n. This gives

$$|e^{a_n} \cdot_* z^{n*}|_* \le C \cdot_* (|z|_*/_*(\alpha -_* \epsilon))^{n*}$$

for all sufficiently large n, with some constant $C > 0$. By choosing $\epsilon > 0$ small enough, we show the convergence of $\sum_{*n=1}^{\infty} e^{a_n} \cdot_* z^{n_*}$. Therefore, $\alpha \leq R$. Let now

$$\beta = \limsup_{n \to \infty} (|e^{a_n}|_*/_* |e^{a_{n+1}}|_*.$$

Suppose that $|z|_* > \beta$ and $\epsilon = |z|_* -_* \beta > 1$. Then, $|e^{a_n}|_* \leq (\beta +_* \epsilon) \cdot_* |e^{a_{n+1}}|_*$ for sufficiently large n. So

$$|e^{a_n} \cdot_* z^{n_*}|_* \geq C \cdot_* (|z|_*^{n_*})/_*(\beta +_* \epsilon)^{n_*} \geq C$$

for some constant $C > 0$, and the series diverges. Therefore, $R \leq \beta$.

Let Ω denotes an open set in $\mathbb{C}_*$.

Definition 7.3 A multiplicative complex-valued function $f \colon \Omega \to \mathbb{C}_*$ is called multiplicative analytic at $z \in \Omega$, if there are coefficients $e_n^a \in \mathbb{C}$ and a radius $e^r > 1$ such that

$$f(z +_* h) = \sum_{*n=0}^{\infty} e^{a_n} \cdot_* h^{n_*}$$

for all $h \in D_r$. Moreover, f is said to be multiplicative analytic on Ω if it is analytic at each $z \in \Omega$. The set of multiplicative analytic functions on Ω is denoted by $\mathcal{C}_*^{\omega}(\Omega)$.

Theorem 7.2 *Let $e^R > 1$ be the convergence radius of the power series*

$$f(z) = \sum_{*n=0}^{\infty} e^{a_n} \cdot_* (z -_* e^c)^{n_*}.$$

Let also $d \in D_R(c)$. Then

$$f(z) = \sum_{*j=0}^{\infty} \left(\sum_{*n=j}^{\infty} \binom{n}{j} e^{a_n} \cdot_* (d -_* e^c)^{(n-j)_*} \right) \cdot_* (z -_* d)^{j_*},$$

where the convergence radius of the power series is at least $e^{R -_* |d -_* c|_*}$. In particular, we have $f \in \mathcal{C}_*^{\omega}(D_R(e^c))$, and the convergence radius of a rearranged power series depends continuously on its centre.

Proof We have

$$(z -_* e^c)^n = (z -_* d +_* d -_* e^c)^n$$

$$= \sum_{*j=0}^{n} \binom{n}{j} (z -_* d)^{j_*} \cdot_* (d -_* e^c)^{(n-j)_*}.$$

Hence,

$$f(z) = \sum_{*n=0}^{\infty} e^{a_n} \cdot_* (z -_* e^c)^{n_*}$$

$$= \sum_{*n=0}^{\infty} \left(e^{a_n} \cdot_* \sum_{*j=0}^{n} \binom{n}{j} (z -_* d)^{j_*} \cdot_* (d -_* e^c)^{(n-j)_*} \right).$$

Let $|z -_* d|_* \leq e^{\rho -_*(d-_*e^c)}$ with $\rho < R$. Then,

$$\sum_{*j=0}^{n} \binom{n}{j} |z -_* d|_*^{j_*} \cdot_* |d -_* e^c|_*^{(n-j)_*} = (|z -_* d|_* +_* |d -_* e^c|_*)^{n_*} \leq \rho^{n_*}$$

and since $e^{a_n} \cdot_* \rho^{n_*} = e^{a_n R^n} \cdot_* \left(e^{\frac{\rho}{R}} \right)^{n_*}$, we obtain that

$$\sum_{*n=0}^{\infty} \left(e^{a_n} \cdot_* \sum_{*j=0}^{n} \binom{n}{j} (z -_* d)^{j_*} \cdot_* (d -_* e^c)^{(n-j)_*} \right)$$

is absolutely convergent on each compact subset of $D_r(d)$. Therefore,

$$\sum_{*n=0}^{\infty} \left(e^{a_n} \sum_{*j=0}^{n} \binom{n}{j} (z -_* d)^{j_*} \cdot_* (d -_* e^c)^{(n-j)_*} \right)$$

$$= \sum_{*j=0}^{\infty} \sum_{*n=j}^{\infty} \binom{n}{j} e^{a_n} \cdot_* (z -_* d)^{j_*} \cdot_* (d -_* e^c)^{(n-j)_*}.$$

Let $e^{R'}$ denote the convergence radius of the rearranged series centred at e^d. Then $e^{R'} \geq e^{R -_* |d -_* e^c|_*}$ or $e^{R -_* R'} \leq |d -_* e^c|_*$. If $|d -_* e^c|_* < e^{\frac{R}{2}}$, and $e^c \in D_{R'}(e^d)$, which means that the above reasoning can be applied with the roles of the two power series interchanged, giving $e^{R' -_* R} \leq |e^c -_* e^d|_*$.

Theorem 7.3 *Let e^R be the convergence radius of the power series (7.1). Then both*

$$g(z) = \sum_{*n=0}^{\infty} n e^{a_n} \cdot_* (z -_* e^c)^{(n-1)_*} \quad and$$

$$F(z) = \sum_{*n=0}^{\infty} (e^{a_n} /_* e^{n+1}) \cdot_* (z -_* e^c)^{(n+1)_*}$$

have convergence radii equal to e^R and

$$f^* = g \quad and \quad F^* = f \quad in \quad D_R(e^c).$$

Proof Let $e^{R'}$ be the convergence radius of g. For $z \neq e^c$, we have

$$g(z) = (e/_*(z -_* e^c)) \cdot_* \sum_{*n=0}^{\infty} e^{na_n} \cdot_* (z -_* e^c)^{n*}.$$

Therefore,

$$(e/_*e^{R'}) = \limsup_{n\to\infty} e^{\sqrt[n]{na_n}} \geq \limsup_{n\to\infty} e^{\sqrt[n]{a_n}} = e^{\frac{1}{R}},$$

i.e., $e^{R'} \leq e^R$. Let $e^r < e^R$. Then for any $e^\epsilon > 1$, there is a constant $C_\epsilon >$ such that

$$e^{n \cdot_* |a_n|_*} \cdot_* r^{n*} \leq e^{C_\epsilon} \cdot_* (e +_* e^\epsilon)^{n*} \cdot_* |a_n|_* r^{n*}$$

$$= e^{C_\epsilon} \cdot_* (e +_* e^\epsilon)^{n*} \cdot_* e^{\left(\frac{r}{R}\right)^n \cdot_* |a_n|_*} \cdot_* R^{n*}.$$

Hence, choosing $e^\epsilon > 1$ small enough, we see that $e^r \leq e^{R'}$. Consequently, $e^R \leq e^{R'}$, whence $R = R'$. Now, we will prove that $f^* = g$ in $D_R(e^c)$, i.e., for each $z \in D_R(e^c)$, we have

$$f(z +_* h) = f(z) +_* g(z) \cdot_* h +_* o(|h|_*).$$

Note that

$$f(z +_* h) -_* f(z) = \sum_{*n=0}^{\infty} e^{a_n} \cdot_* ((z +_* h)^{n*} -_* z^{n*})$$

$$= h \cdot_* \sum_{*n=0}^{\infty} e^{a_n} \cdot_* \sum_{*j=0}^{n-1} (z +_* h)^{j*} \cdot_* z^{(n-1-j)*}$$

$$= h \cdot_* \lambda_z(h).$$

Let $r < R$ be such that $|z|_* < e^r$, and consider all h satisfying $|z +_* h|_* \leq e^r$. Then

$$\sum_{*n=0}^{\infty} |e^{a_n}|_* \sum_{*j=0}^{n-1} |z +_* h|_*^{j*} \cdot_* |z|_*^{(n-1-j)*} \leq \sum_{*n=0}^{\infty} r^{n*} \cdot_* e^{n \cdot_* |a_n|_*} < \infty,$$

so $\lambda_z(h)$ converges locally uniformly in a neighborhood of the multiplicative origin. Hence, λ_z is continuous at 1. Moreover, $\lambda_z(1) = g(z)$. Therefore,

$$\lambda_z(h) = g(z) +_* o(e),$$

with $o(e) \to 1$ as $|h|_* \to \infty$, i.e.,

$$f(z +_* h) = f(z) +_* g(z) \cdot_* h +_* o(|h|_*).$$

The statements about F follow from the above if we replace f with F and g with f.

Definition 7.4 By repeatedly applying Theorem 7.3, we see that the coefficients of the power series f about $e^c \in \Omega$ are given by $e^{a_n} = ((f^{*(n)}(e^c))/_* n!_*$. Therefore, if $f \in C_*^\omega(\Omega)$ and $e^c \in \Omega$, then the following Taylor series converges in a neighborhood of e^c.

$$f(z) = \sum_{*n=0}^{\infty} ((f^{*(n)}(e^c))/_* n!_*) \cdot_* (z -_* e^c)^{n_*}.$$

Definition 7.5 A multiplicative accumulation point of a set $D \subset \mathbb{C}_*$ is a point $z \in \mathbb{C}_*$ such that any neighborhood of z contains a point $w \neq z$ from D.

Definition 7.6 We say that $z \in D$ is a multiplicative isolated point if it is not a multiplicative accumulation point of D.

Definition 7.7 If all points of D are multiplicative isolated, then D is called multiplicative discrete.

Theorem 7.4 *(Identity Theorem) Let Ω be a multiplicative connected open set in $\mathbb{C}_*$ and $f \in C_*^\omega(\Omega)$. Let also at least one of the following statements hold.*

1. *There is $e^b \in \Omega$ such that $f^{*(n)}(e^b) = 1$ for all $n \in \mathbb{N}$.*

2. *The multiplicative zero set of f has a multiplicative accumulation point in Ω.*

Then, $f \equiv 1$ in Ω.

Proof Observe that each of the sets

$$\Sigma_n = \{z \in \Omega : f^{*(n)}(z) = 1\}$$

is relatively closed in Ω. Then $\Sigma = \bigcap_n \Sigma_n$ is also closed. Since $z \in \Sigma$ implies that $f \equiv 1$ in a small disk centred at z by a Taylor series argument, we have that Σ is open.

1. Suppose that there is $e^b \in \Omega$ such that $f^{*(n)}(b) = 1$ for any n. Since $e^b \in \Sigma$, we have that Σ is nonempty. Therefore, $\Sigma = \Omega$.

2. Suppose that the zero set of f has a multiplicative accumulation point e^c. If $e^c \in \Sigma$, then $\Sigma = \Omega$. Let $e^c \notin \Sigma$. Then there is an n such that $f^{*(n)}(e^c) \neq 1$. So, we have that $f(z) = (z -_* e^c)^{n_*} \cdot_* g(z)$ for some continuous function g such that $g(e^c) \neq 1$. Hence, there is a neighborhood of e^c where f has at most one multiplicative zero, which is a contradiction because e^c is a multiplicative accumulation point of the multiplicative zero set of f. Therefore, $e^c \in \Sigma$ and then $\Sigma = \Omega$.

7.2   Autonomous Multiplicative Differential Equations

Here we will give the MDE case of the Cauchy–Kovalevskaya[1] theorem which is simpler and contains half of the main ideas.

Consider the initial value problem:

$$u^* = f(u),$$
$$u(1) = e^e, \tag{7.3}$$

where $f : \mathbb{C}_* \to \mathbb{C}_*$ is a given function multiplicative analytic at 1, u is the unknown function.

Theorem 7.5 *The initial value problem* (7.3) *has a unique solution u that is multiplicative analytic at 1.*

Proof Without loss of generality, we suppose that $e^e = 1$. Otherwise, we change u with $u -_* e^e$ and f with $f(e^e +_* \cdot)$. We repeatedly differentiate the first equation of (7.3) and we get

$$u^{**} \;=\; (f(u))^*$$

$$\;=\; f^*(u) \cdot_* u^*,$$

$$u^{***} \;=\; (f(u))^{**}$$

$$\;=\; f^{**}(u) \cdot_* (u^*)^{2_*} +_* f^*(u) \cdot_* u^{**},$$

$$\vdots$$

$$u^{*(k)} \;=\; (f(u))^{*(k-1)}$$

$$\;=\; q_k(f(u), \dots, f^{*(k-1)}(u), u^*, \dots, u^{*(k-1)}),$$

where q_k is a multivariate polynomial with multiplicative nonnegative integer coefficients. We evaluate this at 1 and we get

$$u^{*(k)}(1) \;=\; q_k(f(1), \dots, f^{*(k-1)}(1), u^*(1), \dots, u^{*(k-1)}(1)).$$

---

[1]Sofia Kovalevskaya (January 15, 1850–February 10, 1891) was the first major Russian female mathematician and responsible for important original contributions to analysis, partial differential equations and mechanics.

Now repeatedly applying the same formula with k having values $k-1$, $k-2$ etc., to eliminate $u^{*(m)}(1)$ from the right-hand side, we obtain

$$u^{*(k)}(1) = Q_k(f(1), \ldots, f^{*(k-1)}(1)), \tag{7.4}$$

where Q_k is another multivariate polynomial having multiplicative nonnegative integer coefficients. This proves uniqueness of multiplicative analytic solutions of (7.3), because (7.4) fixes their Maclaurin series coefficients at 1. Provided that the Maclaurin series

$$u(z) = \sum_{*n=0}^{\infty} ((u^{*(n)}(1))/_* n!_*) \cdot_* z^{n_*}, \tag{7.5}$$

with $u^{*(n)}(1)$ given by (7.4) converges in a neighborhood of 1, the function $v = u^* -_* f(u)$ is multiplicative analytic at 1 and its Maclaurin series is identically multiplicative zero. Hence, by the Identity Theorem, v must vanish wherever it is defined.

Now we will prove that (7.5) converges in a neighborhood of 0. Since the right-hand side f is multiplicative analytic at 1, there exist constants $M > 0$ and $r > 0$ such that

$$((|f^{*(k)}(1)|_*)/_*(k!_*)) \leq (e^M/_* r^{k_*}), \quad k = 0, 1, \ldots.$$

Then the function:

$$F(z) \;=\; (e^M/_*(e -_* z/_* e^r))$$

$$\;=\; e^M +_* e^{\frac{M}{r}} \cdot_* z +_* \cdots +_* e^{\frac{M}{rk}} \cdot_* z^{k_*} +_* \cdots$$

majorizes f at 1. Consider the initial value problem:

$$U^* = F(U),$$
$$U(1) = 1. \tag{7.6}$$

Then, using (7.5), we have

$$U^{*(k)}(1) = Q_k(F(1), \ldots, F^{*(k-1)}(1))$$

and

$$|u^{*(k)}(1)|_* \;=\; |Q_k(f(1), \ldots, f^{*(k-1)}(1))|_*$$

$$\leq\; Q_k(|f(1)|_*, \ldots, |f^{*(k-1)}(1)|_*)$$

$$\leq\; Q_k(|F(1)|_*, \ldots, |F^{*(k-1)}(1)|_*)$$

$$=\; U^{*(k)}(1).$$

Therefore, the solution u of (7.3) is majorized by the solution U of (7.6) at 1. Note that (7.6) is solvable with

$$U(z) = r\left(e -_* (e -_* (e^M \cdot_* z)/_* r)^{\frac{1}{2}}_*\right) = e^M \cdot_* (z/_* e^2) +_* \cdots,$$

whose Taylor series around 1 has multiplicative nonnegative coefficients. Consequently, u is multiplicative analytic at 1.

7.3   Systems of Ordinary Differential Equations

In this section, we will consider the Cauchy–Kovalevskaya theorem for the system

$$u_j^* = f_j(z_1, \ldots, z_n, u_1, \ldots, u_m),$$

$$u_j^*(1) = 1, \quad j = 1, \ldots, m.$$

We could have eliminated the dependence of f on $z_1, \ldots, z_n$ by introducing the new variables $u_{m+1}^* = z_1, \ldots, u_{m+n}^* = z_n$ with the equations $u_{m+k}^* = e$, $k = 1, \ldots, n$. The above system can be written in the following manner:

$$u^* = f(z, u),$$

$$u(1) = 1,$$

(7.7)

where $u = (u_1, \ldots, u_m)$ has values in $\mathbb{C}_*^m$, $z = (z_1, \ldots, z_n)$. In $\mathbb{C}_*^n$, a power series is an expression of the form

$$f(z) = \sum_{*\alpha_1=0}^{\infty} \cdots \sum_{*\alpha_n=0}^{\infty} a_{\alpha_1 \ldots \alpha_n} \cdot_* (z_1 -_* e_1^c)^{\alpha_{1*}} \cdot_* \cdots \cdot_* (z_n -_* e^{c_n})^{\alpha_{n*}},$$

with coefficients $a_{\alpha_1 \ldots \alpha_n} \in \mathbb{C}_*$, and centre $e^c \in \mathbb{C}_*^n$. Introduce the multi-index $\alpha = (\alpha_1, \ldots, \alpha_n) \in \mathbb{N}_0$ and the conventions

$$|\alpha| = \alpha_1 + \cdots + \alpha_n \quad \text{and} \quad z^{\alpha_*} = z_1^{\alpha_{1*}} \cdot_* \cdots \cdot_* z_n^{\alpha_{n*}} \quad \text{for} \quad z \in \mathbb{C}_*^n.$$

Then the above series can also be written as follows:

$$f(z) = \sum_{*|\alpha| \geq 0} a_\alpha \cdot_* (z -_* e^c)^{\alpha_*}.$$

(7.8)

If the series (7.8) converges for some z, then there is a constant $0 < M < \infty$ such that

$$|a_\alpha|_* \cdot_* |z_1 -_* e^{c_1}|_*^{\alpha_{1*}} \ldots |z_n -_* e^{c_n}|_*^{\alpha_{n*}} \leq e^M \quad \text{for all} \quad \alpha.$$

In particular, if this series converges in a neighborhood of e^c, then there are constants $0 < M < \infty$ and $r > 0$, such that $|a_\alpha|_* \leq e^{Mr^{-*|\alpha|}}$ for all α. On the other hand, if $r \in \mathbb{R}_*^n$ and $M < \infty$ satisfy $|a_\alpha|_* \cdot_* r_1^{\alpha_1*} \ldots r_n^{\alpha_n*} \leq e^M$ for all α, then the series converges absolutely for all $z \in \mathbb{C}_*^n$ satisfying $|z_i -_* e^{c_i}|_* < r_i$ for each $i \subset \{1, \ldots, n\}$.

Definition 7.8 Let Ω be an open subset of $\mathbb{C}_*^n$. A complex-valued function $f \colon \Omega \to \mathbb{C}_*$ is called multiplicative analytic at $e^c \in \Omega$ if there are constants $a_\alpha \in \mathbb{C}_*$, $\alpha \in \mathbb{N}_0^n$, such that the power series (7.8) converges in a neighborhood of e^c. Moreover, f is said to be multiplicative analytic in Ω if it is multiplicative analytic at each $e^c \in \Omega$. The set of multiplicative analytic functions on Ω is denoted by $\mathcal{C}_*^\omega(\Omega)$.

As in the single-variable case, one can show that if f is multiplicative analytic at e^c, then the series (7.8) is its multivariate Taylor series, i.e., the coefficients are given by

$$a_\alpha = (\partial_*^\alpha f(e^c)/_* \alpha!_*)$$

$$= (\partial_{*1}^{\alpha_1} \ldots \partial_{*n}^{\alpha_n} f(e^c))/_* (\alpha_1!_* \ldots \alpha_n!_*),$$

where we have introduced the conventions $\alpha!_* = \alpha_1!_* \cdot_* \ldots \cdot_* \alpha_n!_*$ and $\partial_{*j}^{\alpha_j} f = \partial_{*z_j}^{\alpha_j} f$, $j = 1, \ldots, n$.

Theorem 7.6 *(Identity Theorem) Let $f \in \mathcal{C}_*^\omega(\Omega)$ with Ω a multiplicative connected open set in $\mathbb{C}_*^n$, and with some $e^b \in \Omega$, let $\partial_*^\alpha f(e^b) = 1$ for all α. Then $f \equiv 1$ in Ω.*

Proof Let

$$\Sigma_\alpha = \{z \in \Omega : \partial_*^\alpha f(z) = 1\}.$$

Then Σ_α is relatively closed in Ω. Hence, $\Sigma = \bigcap_\alpha \Sigma_\alpha$ is also closed. Since $z \in \Sigma$ implies that $f \equiv 1$ in a neighborhood of z by a Taylor series argument, we have that Σ is also open. Because $e^b \in \Sigma$, we have that Σ is nonempty. Consequently, $\Sigma = \Omega$. This completes the proof.

Theorem 7.7 *Consider the following Cauchy problem:*

$$u_j^* = f_j(z, u),$$

$$u_j^*(1) = e_{*j}, \quad j = 1, \ldots, m,$$

where $f_j \colon \mathbb{C}_*^{n+m} \to \mathbb{C}_*$ are multiplicative analytic at 1 for each $j = 1, \ldots, m$. Then there exists a unique solution u that is multiplicative analytic at 1.

Proof Without loss of generality, we suppose that $e_{*j} = 1$, $j = 1, \ldots, m$. First of all, we will determine the higher derivatives of u. For this aim, we multiplicative differentiate the equation $u_j^* = f_j(z, u)$ and we get

$$u_j^{**} = (f_j(z, u))^*$$

$$= \partial_{*z} f_j +_* \sum_{*i=1}^{m} \partial_{*u_i} f_j \cdot_* u_i^*,$$

$$u_j^{***} = (f_j(z, u))^{**}$$

$$= \partial_{*z}^2 f_j +_* \sum_{*l=1}^{m} \sum_{*i=1}^{m} \partial_{*u_i} \partial_{*u_l} f_j \cdot u_i^* \cdot_* u_l^* +_* \sum_{*i=1}^{n} \partial_{*u_i} f_j \cdot_* u_i^*,$$

$$\vdots$$

$$u_j^{*(k)} = q_k \left(\partial_*^\beta f_j(z, u), u^{*(l)} \right),$$

where q_k is a multivariate polynomial with multiplicative nonnegative coefficients and the arguments of q_k are all $\partial_*^\beta f_j(z, u)$ with $|\beta| \leq k - 1$, and all components of all $u_*^{(l)}$ with $l \leq k - 1$. We evaluate this at $z = 1$, and we use $u(1) = 1$, to get

$$u_j^{*(k)}(1) = q_k \left(\partial_*^\beta f_j(1), u^{*(l)}(1) \right)$$

$$= Q_{jk} \left(\partial_*^\beta f(1) \right),$$

where $Q_{jk} \left(\partial_j^\beta f(1) \right)$ is a multivariate polynomial having multiplicative non-negative coefficients. Note that the arguments of Q_{jk} are all components of all $\partial_*^\beta f(1)$ with $|\beta| \leq k - 1$. Since f is componentwise multiplicative analytic at 0, there exist constants $M > 0$ and $r > 0$ such that

$$(|\partial_*^\alpha f_j(1)|_*)/_* \alpha!_* \leq (e^M/_* r^{|\alpha|_*}) \quad \text{for all} \quad \alpha, \quad \text{and all} \quad j.$$

Note that

$$F_j(z, u) = e^M/_*(e -_* z/_* r)(e -_* (u_1^* +_* \cdots +_* u_m^*)/_* r)$$

majorizes f_j at 1, $j = 1, \ldots, m$. Consider the system

$$U_j^* = F_j(z, U),$$

$$U_j^*(1) = 1, \quad j = 1, \ldots, m.$$

$$(7.9)$$

Using the multiplicative positivity of the coefficients of Q_{jk}, we get

$$
\begin{aligned}
\left|\partial_*^\alpha u_j^*(1)\right|_* &= \left|Q_{jk}\left(\partial_*^\beta f(1)\right)\right|_* \\
&\leq Q_{jk}\left(\left|\partial_*^\beta f(1)\right|_*\right) \\
&\leq Q_{jk}\left(\partial_*^\beta F(1)\right) \\
&= \partial_*^\alpha U_j^*(1),
\end{aligned}
$$

i.e., U_j^* majorizes u_j^* at 1. Observe that

$$
U_1^*(z) = \cdot_* = U_m^*(z) = e^{\frac{r}{m}} \cdot_* \left(e -_* (e +_* e^{2mM} \cdot_* \log_*(e -_* (z/_* r)))^{\frac{1}{2}}_*\right)
$$

solves (7.9), which is multiplicative analytic at 0. Consequently, u is multiplicative analytic at 1.

7.4 Partial Differential Equations

Theorem 7.8 *(Cauchy–Kovalevskaya Theorem) Consider the following Cauchy problem:*

$$
\partial_{*n} u_j^* = f_j(z, u, \partial_{*1} u, \dots, \partial_{*n-1} u), \quad u_j^*(\zeta, 1) = 1, \quad \zeta \in \mathbb{C}_*^{n-1}, \quad j = 1, \dots, m.
\tag{7.10}
$$

Let $f_j \in \mathbb{C}_*^{n+m+(n-1)\times m} \to \mathbb{C}_*$ be multiplicative analytic at 1 for all $j = 1, \dots, m$. Then there exists a unique solution u that is multiplicative analytic at 1.

Proof Without loss of generality, we can assume that $f_j(1) = 1$ for all $j = 1, \dots, m$. Otherwise, we replace $u(z)$ with $u(z) -_* z_n \cdot_* \partial_{*n} u(1)$. With p_{ik} we denote the multiplicative slot of f_j that takes $\partial_i u_k^*$ as its arguments, i.e, $f_j = f_j(z, u, p)$ with $z \in \mathbb{C}_*^n$, $u \in \mathbb{C}_*^m$, $p \in \mathbb{C}_*^{(n-1)\times m}$, $j = 1, \dots, m$. Because $u_j^*(\zeta, 1) = 1$, $\zeta \in \mathbb{C}_*^{n-1}$, $j = 1, \dots, m$, we have

$$
\partial_*^\alpha u(1) = 1 \quad \text{if} \quad \alpha_n = 0.
$$

The multiplicative derivatives $\partial_*^\alpha u$ with $\alpha_n > 0$ can be found by multiplicative differentiating equation (7.10). We get

$$
\partial_{*k}\partial_{*n} u_j = \partial_{*z_k} f_j +_* \sum_{*q=1}^m \partial_{*u_q} f_j \cdot_* \partial_{*z_k} u_q +_* \sum_{*i=1}^{n-1}\sum_{*q=1}^m \partial_{*p_{iq}} f_j \cdot_* \partial_{*z_i}\partial_{*z_k} u_q,
$$

in general, for α with $\alpha_n > 0$, we have

$$\partial_*^\alpha u_j = q_\alpha \left(\partial_*^\beta f_j, \partial_*^\gamma u\right), \quad j = 1, \ldots, m,$$

where q_α is a polynomial with multiplicative nonnegative coefficients, depending on $\partial_*^\beta f_j$ with $|\beta| \leq |\alpha| - 1$ and $\partial_*^\gamma u$ with $|\gamma| \leq |\alpha|$ and $\gamma_n \leq \alpha_n - 1$. As in the previous sections we can eliminate the terms $\partial_*^\gamma u$ and evaluate the result at 1 to get

$$\partial_*^\alpha u_j(1) = q_\alpha \left(\partial_*^\beta f_j(1), \partial_*^\gamma u(1)\right) = Q_{j\alpha} \left(\partial_*^\beta f(1)\right), \quad j = 1, \ldots, m,$$

where $Q_{j\alpha}$ is a polynomial with multiplicative nonnegative coefficients depending on $\partial_*^\beta f_j$ with $|\beta| \leq |\alpha| - 1$. Since f is componentwise multiplicative analytic at 1, there exist constants $M > 0$ and $r > 0$ such that

$$|\partial_*^\alpha f_j(1)|_* /_* \alpha!_* \leq e^{\frac{M}{r^{|\alpha|}}} \quad \text{for} \quad \text{all} \quad \alpha \quad \text{and} \quad \text{all} \quad j.$$

Note that

$$F_j(z, u, p) = e^M /_* (e -_* (z_1 +_* \cdots +_* z_{n-*1} +_* (z_n/_*\rho)$$

$$+_* u_1^* +_* \cdots +_* u_m^*)/_* r) \cdot_* \left(e -_* (e/_* r)\sum_{*i,k} p_{ik}\right) -_* e^M,$$

where $\rho \in (1, e]$ is a constant which will be determined below, majorizes f_j at 1, $j = 1, \ldots, m$. Consider the system

$$U_j^* = F_j(z, U),$$

$$U_j^*(1) = 1, \quad j = 1, \ldots, m. \tag{7.11}$$

Using the positivity of the coefficients of $Q_{j\alpha}$, we get

$$\begin{aligned} |\partial_*^\alpha u_j(1)|_* &= \left|Q_{j\alpha}\left(\partial_*^\beta f(1)\right)\right|_* \\ &\leq Q_{j\alpha}\left(\left|\partial_*^\beta f(1)\right|_*\right) \\ &\leq Q_{j\alpha}\left(\left|\partial_*^\beta F(1)\right|_*\right) \\ &= \partial_*^\alpha U_j(1), \end{aligned}$$

i.e., U_j majorizes u_j at 1, $j = 1, \ldots, m$. Now we will prove that (7.11) has a solution that is multiplicative analytic at 0. Put

$$\begin{aligned} s &= z_1 +_* \cdots +_* z_{n-1}, \\ t &= z_n, \\ v &= U_1^* = \cdots = U_m^*, \end{aligned}$$

to get

$$\partial_{*t} v = e^M /_* (e -_* ((s +_* t)/_* \rho +_* m \cdot_* v)/_* r)$$
$$\cdot_* \left(e -_* (e^{(n-1)m\rho/_* r}) \cdot_* \partial_{*s} v \right) -_* e^M.$$

We define the new variable $\sigma = t +_* \rho \cdot_* s$ and assume that v depends only on σ. Then,

$$\partial_{*\sigma} v = e^M /_* (e -_* ((e^\sigma/_* e^\rho) +_* m \cdot_* v)/_* r) \cdot_* \left(e -_* e^{\frac{(n-1)m\rho}{r}} \cdot_* \partial_{*\sigma} v \right) -_* e^M,$$

whereupon

$$\left(e -_* e^{\frac{(n-*1)mM\rho}{r}} \right) \cdot_* \partial_{*\sigma} v -_* e^{\frac{(n-1)m\rho}{r}} \cdot_* (\partial_{*\sigma} v)^{2*}$$

$$= e^{\frac{M}{e -_* (\sigma/_* \rho +_* m \cdot_* v)/_* r}} -_* e^M. \tag{7.12}$$

We choose $\rho \in (0, 1]$ so small that

$$e^{e -_* e^{((n-1)mM\rho)/_* r}} > 1.$$

Then equation (7.12) can be solved for $\partial_{*\sigma} v$ in the power series

$$\partial_{*\sigma} v = e^{c_1} \cdot_* \left(e^{\frac{\sigma}{\rho}} +_* m \cdot_* v \right) +_* e^{c_2} \cdot_* \left(e^{\frac{\sigma}{\rho}} +_* m \cdot_* v \right)^{2*} +_* \cdots$$

convergent for some $e^{\frac{\sigma}{\rho}} +_* m \cdot_* v \neq 1$, with multiplicative nonnegative coefficients e^{c_k}. In other words, there is a function g multiplicative analytic at 1, with multiplicative nonnegative Maclaurin series coefficients and with $g(1) = 1$, such that

$$\partial_{*\sigma} v = g \cdot_* \left(e^{\frac{\sigma}{\rho}} +_* m \cdot_* v \right). \tag{7.13}$$

Now, we can apply the Cauchy–Kovalevskaya theorem for multiplicative analytic MDEs from the previous sections of this chapter to conclude that equation (7.13) has a solution v multiplicative analytic at 1, satisfying $v(1) = 1$, whose Maclaurin series coefficients are multiplicative nonnegative. Hence, the vector function U with components

$$U_j(z) = v \cdot_* (\rho \cdot_* (z_1 +_* \cdots +_* z_{n-1}) +_* z_n), \quad j = 1, \ldots, m,$$

solves (7.11). Since the Maclaurin series coefficients of v are multiplicative nonnegative, the same holds for U_j implying that $U_j\big|_{z_n=1}$ majorizes 1 at 1.

References

[1] Arsenin, V. Ya. *Basic Equations and Special Functions of Mathematical Physics*, Iliffe, 1968.

[2] Asmar, N. *Partial Differential Equations with Fourier Series and Boundary Value Problems*, Pearson Education, Inc., 2005.

[3] Bitsadse, A. *Equations of Mathematical Physics*, Mir Publishers, Moscow, 1980. (in English).

[4] Copson, E. *Partial Differential Equations*, Cambridge University Press, Cambridge, 1975.

[5] Courant, R. *Methods of Mathematical Physics*, Vol. II, Wiley-Interscience, New York, 1962.

[6] Courant, R. *Course of Differential and Integral Calculus*, Vols. I–III, Pergamon, New York, 1965.

[7] Courant, R. *Calculus of Variations*, New York, 1962.

[8] Epstein, B. *Partial Differential Equations: An Introduction*, Robert E. Krieger Publishing Company, 1975.

[9] Evans, L. *Partial Differential Equations*, AMS, 1997.

[10] Folland, G. *Introduction to Partial Differential Equations*, Princeton University Press, Princeton, 1995.

[11] Jeffrey, A. *Applied Partial Differential Equations: An Introduction*, Academic Press, 2003.

[12] John, F. *Partial Differential Equations*, reprint of the fourth edition, *Applied Mathematical Sciences Vol. 1.* Springer, Berlin, 1991.

[13] Jost, J. *Partial Differential Equations*, Springer, New York, 2002.

[14] Miranda, K. *Lectures on Partial Differential Equations*, Wiley-Interscience, New York, 1954.

[15] Petrovsky, I. G. *Lectures on Partial Differential Equations*, Dover Publications Inc, 1991.

[16] Pinchover, Y., J. Rubinstein. *An Introduction to Partial Differential Equations*, Cambridge University Press, Cambridge, 2005.

[17] Rauch, J. *Partial Differential Equations*, Springer, 1996.

[18] Renardy, M., R. Rogers. *An Introduction to Partial Differential Equations*, Springer, New York, 2004.

[19] Vladimirov, V. *Equations of Mathematical Physics*, Marcel Dekker, Inc., New York, 1971.

[20] Webster, A. *Partial Differential Equations of Mathematical Physics*, Dover Publications, Inc., 1955.

Index